Proceedings of the 14th International Symposium on

BIOLUMINESCENCE AND CHEMILUMINESCENCE

Chemistry, Biology and Applications

San Diego, USA
15 – 19 October 2006

Proceedings of the 14th International Symposium on

BIOLUMINESCENCE AND CHEMILUMINESCENCE

Chemistry, Biology and Applications

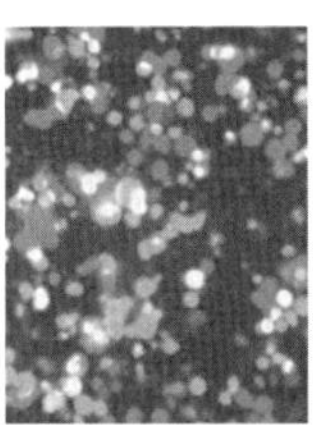
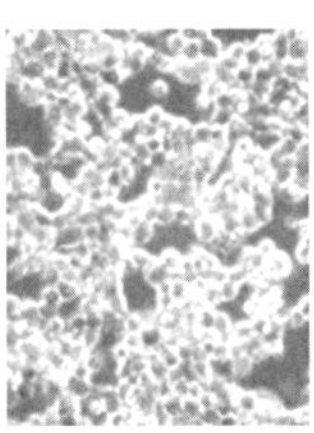

edited by

Aladar A Szalay
University of Würzburg, Germany

Philip J Hill
University of Nottingham, UK

Larry J Kricka
University of Pennsylvania Medical Center, USA

Philip E Stanley
Cambridge Research & Technology Transfer Ltd, UK

World Scientific

NEW JERSEY • LONDON • SINGAPORE • BEIJING • SHANGHAI • HONG KONG • TAIPEI • CHENNAI

Published by

World Scientific Publishing Co. Pte. Ltd.

5 Toh Tuck Link, Singapore 596224

USA office: 27 Warren Street, Suite 401-402, Hackensack, NJ 07601

UK office: 57 Shelton Street, Covent Garden, London WC2H 9HE

British Library Cataloguing-in-Publication Data
A catalogue record for this book is available from the British Library.

BIOLUMINESCENCE AND CHEMILUMINESCENCE
Chemistry, Biology and Applications

ISBN-13 978-981-270-816-8
ISBN-10 981-270-816-2

Printed in Singapore by World Scientific Printers (S) Pte Ltd

PREFACE

These are the Proceedings of the 14th Symposium on Bioluminescence and Chemiluminescence held at Paradise Point Resort & Spa from October 15-19, 2006. This series of symposia started in Brussels in 1978, and a list of the other Proceedings volumes appears at the end of this Preface. As in previous symposia, participants came from far and wide and in all 18 countries were represented.

The Organizing Secretariat was fortunate to have the continued association with the International Society for Bioluminescence & Chemiluminescence. The organizers are thankful for the kind support of the society. We also thank John Wiley & Sons for publishing the regular abstracts in the journal *Luminescence* Vol. 21(5) 2006.

Editorial Note

This volume was compiled without peer review from camera-ready manuscripts of lectures and posters presented at the Symposium. The Editors have, in the interest of rapid publication, made only minor stylistic changes. They take no responsibility for scientific or priority matters.

The Editors: AA Szalay, PJ Hill, LJ Kricka, PE Stanley

THE MARLENE DELUCA PRIZE

The Marlene DeLuca prizes were again generously given by Dr Fritz Berthold, together with Berthold Technologies. Dr. Berthold has provided these prizes at each symposium since the 1988 Symposium in Florence. The prize can be awarded to symposium participants under the age of 35 at the day before the starting date of the symposium. The prize is given in memory of Dr. Marlene DeLuca who made major contributions to the science of bioluminescence (*see* Stanley PE. Dedication to Marlene DeLuca: *J Biolumin Chemilumin* 4;1989:7-11 (includes list of her papers). Similarly to previous years' selections, the President of the International Society, Professor Aladar A Szalay (University of Würzburg, Würzburg, Germany), assembled a five-member selection committee from the society to choose the three candidates based on their presentations.

The 2006 prize winners were:

Yoriko Ando (University of Tokyo, Chiba, Japan) *Quantitative luminescence, yield spectra of firefly bioluminescence*

Jenny Krönström (Department of Zoophysiology, Göteborg University, Göteborg, Sweden) *Do muscular sphincters control oxygen supply to photocytes in Northern Krill,* Meganyctiphanes norvegica*?*

Kazuki Niwa (National Institute of Advanced Industrial Science and Technology, Osaka, Japan) *Biosynthesis of firefly D-luciferin*

INTERNATIONAL SOCIETY FOR BIOLUMINESCENCE AND CHEMILUMINESCENCE

2004-2006 ISBC COUNCIL
Council Members
A.A. Szalay (President), J. Lee (Past President), B Branchini (President-Elect and Past Treasurer & Membership Secretary), P. Pasini (Secretary), E. Hawkins (Treasurer & Membership Secretary), LJ Kricka (Publication Officer)
Councilors
H. Akhavan-Tafti, L. Brovko, S. Haddock, V. Kratasyuk, M. Maeda, O. Nozaki, A. Roda, B. Zomer
Scientific Advisory Board
P. De Sole, S. Girotti, D. Shah, P. Stanley, A. Tsuji, V. Viviani, Xiaolin Yang

2006-2008 ISBC COUNCIL
Council Members
B. Branchini (President), A. A. Szalay (Past President), M. Aizawa (President Elect), Y. Ohmiya (Secretary), P. Pasini (Past Secretary), E. Hawkins (Treasurer & Membership Secretary), L. Kricka (Publication Officer).
Councilors
H. Akhavan-Tafti, L. Brovko, R. Hart, P. Hill, O. Nozaki, A. Roda, E. Widder, K. Wood

SYMPOSIUM ORGANIZING COMMITTEE

CHAIRMAN
Dr. Aladar A. Szalay

SYMPOSIUM ORGANIZERS
Dr. Aladar A. Szalay, Dr. Phil Hill (co-organizer)

PROGRAM ORGANIZERS
Dr. Phil Hill, Dr. Aladar A. Szalay, members of the program advisory committee

SECRETARIAT
Alison Estrada (coordinator), Valere Rollings (secretary), Andrea Feathers (program coordinator), Tom Hagood (audio/visual coordinator)

MANUSCRIPT EDITORS
Professor L. Kricka, Dr. Phil Stanley, Andrea Feathers

ACKNOWLEDGEMENTS

We wish to express our sincere appreciation to the following for their generous support of this symposium and to the ISBC for the loan of deposit for the conference venue.

SPONSORS/EXHIBITORS

Genelux Corporation; Promega Corporation; American Gene Therapy, Inc.; Applied Biosytems; Berthold Detection Systems GmbH; Berthold Technologies GmbH & Co. KG; Kodak Molecular Imaging Systems; Teledyne Benthos, Inc.; Turner BioSystems; Xenogen Corporation.

NEXT SYMPOSIUM

The 15th International Symposium on Bioluminescence and Chemiluminescence is proposed to be held in 2008 in China. Dr. Xiaolin Yang (People's Hospital of Peking University, Beijing, China, xiaolin.yang@yuande.com) volunteered to organize the symposium as symposium chairman. Details of the organization of the 15th BL&CL Symposium will be posted on the society website, http://www.isbc.unibo.it.

PROCEEDINGS OF PREVIOUS SYMPOSIUMS

13th 2004 Yokohama, Japan

Bioluminescence & Chemiluminescence: Progress and Perspectives. Editors: Tsuji A, Matsumoto M, Maeda M, Kricka LJ, Stanley PE. Singapore: World Scientific 2002. pp. 520. ISBN 981-238-156-2.

12th 2002 Cambridge, UK

Bioluminescence & Chemiluminescence: Progress & Current Applications. Editors: Stanley PE, Kricka LJ. Singapore: World Scientific 2002. pp. 520. ISBN 981-238-156-2.

11th 2000 Monterey, CA, USA

Proceedings of the 11th International Symposium on Bioluminescence & Chemiluminescence. Editors: Case JF, Herring PJ, Robison BH, Haddock SHD, Kricka LJ, Stanley PE. Singapore: World Scientific 2001. pp. 517. ISBN 981-02-4679-X.

10th 1998 Bologna, Italy

Bioluminescence and Chemiluminescence: Perspectives for the 21st Century. Editors: Roda A, Pazzagli M, Kricka LJ, Stanley PE. Chichester: Wiley 1999. pp. 628. ISBN: 0-471-98733-6.

9th 1996 Woods Hole, MA, USA

Bioluminescence and Chemiluminescence: Molecular Reporting with Photons.

Editors: Hastings JW, Kricka LJ, Stanley PE. Chichester: Wiley 1997. pp. 568. ISBN: 0-471-97502-8.

8th 1994 Cambridge, UK

Bioluminescence and Chemiluminescence: Fundamentals and Applied Aspects. Editors: Campbell AK, Kricka LJ, Stanley PE. Chichester: Wiley 1994. pp. 672. ISBN: 0-471-95548-5.

7th 1993 Banff, Canada

Bioluminescence and Chemiluminescence: Status Report. Editors: Szalay AA, Kricka LJ, Stanley PE. Chichester: Wiley. 1993, pp. 548. ISBN: 0-471-94164-6.

6th 1990 Cambridge, UK

Bioluminescence and Chemiluminescence: Current Status. Editors: Stanley PE, Kricka LJ. Chichester: Wiley 1991. pp. 570. ISBN: 0-471-92993-X.

5th 1988 Florence, Italy

Bioluminescence and Chemiluminescence: Studies and Applications in Biology and Medicine. Editors: Pazzagli M, Cadenas E, Kricka LJ, Roda A, Stanley PE. Chichester: Wiley 1989. pp. 646. (published as volume 4, issue 1 of the *Journal of Bioluminescence and Chemiluminescence*, 1989). ISBN: 0-471-92264-1.

4th 1986 Freiburg, Germany

Bioluminescence and Chemiluminescence: New Perspectives. Editors: Schölmerich J, Andreesen R, Kapp A, Ernst M, Woods WG. Chichester: Wiley 1987. pp. 600. ISBN: 0-471-91470-3.

3rd 1984 Birmingham, UK

Analytical Applications of Bioluminescence and Chemiluminescence. Editors: Kricka LJ, Stanley PE, Thorpe GHG, Whitehead TP. London: Academic Press 1984. pp. 602. ISBN: 0-12-426290-2.

2nd 1980 San Diego, CA, USA

Bioluminescence and Chemiluminescence: Basic Chemistry and Analytical Applications. Editors: DeLuca MA, McElroy WD. New York: Academic Press 1981. pp.782. ISBN: 0-12-208820-4.

1st 1978 Brussels, Belgium

International Symposium on Analytical Applications of Bioluminescence and Chemiluminescence. Proceedings 1978. Editors: Schram E, Stanley PE. Westlake Village, CA: State Printing & Publishing, Inc., 1979, pp. 696. (Privately published).

INTRODUCTION

On behalf of the Organizing Committee of 14th International Symposium on Bioluminescence & Chemiluminescence (IS BL and CL), held October 15-19, 2006, I would like to thank the International Society of Bioluminescence and Chemiluminescence (ISBC) for their trust and support to host this beautiful event. The symposium brought experts from different areas of the world to Paradise Point Resort & Spa in San Diego, California. The first symposium was held in 1978 in Brussels, Belgium, and has subsequently been held every two years in Europe, America or Asia. After unforeseen complications in Europe, we were pleased to be able to create an opportunity for scientists interested in bioluminescence and chemiluminescence to meet and exchange new ideas in the magnificent city of San Diego, which is considered the center of biomedical science and industry in the southwestern United States.

After opening remarks from the President of the Symposium, Dr. Keith Wood from Promega Corporation delivered the keynote address entitled "Switchable luciferases: the third wave of bioluminescence technology." The Science and Technology Advance Lecture, titled "Subcellular imaging *In vivo:* The new cell biology," was presented by Dr. Robert M. Hoffman, from Anti-Cancer, Inc., and the University of California, San Diego. We were fortunate to have oral and poster presentations given by scientists from 18 countries, as well as active participation from industrial exhibitors. Like previous symposia, the sessions included luciferases, instrumentation, bioluminescent and chemiluminescent assays, multicolor assays, optical imaging, physiology, biochemistry, detection of biological light emission in bioluminescent oceans, as well as its utility in space exploration.

The participants very much enjoyed the lovely gala dinner overlooking Mission Bay and a harp concert by artist Alfredo Rolando Ortiz. The highlight of the festive dinner was the Marlene DeLuca Prize award ceremony, which has become a tradition for honoring outstanding scientific achievements. The President and Dr. Fritz Berthold handed over the prizes to three young scientists under the age of 35: Dr. Yoriko Ando (University of Tokyo, Chiba, Japan), Dr. Jenny Krönström (Department of Zoophysiology, Göteborg University, Göteborg, Sweden) and Dr. Kazuki Niwa (National Institute of Advanced Industrial Science and Technology, Osaka, Japan). In addition to the very capable young awardees, it was refreshing to see the active participation of many young men and women from different scientific disciplines. I predict that this new generation of young scientists, if nurtured, will soon become a well-recognized, innovative scientific force in the footsteps of the greatest pioneers in the BL/CL sciences, many of whom were unfortunately unable to attend the symposium in San Diego, in 2006.

In 1986, exactly 20 years ago, at the 4th International Symposium on Bioluminescence & Chemiluminescence, convened in Freiburg, Germany, I presented, for the first time, visualization of nitrogenase activation in live root nodules (bacterium-caused plant tumors) infected by stably transformed (genomic insertion), *nif* promoter luxAB gene containing Rhizobia. The visualization experiments occurred two years prior to the symposium in my laboratory at Cornell

University (Ithaca, New York) by low light imaging using the Hamamatsu Argus 100 dual microchannel plate camera.

Since that time, numerous publications on imaging of live organisms have appeared from our and other laboratories using many different cells, tissues and organisms with luciferases alone, with fluorescent proteins alone or with separate luciferase and fluorescence protein combinations, as well as with luciferase-fluorescence protein fusion derivatives.

In addition to excellent presentations in the fundamental science session of the 14th International Symposium, the highlight of the symposium was a whole body live imaging session using light-emitting bacteria or viruses as vectors for detection of tumors and metastases. These unexpected findings presented by myself in Yokohama (13th International Symposium on Bioluminescence & Chemiluminescence) are now confirmed by participants from several laboratories, four of whom gave fascinating presentations during the optical imaging session. This optical imaging research may one day lead to the monitoring of tumor detection, progression and therapy in man. The combined scientific contributions of the bioluminescence/chemiluminescence research community to live imaging by light-emitting vectors is beginning to exceed the *in vitro* assay or diagnostic field and is slowly expanding its utilization into live animal (veterinary) applications and even into human clinical trials.

The ability to monitor host-pathogen interactions in live animals based on low light imaging, especially in tumors and metastases, will set new horizons for understanding the role of the host immune system in eradication of tumors and, thus, reaches far beyond the capability of cell culture infection assays. These developments in imaging over the past 20 years demonstrate once again the need for longterm support from funding agencies such as the National Institutes of Health, especially in novel and innovative research.

The organizers and I are grateful to all the generous donors for their support. A special thanks is owed to the management of Genelux Corporation, which contributed the majority of the outside financial support, and all my competent and friendly staff who aided the participants of the 14th International Symposium without pay. Lastly, I would like to thank Dr. Phil Hill, all of my colleagues in the ISBC, the chairpersons and organizers and the graduate students for their selflessness, unconditional and generous help. Without them, none of this would have happened in San Diego.

Cordially,
Aladar A. Szalay, PhD
The President of 14th International Symposium on Bioluminescence and Chemiluminescence

CONTENTS

Preface v

Introduction to the 14th Symposium ix

PART 1. LUCIFERASES

Refolding of the recombinant luciferases of *Metridia longa* 3
Borisova VV, Frank LA, Markova SV, Golz S and Vysotski ES

The main function of HIS175, TRP179, and TYR190 residues of the obelin binding site is to stabilize the hydroperoxycoelenterazine intermediate 7
Eremeeva EV, Markova SV, Frank LA and Vysotski ES

Mechanistic study on the *Cypridina (Vargula)* bioluminescence reaction 11
Hirano T, Takahashi Y, Kondo H, Ikeda H, Maki S and Niwa H

Stabilization of *Luciola mingrelica* firefly luciferase by mutation of non-conservative cysteine residues 15
Lomakina GY, Legotsky SA, Tishkov VI and Ugarova NN

Firefly luciferase arose from fatty acyl-CoA synthetase by gene duplication and gene recruitment 19
Oba Y and Inouye S

Bioluminescence in click beetles (family Elateridae): Molecular phylogenetic analysis, inferred from 28S ribosomal DNA 23
Sagegami-Oba R, Oba Y and Ôhira H

Effect of microenvironmental changes on kinetic parameters of steady-state enzyme-induced bacterial bioluminescent reaction 27
Sukovataya IE, Tyulkova NA and Kaykova EV

Thermal stability and absorption spectra of bacterial luciferases from *Photobacterium leiognathi* in water-organic solvents 31
Sukovataya IE, Tyulkova NA and Kaykova EV

Characterization of a fish symbiont of *Photobacterium phosphoreum* with altered spectral properties 35
Whitaker S, Herring PJ, Campbell AK, Hill PJ and Rees CED

PART 2. INSTRUMENTATION

Luminometers for industrial applications of rapid microbiology 41
Berthold F, Hafner K, Schreibers M and Tarkkanen V

Application of a luminescence microscope with novel optical system for detection of the gene expression pattern of individual cells 45
Hatta-Ohashi Y, Hayasaka N, Takahasi T, Shigeyoshi Y and Suzuki H

Novel tools for quantification of luminescence 49
Hickey PC

Luminescence microscope for reporter assay of single live cells 53
Suzuki H, Dosaka S, Ohashi-Hatta Y and Sugiyama T

PART 3. ASSAYS

Elevation of choline concentration in cardiac troponin-I positive human plasma 59
Adamczyk M, Brashear RJ and Mattingly PG

High throughput detection of hydrogen peroxide: Validation of homogeneous chemiluminescent assays for choline in human plasma and whole blood 63
Adamczyk M, Brashear RJ, Mattingly PG and Tsatsos PH

Development of bioluminescent pyrophosphate assay using pyruvate phosphate dikinase and its application to SNPs analysis 67
Arakawa H, Munakata E, Obinata R, Maeda M, Suzuki S, Kamahori M and Kambara H

Bioluminescent method for detection of proteinase activity in UHT milk 71
Brovko LY, Hajdok C, Tapp C, Ramsay C, Driscoll M and Griffiths MW

Detection and control of *Salmonella typhimurium* growth using specific bacteriophage 75
Brovko LY, Romanova NA, Allain B, Mandeville R and Griffiths MW

Improved sensitivity method for rapid hygiene monitoring using ATP bioluminescence 79
Driscoll M, Ramsay C and Watkin J

Determination of antioxidant capacity and reactivity using hemin-catalyzed luminol chemiluminescence 83
Eckert CR, Esteves CH, Bastos EL, Romoff P and Baader WJ

Applications of adenylate kinase detection using bioluminescence 87
Foote N, Green M and Gregory S

Standardization of bioluminescent raw milk quality assay using Lumtek test-systems 91
Frundzhyan VG, Ugarova NN, Makarushin AA and Gavrilov GB

BART – Bioluminescent alternative to real-time PCR 95
Gandelman O, Church VL, Moore CA, Carne C, Jalal H, Murray JAH and Tisi LC

Advancing the development of dual-luciferase assays 99
Hawkins EM, Garvin D, Paguio A, Stecha P, Swanson B, Fan F and Wood KV

Simultaneous bioluminescent immunoassay using aequorin labeled Fab fragment and biotinylated firefly luciferase 103
Ito K, Nishimura W, Maeda M, Gomi K, Inouye S and Arakawa H

Designing enhanced thermostable luciferase for proteolytic assay 107
Jathoul AP, Gandelman OA, Law GHE, Murray JAH and Tisi LC

Biosensors for the non-invasive evaluation of bacterial quorum sensing in GI disorders 111
Kumari A, Pasini P, Deo SK, Flomenhoft D, Shashidhar H and Daunert S

Improved Ca^{2+}-activated photoproteins for HTS applications 115
Mastroianni N, Flak T and Corazza S

Development of a multiplexed bioluminescent cell-based assay with the *luc* gene from *Luciola italica* for high throughput screening of cholesterol-lowering drugs 119
Michelini E, Southworth TL, Ablamski D, Branchini BR and Roda A

Novel BRET-based biosensors with a new bioluminescent donor, *Gaussia* luciferase, for estrogen receptor ligands 123
Michelini E, Magliulo M, Baraldini M and Roda A

The use of lux genes for monitoring uptake of *Pseudomonas aeruginosa* by respiratory epithelial cells 127
Robinson GM, Jackson SK, Salisbury VC and Nelson SM

Bioluminescence characteristics of an Obelin mutant in varying solvent conditions 131
Rowe L, Rothert A, Deo SK, Ensor M and Daunert S

A simple chemiluminescence flow injection analysis for the determination of sulphide in aqueous samples 135
Satienperakul S, Liawruangrath S and Cardwell TJ

Using recombinant antigens for the detection of antibodies to *Trypanosomas cruzi* on a fully automated chemiluminescence analyzer 139
Shah D, Chang C-D, Cheng K, Jiang L, Salbilla V, Haller A and Schochetman G

Quantitative bioluminescent RNA assay 143
Sun Y, Jacobson KB and Golovlev V

A novel multi-channel chemiluminescence-immunoassay to detect HIV antibodies as a more specific supplemental test 147
Sun X and Yang X

Time-resolved measurement of firefly bioluminescence using photolysis of caged-ATP 151
Tanaka M, Igarashi R and Wada N

Red-chemiluminescent probes for detecting superoxide anions 155
Teranishi K

Aequorin bioluminescence dissected by random mutagenesis 159
Tricoire L, Tsuzuki K and Lambolez B

Influence of inorganic sulfur compounds on bioluminescent system NAD(P)H: FMN-oxidoreductase–luciferase 163
Vetrova EV and Prokhorova EV

Development of the *Cypridina* bioluminescent system for high throughput screening assays 167
Wu C, Kawasaki K, Ohgiya S and Ohmiya Y

Quantitative Western blotting using a GZ-11 based chemiluminogenic signal reagent 171
Zomer G, Smitsman C, Arts R, Hamzink M and Kooijman M

PART 4. OPTICAL IMAGING AND MULTICOLOR ASSAYS

Demonstration of tri-colored reporter assay using a filter luminometer "Sirius C" 177
Asai T, Berthold A and Ohmiya Y

Real-time monitoring of vaccinia virus infection in cultured cells and in living mice using light-emitting proteins 181
Chen N, Yu YA, Zhang Q, Raab V and Szalay AA

Use of bioluminescent reporter bacteria to study invasion and survival within mammalian cell lines 185
Cutter KL, Nelson SM, Greenman J, Robinson GM and Salisbury VC

Network activities of the cerebral cortex monitored by bioluminescent aequorin imaging 189
Drobac E, Tricoire L and Lambolez B

Bivalent fluorescent reporters for gene expression studies in Gram-positive bacteria 193
Gaddipati SR, Perehinec TM, Qazi SNA, Rees CED and Hill PJ

Four-color labeling of cell culture and tumors of live mice upon infection with: *GFP-Ruc* and *RFP-CBG99* expressing *Vaccinia* virus strains 197
Raab V, Horbaschek C, Chen N, Zhang Q, Yu YA, Seubert C, Geissinger U, Worschech A, Tietze CJ, Zellner E, Le TH, Grummt F, Stritzker J and Szalay AA

Quantitative immunohistochemistry and *in situ* hybridization with luminescence detection: a new potential diagnostic tool 201
Roda A, Guardigli M, Mirasoli M, Ambretti S, Venturoli S, Musiani M, Casanova S and Roda E

Evaluation of an *in vivo* gene induction system in infected tumor-bearing mice 205
Stritzker J, Weibel S, Hill PJ and Szalay AA

Examinations of bacterium-mediated detection of tumors in mice models 209
Yu YA and Szalay AA

PART 5. BIOCHEMISTRY/LUMINESCENCE

Direct observation of a radical-ion intermediate in the chemiexcitation step of peroxyoxalate chemiluminescence 215
Bastos EL, Ciscato LFML, Bartoloni FH and Baader WJ

Kinetic observation of the chemiexcitation step in peroxyoxalate chemiluminescence 219
Ciscato LFML, Bartoloni FH and Baader WJ

Synthesis and studies of fenchyl-derived 1,2-dioxetanes 223
Ciscato LFML, Weiss D, Flammersheim HJ and Baader WJ

Enzyme embedded organic monolith sensor for flow injection — chemiluminescence 227
Nozaki O, Munesue M and Kawamoto H

Efficiency of electron-transfer induced chemiexcitation: A comparison of inter- and intramolecular processes 231
Oliveira MA and Baader WJ

Study of metal binding to monomeric red fluorescent protein, DsRed-monomer 235
Rahimi Y, Shrestha S and Deo SK

Time-dependent spectral change of chemiluminescence from firefly luciferin in deoxygenated dimethyl sulfoxide 239
Shibata R, Yoshida Y and Wada N

A quantum-chemical approach to the amino analogs of firefly luciferin 243
Wada N, Fujii H and Sakai H

Firefly-luciferin and its analogs: a source of new luminescence dyes and ligands 247
Weiss D, Trupp S, Ciscato LFML, Beckert R and Anders E

Chemiluminescence enhancement effect of trace metal elements 251
Yoshiki Y

PART 6. BLCL OCEANS, LAND & SPACE

Bioluminescence in the spaceflight and life science training program at the Kennedy Space Center 257
Kratasyuk VA, Gusev SM, Remmel NN, Osipenko OA, Esimbekova EN, Shoeman DM, Dreschel TW, Chetirkin PV, Correll MJ, Bucklin RA and Rygalov VE

A comparison of bioluminescence and chlorophyll fields of the world ocean 261
Piontkovski SA, Tokarev YN and Levin LA

Brazilian species of bioluminescence fungi 265
Stevani CV, de Oliveira AG, Mendes LF and Domingues O

Molecular phylogeny of cypridinid ostracodes and the evolution of bioluminescence 269
Torres E and Gonzalez VL

Index 273

PART 1

LUCIFERASES

REFOLDING OF THE RECOMBINANT LUCIFERASES OF *METRIDIA LONGA*

VV BORISOVA,[1] LA FRANK,[1] SV MARKOVA,[1] S GOLZ,[2] ES VYSOTSKI[1,3]
[1] Photobiology Lab, Institute of Biophysics SB RAS, Krasnoyarsk 660036, Russia;
[2] Global Drug Discovery – Target Research, Bayer HealthCare AG, 42096 Wuppertal, Germany; [3] Dept of Molecular Biology and Biochemistry, University of Georgia, Athens, GA 30602, USA

INTRODUCTION

Metridia longa is a marine copepod from which a blue bioluminescence originates as a secretion from epidermal glands in response to various stimuli. Its bioluminescence is conditioned by secreted coelenterazine-dependent luciferases. Using functional screening the three cDNAs encoding different luciferases were cloned from the expression cDNA library of *Metridia*. One of them, MLuc164, was successfully applied as a secreted reporter enzyme in mammalian cells.[1] Two luciferases, MLuc164 and MLuc39, without signal peptides for secretion were expressed in *E. coli*. In *E. coli* cells, most of the synthesized protein is accumulated in insoluble inclusion bodies. Here we report the results of testing various approaches for solubilization and refolding of these luciferases. Despite high identity of luciferases, MLuc39 yielded more active monomeric proteins.

MATERIALS AND METHODS

Reagents. Coelenterazine was from PJK GmbH (Kleinblittersdorf, Germany). Dithiothreitol (DTT) and 5,5'-dithiobis-(2-nitrobenzoic acid) (DTNB) were from Aldrich and Serva respectively. Oxidized (GSSG) and reduced (GSH) glutathione were from Sigma. All other reagents were of highest quality commercial grade.

Bacterial expression. *E. coli* cells BL21 CodonPlus (DE3)-RIPL (Stratagene) transformed with the plasmids pET22-MLuc164 and pET22-MLuc39 were cultivated with vigorous shaking at 37°C in LB medium containing 200 μg/mL ampicillin. When the culture reached an OD (600 nm) of 0.8-1.0 the luciferase synthesis was induced with 100 mg/L IPTG. After induction, the cultivation was continued for 3 h. Cells were harvested by centrifugation at 4°C.

Isolation and purification. The cell paste was resuspended in 20 mM Tris-HCl, pH 7.0 (1:5, w/v), disrupted by sonication (20 sec × 6) at 0°C and centrifuged. The pellet was sequentially washed with 0.9% NaCl, 0.5% Tween-20 (× 3), and 20 mM Tris-HCl pH 7.0. All the washing procedures were followed by centrifugation at 4°C. The final pellet was extracted either with 6 M urea, 20 mM Tris-HCl pH 8.8 or 6 M guanidine HCl. The 6 M urea extracts of the luciferases were purified on DEAE Sepharose Fast Flow (Pharmacia) column preliminary equilibrated with 20 mM Tris-HCl pH 8.8. The proteins were eluted with linear salt gradient: 0–0.5 M NaCl in the same buffer, then dialyzed overnight at 4°C against 0.1 M NaCl, 20 mM Tris-

HCl pH 8.8. Luciferases were concentrated and subjected to gel filtration on a column of Bio-Gel P100 equilibrated with 0.5 M NaCl, 50 mM Tris-HCl pH 8.8.
Refolding experiments. The inclusion bodies were dissolved in 6 M guanidine HCl with or without DTT and then diluted 100 times into refolding solution: No. 1, 20 mM Tris-HCl pH 8.8; No. 2, 0.1 M NaCl, 20 mM Tris-HCl pH 8.8; No. 3, 2 M NaCl, 20 mM Tris-HCl pH 8.8; No. 4, 1% Tween-20, 20 mM Tris-HCl pH 8.8; No. 5, 0.1% Tween-20, 20 mM Tris-HCl pH 8.8; No. 6, 5 mM GSSG, 0.5 mM GSH, 20 mM Tris-HCl pH 8.8; No. 7, 0.1 M NaCl, 20 mM Tris-HCl pH 7.0; No. 8, 0.1 M NaCl, 20 mM Tris-HCl pH 6.0; No. 9, 0.1 M NaCl, 5 mM GSSG, 0.5 mM GSH, 20 mM Tris-HCl pH 8.8; No. 10, 0.1 M NaCl, 5 mM GSSG, 0.5 mM GSH, 20 mM Tris-HCl pH 7.0; No. 11, 0.1 M NaCl, 5 mM GSSG, 0.5 mM GSH, 20 mM Tris-HCl pH 6.0 (Table 1). Samples were concentrated and analyzed by native PAGE.

Table 1. Distribution of bioluminescent activity and protein between fractions at gel filtration on Bio-Gel P100

	polymer		monomer	
	activity %	protein %	activity %	protein %
MLuc164	3.9	63.0	96.1	37.0
MLuc39	2.5	43.8	97.5	56.3

Native PAGE. Samples were run on a 12.5% polyacrylamide gel. Protein solutions were mixed 2:1 with 50 mM Tris-HCl pH 7.0 buffer including 0.1% bromophenol blue, 10% glycerol, and loaded directly onto gel without boiling. A low range prestained molecular weight marker (Bio-Rad) was used as size standards.
Quantitation of SH-groups. The SH-groups were determined with DTNB (Ellman's reagent).[2] The luciferase was treated with 20 molar DTNB excess in 20 mM Tris-HCl pH 8.0 and incubated at room temperature for 10 min prior to measurement of absorbance at 412 nm. The concentration of 5-thio-2-nitrobenzoate (TNB) was determined using an extinction coefficient of 14.140 $M^{-1}cm^{-1}$ at pH 8.0.
Bioluminescence assay. The luciferase activity was measured with a BLM 8801 luminometer (SKTB "Nauka", Russia) by injection of 5 μL of 10^{-5} M coelenterazine in methanol into cuvette containing 5 μL of luciferase sample and 500 μL of buffer (50 mM Tris-HCl pH7.5, 0.1 M NaCl, 10 mM $MgSO_4$, 0.01% gelatin).

RESULTS AND DISCUSSION

Both Mlucs, expressed in high yields, accumulated in insoluble inclusion bodies, and both luciferases, after purification on DEAE Sepharose, had high purity by SDS-PAGE (Fig. 1), but gel filtration of these revealed two peaks that respectively corresponded to proteins with the molecular masses over 100 kDa (polymeric fraction) and 27.4 kDa for MLuc39 or 30.4 kDa for MLuc164 (monomeric fraction).

Most of a protein is in a polymeric fraction, and the corresponding bioluminescent activity is much lower than that of monomeric fraction (Table 1).

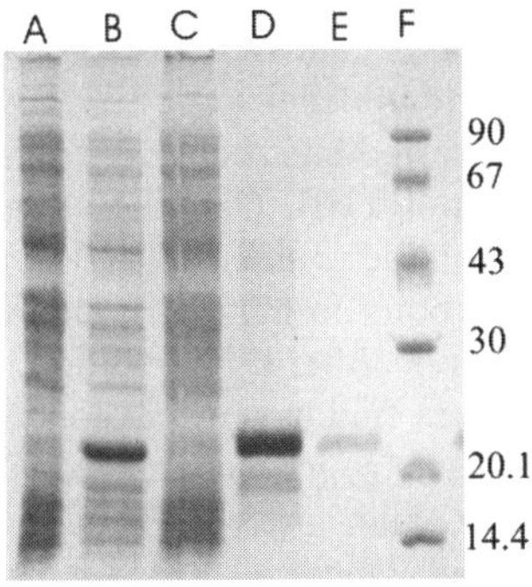

Fig. 1. SDS-PAGE of MLuc39 samples. Lanes: A, B – *E. coli* cells before and after induction; C – cells lysate; D – inclusion bodies in 6 M urea; E –after DEAE Sepharose; F – molecular weight markers

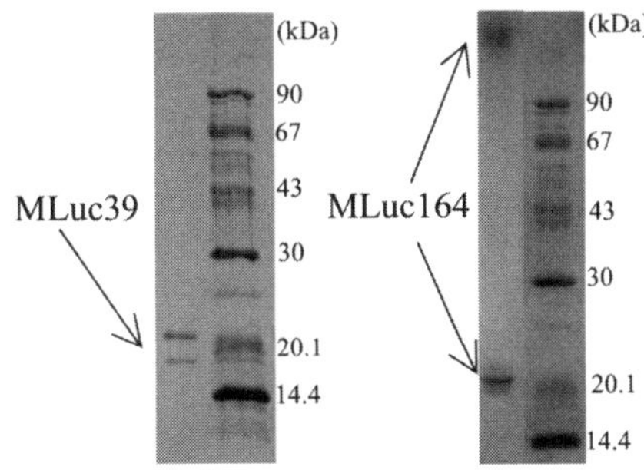

Fig. 2. Native PAGE analysis of MLuc39 and MLuc164 after refolding at No. 9 and No. 6 conditions

Table 2. Bioluminescent activity of MLuc164 and MLuc39 depending on refolding conditions, with maximum of bioluminescent activity marked in bold

MLuc164			MLuc39		
refolding solution	RLU		refolding solution	RLU	
	– DTT	+ DTT		– DTT	+ DTT
No. 1	550	148	No. 2	288	360
No. 2	200	**973***	No. 7	61	4
No. 3	290	48	No. 8	18	0.3
No. 4	2012	381	No. 9	**640**	**480**
No. 5	1000	677	No. 10	0.64	0.016
No. 6	**3250**	719	No. 11	0.48	0.032

It is well known that, for the majority of the proteins, a removal of the denaturant leads rather to aggregation than to the correct protein folding. Although in recent years various strategies for overcoming aggregation have been developed, the folding conditions for each protein are unique and their finding is still empirical.[3]

MLuc164 and MLuc39 are single chain polypeptides with molecular masses 22 kDa and 21 kDa respectively. They display high degree of identity (82 %), and both contain 10 cysteine residues. A lot of cysteine residues suggest these luciferases to have some disulfide bonds, which might be responsible for bioluminescent activity or the stability of a protein conformation. The quantity of disulphide bonds in monomeric luciferases (after gel filtration on Bio-Gel P100) was estimated by comparing the number of free SH-groups in the unfolded proteins produced with 6 M guanidine HCl with or without DTT. Both luciferases contain no less than four disulphide bonds. However at the surface of monomeric molecules, free SH-groups are lacking. This shows that two SH-groups might be buried into molecule core and, therefore, they will be inaccessible for modification with DTNB. The DTT adding to luciferases turns those into monomeric forms. It might be evidence that *Metridia* luciferases aggregate through intermolecular disulfide bond formation. It should be also noted that both luciferases lose bioluminescent activity after DTT addition.

All in all, eleven different conditions were tested to find the correct conditions for refolding of luciferases (Table 2). Refolding accuracy was monitored by bioluminescent activity, free SH-groups quantity and native PAGE data. For Mluc164 proper refolding condition was not found. Under all tested refolding conditions MLuc164 is a mixture of polymeric and monomeric forms (Fig. 2). That's why the number of free SH-groups in the samples was not determined. However, the largest bioluminescent activity was found at refolding under alkaline condition with GSSG/GSH (Table 1, No. 6). In contrast to MLuc164, MLuc39 is a monomer under all examined refolding conditions (Fig. 2). Nevertheless, the most of bioluminescent activity was found at refolding under alkaline conditions with GSSG/GSH adding (Table 1, No. 9). At that free SH-groups at the molecule surface are not determined, though at another refolding conditions the ones were found. It is interesting that the number of free SH-groups correlates with bioluminescent activity; the increase of the number of free SH-groups is accompanied by the decrease of bioluminescent activity. Although both *Metridia* luciferases display a high degree of identity it was surprisingly to find that at tested refolding conditions MLuc39 reveals the better yield of the active monomeric protein than MLuc164.

ACKNOWLEDGEMENTS

Supported by RFBR grant 05-04-48271, the RAS program for "Molecular and Cellular Biology," the SB RAS Lavrentiev grant for Young Scientists, and Bayer HealthCare AG (Germany).

REFERENCES

1. Markova SV, Golz S, Frank LA, Kalthof B, Vysotski ES. Cloning and expression of cDNA for luciferase from the marine copepod *Metridia longa*. J Biol Chem 2004;279:3212-7.
2. Creighton TE. Disulfide bonds between cysteine residues. In: Creighton TE. eds. Protein structure: a practical approach. Oxford: IRL Press, 1995:156-8.
3. Singh SM, Panda AK. Solubilization and refolding of bacterial inclusion body proteins. J Biosci Bioeng 2005;99:303-10.

THE MAIN FUNCTION OF HIS175, TRP179, AND TYR190 RESIDUES OF THE OBELIN BINDING SITE IS TO STABILIZE THE HYDROPEROXYCOELENTERAZINE INTERMEDIATE

EV EREMEEVA,[1] SV MARKOVA,[1] LA FRANK,[1] ES VYSOTSKI[1,2]
[1] *Photobiology Lab, Institute of Biophysics, SB RAS, Krasnoyarsk 660036, Russia;*
[2] *Dept of Molecular Biology and Biochemistry, University of Georgia, Athens, GA 30602, USA*

INTRODUCTION

Obelin is a Ca^{2+}-regulated photoprotein isolated from marine bioluminescent hydroid *Obelia longissima*. It consists of a single polypeptide chain (~ 22 kDa) with a hydrophobic cavity in which is bound the "pre-activated" coelenterazine, 2-hydroperoxycoelenterazine. The bioluminescent reaction is triggered by Ca^{2+} and results in light emission with λ_{max} = 485 nm. The spatial structures of obelin from *O. longissima* have been determined with resolution at 1.7 Å and 1.1 Å.[1,2] The side chains of His175, Trp179, and Tyr190 form hydrogen bonds with peroxy and carbonyl groups of a substrate (Fig. 1). The 2-hydroperoxycoelenterazine is stable in the hydrophobic environment of the binding site; it seems to be stabilized by the H-bond network. To determine the function of these hydrogen bonds we substituted His175, Trp179, and Tyr190 by residues with different donor-acceptor hydrogen bond properties of side chains.

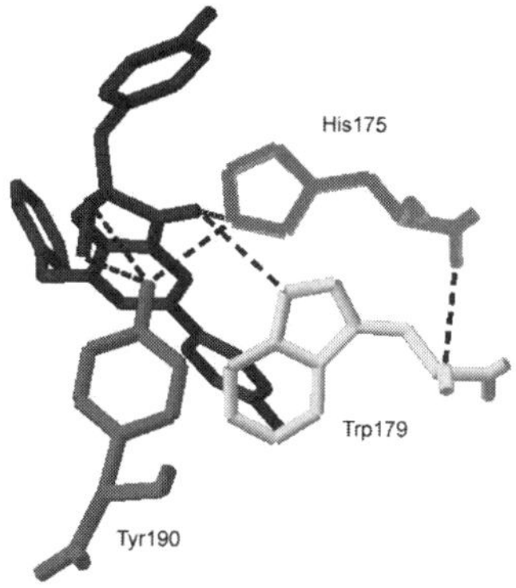

Fig. 1. Hydrogen bonds formed by atoms of 2-hydroperoxycoelenterazine with His175, Trp179, and Tyr 190 at the binding pocket of obelin. The 2-hydroperoxycoelenterazine molecule is shown in black

MATERIALS AND METHODS

Molecular biology. Site-directed mutagenesis was done on the template pET19-OL8 *E. coli* expression plasmid carrying the *O. longissima* wild-type apoobelin gene.[3] Mutations resulting in the amino acid change were carried out using QuickChange site-directed mutagenesis kit (Stratagene) according to the kit's protocol. The plasmids harboring mutations were verified by DNA sequencing. For protein production, the transformed *E. coli* BL21-Gold was cultivated with vigorous shaking at 37°C in LB medium containing ampicillin and induced with 1 mM IPTG when the culture reached an OD_{590} of 0.6-0.8. After addition of IPTG, the cultivation was continued for 3 h.

Protein purification. Obelin mutants were purified and charged with coelenterazine as previously reported for recombinant WT obelin.[4,5] At the last step of protein purification by ion-exchange chromatography on Mono Q column (Amersham Bioscience) obelin mutants were clearly separated from the uncharged proteins. The final mutant proteins were homogeneous according to SDS-PAGE.

Bioluminescence assay. Bioluminescence (6 sec interval) was measured using a Luminoskan Ascent (Thermo Electron, Finland) by injection of 50 μL 100 mM $CaCl_2$, 100 mM Tris-HCl, pH 8.8, into the well containing 100 μL of 5 mM EDTA, 100 mM Tris-HCl pH 8.8 and the photoprotein aliquot.

RESULTS AND DISCUSSION

The bioluminescent activities of mutant obelins with His175 substituted by Gln, Asn, Ala, Glu, Asp, Arg, Lys, Phe; Trp179 substituted by Ala, Tyr, Arg; and Tyr190 substituted by Arg, Lys, Phe, Glu are represented in Table 1. All mutants reveal low (or no) photoprotein bioluminescent activity except for W179Y and Y190F. These mutants retain 23% and 14% of WT obelin activity, respectively.

Table 1. Bioluminescence of obelin mutants relative to WT obelin.

Obelin mutant	Bioluminescent activity [a] (%)	Obelin mutant	Bioluminescent activity (%)
H175Q	0.80	W179A	0.02
H175N	1.70	W179Y	23.00
H175A	0.50	W179R	0.02
H175E	0.20		
H175D	0.01	Y190R	0.01
H175R	0.02	Y190K	0.03
H175Y	0.30	Y190F	14.30
H175F	0.30	Y190E	1.90

All mutants display a luciferase-like bioluminescence activity in their incubation with coelenterazine in Ca^{2+}-free conditions. However, when Ca^{2+} is injected into the solution containing photoprotein and coelenterazine, a flash-like bioluminescent activity of photoprotein was also observed (Fig. 2). It is important to notice that photoprotein activities of mutants were substantially higher than their luciferase-like activities. It can be assumed from this fact that substitution of one of these amino acid residues leads to the formation of unstable photoprotein complex which can be destroyed slowly through Ca^{2+}-independent bioluminescence.

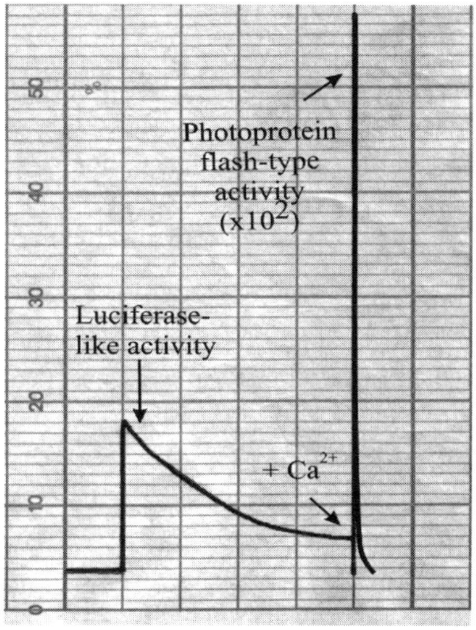

Fig. 2. Simultaneous luciferase-like and photoprotein bioluminescent activities of produced mutants by the example of Y190E

Substitutions of His175 to amino acids with different side chains lead to complete loss of the bioluminescent activity. His175 is apparently necessary for the stability of 2-hydroperocoelenterazine due to its H-bonds with coelenterazine and Tyr190 and may play a key role in the oxidative decarboxylation at the initial step of bioluminescent reaction. The supposition that a His175 residue is essential for bioluminescent activity is supported by the observation that substitution of His169 (corresponding to His175 in obelin) to Ala, Trp or Phe results in complete loss of the activity whereas mutants with modification of the remaining four His residues have varying bioluminescent activity.[6]

The obelin mutant with substitution of W179 to Tyr had good bioluminescent activity (23%) whereas W179A and W179R have low activities. Since the Tyr side chain is approximately the same size as Trp, Trp179 is inferred to take part in a spatial orientation of coelenterazine molecule in the binding site. A proton-relay hypothesis has been formulated to explain how Ca^{2+} binding triggers the bioluminescent reaction.[7] According to this hypothesis, H-bonds of Tyr190 to His175 and to 2-hydroperoxycoelenterazine are indispensable. The substitution of Tyr190 to non-H-binding amino acids should lead to disappearance of H-bonding and consequently make the proton transfer impossible. Hence it was surprising to find that Y190F mutant retains 14.3% of WT obelin bioluminescent activity. Some other amino acid residues in the active site make up the deficiency of this bond.

However this result may be not simply interpretable without knowledge of the spatial structure of this mutant.

CONCLUSIONS

The abolished (or noticeably decreased) bioluminescent activity of obelin mutants by modifying the His175, Trp 179, and Tyr 190 indicate that the function of these amino acid residues is to stabilize the 2-hydroperoxycoelenterazine intermediate rather than to participate in catalyzing the bioluminescent reaction because all mutants revealed luciferase-like bioluminescent activity. Further structural and spectroscopic studies of obelin mutants are planned to follow up this idea.

ACKNOWLEDGEMENTS

This work was supported by grant 05-04-48271 of the Russian Foundation for Basic Research, the "Molecular and Cellular Biology" program of the Russian Academy of Sciences, and by Lavrentiev grant for Young scientists of the Siberian Branch of the Russian Academy of Sciences.

REFERENCES

1. Liu ZJ, Vysotski ES, Chen CJ, Rose JP, Lee J, Wang BC. Structure of Ca^{2+}-regulated photoprotein obelin at 1.7: Å resolution determined directly from its sulfur substructure. Protein Sci 2000;9:2085-93.
2. Liu ZJ, Vysotski ES, Deng L, Lee J, Rose J, Wang BC. Atomic resolution structure of obelin: soaking with calcium enhances electron density of the second oxygen atom substituted at the C2-position of coelenterazine. Biochem Biophys Res Commun 2003;311:433-9.
3. Markova SV, Vysotski ES, Lee J. Obelin hyperexpression in *Escherichia coli*, purification and characterization. In: Case JF, Herring PJ, Robinson BH, Haddock SHD, Kricka LJ, Stanley PE. eds. Bioluminescence and Chemiluminescence. Singapore: World Scientific, 2001:115-8.
4. Illarionov BA, Frank LA, Illarionova VA, Bondar VS, Vysotski ES, Blinks JR. Recombinant obelin: cloning and expression of cDNA, purification and characterization as a calcium indicator. Methods Enzymol 2000;227:223-49.
5. Vysotski ES, Liu ZJ, Rose J, Wang BC, Lee J. Preparation and preliminary study of crystals of the recombinant calcium-regulated photoprotein obelin from the bioluminescent hydroid *Obelia longissima.* Acta Crystallogr, Sect D: Biol Crystallogr 1999;55:1965-6.
6. Ohmiya Y, Tsuji FI. Bioluminescence of the Ca^{2+} binding photoprotein, aequorin, after histidine modification. FEBS Lett 1993;320:267-70.
7. Vysotski ES, Lee J. Ca^{2+}-regulated photoproteins: Structural insight into the bioluminescence mechanism. Acc Chem Res 2004;37:405-15.

MECHANISTIC STUDY ON THE *CYPRIDINA (VARGULA)* BIOLUMINESCENCE REACTION

T HIRANO,[1] Y TAKAHASHI,[1] H KONDO,[1] H IKEDA,[2] S MAKI,[1] H NIWA[1]

[1] Dept of Applied Physics and Chemistry, The University of Electro-Communications, Chofu, Tokyo 182-8585, Japan; [2] Dept of Applied Chemistry, Graduate School of Engineering, Osaka Prefecture University, 1-1 Gakuencho, Nakaku, Sakai, Osaka 599-8531, Japan

INTRODUCTION

The luciferin–luciferase (L-L) reaction of the luminous ostracod *Cypridina* (now *Vargula*) produces blue light with a high quantum yield (Φ_{BL} ca. 0.3).[1-3] In the L-L reaction, Cypridina luciferin (LH) is oxidized by triplet molecular oxygen ($^3O_2$) to give the singlet-excited state of the neutral amide form of oxyluciferin and CO_2. The reaction consists of two important parts; (i) oxygenation to make a high-energy dioxetanone intermediate and (ii) chemiexcitation from the dioxetanone to the singlet excited oxyluciferin with a high efficiency (Φ_S). To obtain mechanistic information on the processes (i) and (ii), we have investigated the chemiluminescent properties of a series of substituted imidazopyrazinones (1H) as luciferin models.[4,5] We report here the characteristics of these elementary processes and the important roles of the indolyl group of LH in determining the bioluminescent property of the *Cypridina* bioluminescence system.

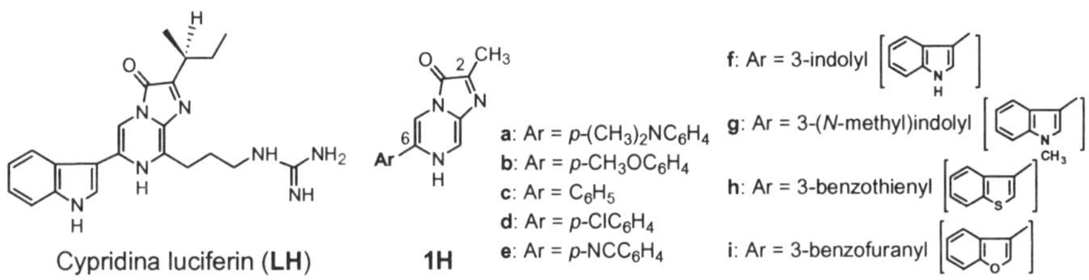

METHODS

Chemiluminescence reactions were monitored by detecting intensity of the total emitted light with a luminometer (Hamamatsu R5929 photomultiplier tube) at 25 ± 1 °C. Chemiluminescence spectra were recorded on an ATTO AB-1850 spectrometer. Chemiluminescence quantum yields (Φ_{CL}) were determined as quantum yields relative to the Φ_{CL} value (0.013) of luminol in DMSO containing *t*-BuOK/*t*-BuOH under air. The yields of amidopyrazine products were determined by GC analyses.

RESULTS AND DISCUSSION

Single electron transfer (SET)-oxygenation mechanism. The rate-determining step in the chemiluminescence reaction of an imidazopyrazinone is the reaction of the imidazopyrazinone anion with $^3O_2$.[4,6] To gain insight into this step, we investigated chemiluminescence reactions of 1H in CH_3CN and in DMSO. Results indicate that the reaction rate is controlled by the single electron transfer (SET) step from anion 1^- to $^3O_2$ and the successive radical coupling between $1^{\bullet}$ and $O_2^{\bullet -}$ gives peroxide 2^- (Scheme 1).[4] We found that the Hammett plot, $\log[k_{obsd}(1H)/k_{obsd}(1Hc)]$ versus σ_p, shows good linearity. The chemiluminescence of 1Ha–e in DMSO containing 1,1,3,3-tetramethylguanidine (TMG, used as a base) under air gave the plot, $\log[k_{obsd}(1H)/k_{obsd}(1Hc)] = -0.37\sigma_p - 0.03$ (r = 0.994). We extended the application of the linear Hammett plot to evaluate the electron donating ability of Ar groups at C6, to give the σ_{Ar} constant. For 1Ha–e we defined $\sigma_{Ar} = \sigma_p$. The σ_{Ar} values for 3-indolyl, 3-(*N*-methyl)indolyl, 3-benzothienyl, and 3-benzofuranyl groups in 1Hf–i were estimated as - 0.92, - 0.67, +0.29, and +0.45, respectively. These values indicate that 3-indolyl and 3-(*N*-methyl)indolyl groups have electron-donating abilities similar to that of the *p*-(dimethylamino)phenyl group, while 3-benzothienyl and 3-benzofuranyl groups have less electron-donating abilities similar to that of the *p*-chlorophenyl group. Therefore, the indolyl group in LH acts as an electron-donating group for accelerating the SET-oxygenation process.

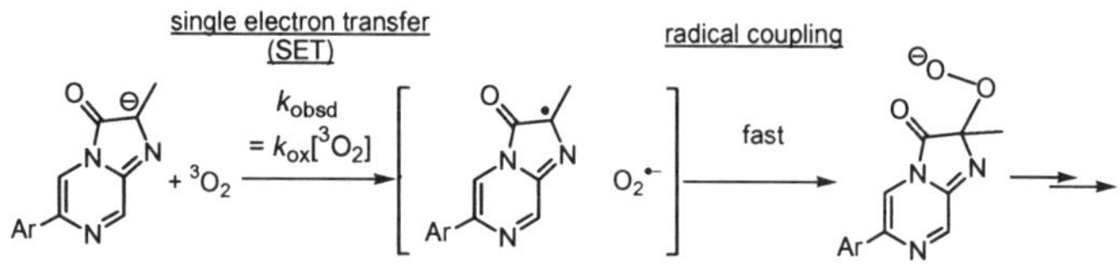

Scheme 1. SET-oxygenation process of imidazopyrazinone anion $\mathbf{1}^-$ with $^3O_2$

Factors for increasing Φ_S in the chemiexcitation process. The Φ_{CL} value of LH or a luciferin derivative depends on the reaction conditions. Especially, Goto *et al.* found that LH shows high Φ_{CL} (10% of Φ_{BL}) in diglyme containing an acetate buffer.[6] This finding led them to predict that the chemiluminescence reaction in diglyme/acetate buffer proceeds via the same pathway as that of the bioluminescence reaction. To clarify the reason for the high Φ_{BL}, we have investigated the chemiluminescent properties of 1H in DMSO/TMG and in diglyme/acetate buffer. Observed Φ_{CL} values are plotted against the σ_{Ar} constants for 1H (Fig. 1). The Φ_{CL} values in DMSO/TMG show the small change (ca. 10^{-3}) on varying the Ar group. On the other hand, the Φ_{CL} values in diglyme/acetate buffer for 1Ha, 1Hf, and 1Hg having electron-donating Ar groups ($\sigma_{Ar} < -0.6$) are high

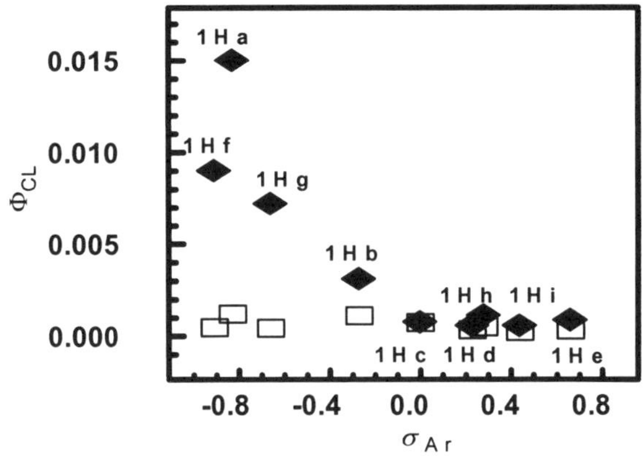

Fig. 1. Plots of the Φ_{CL} values in DMSO/TMG (□) and in diglyme/acetate buffer (♦) against the σ_{Ar} constants

(>0.007), while the Φ_{CL} values of the other derivatives (except 1Hb) are *ca.* 10^{-3} and the Φ_{CL} value of 1Hb falls between those of 1Ha and 1Hc. The Φ_{CL} value of the indolyl derivative 1Hf is similar to that of LH (ca. 10^{-2}). This indicates that the electron-donating indolyl group in LH is a contributor to the high Φ_{CL} value.

Chemiluminescence spectra of 1Ha, 1Hf, and 1Hg having electron-donating Ar groups in diglyme/acetate buffer indicate that light emission occurs from the singlet-excited state of neutral acetamidopyrazine ($^14H^*$) (Scheme 2). This result is reasonably explained by participation of a protonation of 3^- to give neutral dioxetanone 3H and of a chemiexcitation process from 3H to $^14H^*$ with a high Φ_S. Therefore, two factors are important in obtaining a high Φ_{CL}: the chemiexcitation process from 3H to $^14H^*$ and the electron-donating ability of the Ar group in 3H ($\sigma_{Ar} < -0.6$).

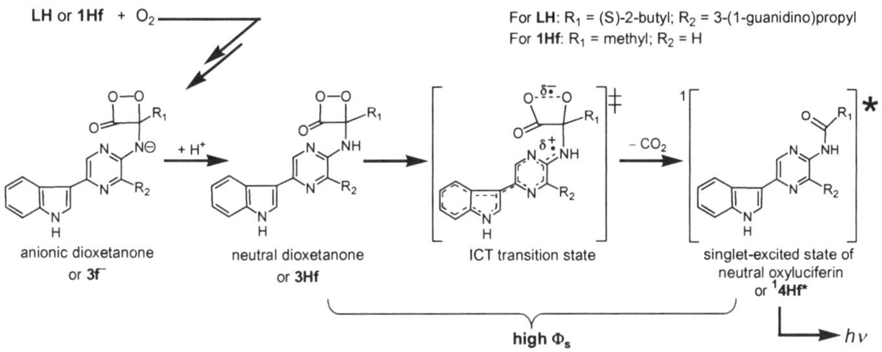

Scheme 2. Reaction mechanism for chemiluminescence of LH and 1Hf via the neutral dioxetanone intermediate

Nakamura *et al.* studied substituent effects of the Ar group on the bio- and chemiluminescent properties of LH using racemic LH and its derivatives.[7] The Φ_{BL} and Φ_{CL} ratios observed for X = NH, S, and O are 100:11:6 and 100:7:5, respectively. The Φ_{CL} ratio (100:6:12) observed for 1Hf, 1Hh, and 1Hi in diglyme/acetate buffer is also similar to the Φ_{BL} and Φ_{CL} ratios for racemic LH and its derivatives, indicating that Nakamura's findings are attributable to the electronic effects of the Ar group on the bioluminescent properties of LH. Therefore, we can apply the factors that contribute to high Φ_{CL} for the chemiluminescence of 1H to the Cypridina bioluminescence mechanism. The singlet-excited state of neutral oxyluciferin is generated with high Φ_S by the thermal decomposition of the neutral dioxetanone (Scheme 2), with the indolyl group acting as an electron-donating group. The decomposition of the neutral dioxetanone proceeds through the intramolecular charge transfer (ICT) transition state in the manner predicted theoretically[8] as the charge transfer–induced luminescence (CTIL) mechanism for high Φ_S.

REFERENCES

1. Goto T. Chemistry of bioluminescence. Pure Appl Chem 1968; 17: 421-41.
2. Johnson FH, Shimomura O. Introduction to the *Cypridina* system. Methods Enzymol 1978; 57: 331-64.
3. Tsuji FI. Cypridina luciferin and luciferase. Methods Enzymol 1978;57:364-72.
4. Kondo H, Igarashi T, Maki S, Niwa H, Ikeda H, Hirano T. Substituent effects on the kinetics for the chemiluminescence reaction of 6-arylimidazo[1,2-*a*] pyrazin-3(*7H*)-ones (Cypridina luciferin analogues): support for the SET-oxygenation mechanism with triplet molecular oxygen. Tetrahedron Lett 2005; 46: 7701-4.
5. Takahashi Y, Kondo H, Maki S, Niwa H, Ikeda H, Hirano T. Chemiluminescence of 6-aryl-2-methylimidazo[1,2-*a*]pyrazin-3(*7H*)-ones in DMSO/TMG and in diglyme/acetate buffer: support for the chemiexcitation process to generate the singlet-excited state of neutral oxyluciferin in high quantum yield in the *Cypridina* (*Vargula*) bioluminescence mechanism. Tetrahedron Lett 2006; 47: 6057-61.
6. Goto T, Inoue S, Sugiura S. Cypridina bioluminescence. IV. Synthesis and chemiluminescence of 3,7-dihydroimidazo[1,2-a]pyrazin-3-one and its 2-methyl derivative. Tetrahedron Lett 1968: 3873-6.
7. Nakamura H, Aizawa M, Takeuchi D, Murai A, Shimomura O. Convergent and short-step syntheses of dl-Cypridina luciferin and its analogues based on Pd-mediated cross couplings. Tetrahedron Lett 2000; 41: 2185-8.
8. Isobe H, Okamura M, Kuramitsu S, Yamaguchi K. Mechanistic insights in charge-transfer-induced luminescence of 1,2-dioxetanones with a substituent of low oxidation potential. J Am Chem Soc 2005;127:8667-79.

STABILIZATION OF *LUCIOLA MINGRELICA* FIREFLY LUCIFERASE BY MUTATION OF NON-CONSERVATIVE CYSTEINE RESIDUES

GY LOMAKINA, SA LEGOTSKY, VI TISHKOV, NN UGAROVA
Chemistry Faculty, Lomonossov Moscow State University, Moscow 119992, Russia,
E-mail: unn@enz.chem.msu.ru

INTRODUCTION

Cysteine residues in firefly luciferases do not belong to the active center;[1] however, they play an important role in stabilization of the active conformation of the enzyme that is proved by the stabilizing effect of dithiothreitol on the luciferase.[2] All known luciferases contain three conservative Cys residues (Cys 82, 260, and 393, numeration according to the *Luciola mingrelica* luciferase). The *Photinus pyralis* luciferase also contains only one non-conservative Cys residue (Cys 216), *Luciola* luciferases have from three to five, and beetle luciferases have 10 non-conservative Cys residues.[3] The higher the Cys residue content, the less stable the luciferase.

Substitution of the conservative Cys 82, 260, and 393 residues, which are located inside the protein globule, to Ala residue in the *L. mingrelica* luciferase had no effect on the kinetic characteristics of the enzyme. The Cys82Ala and Cys260Ala mutations did not change the stability of the luciferase, and thermostability of the luciferase carrying Cys393Ala mutation was somewhat higher than that of the wild-type recombinant enzyme.[2] According to the X-ray structural data, non-conservative Cys residues in the luciferases are located either on the surface or near the surface of the proteins, therefore, Cys residues are accessible for solvent molecules including dissolved oxidizers.[4] Thus, when selecting centers for mutations, preference was given to the surface-located Cys residues. The analysis of three-dimensional model of *L. mingrelica* firefly luciferase[5] indicated that Cys146 is located on the surface of the globule and Cys86 is in close vicinity to the surface. These particular residues were chosen by us for mutagenesis. In order to make a correct choice of replacing residue, we have studied the microenvironment of Cys86 and Cys146 residues in the *L. mingrelica* luciferase molecule within 5 Å radius of the sulfur atom using a Swiss PDB-Viewer 3 program.

An additional criterion for the selection of amino acids for the replacement of Cys residues was sequence homology of different luciferases (Table 1). By the comparison of amino acid sequences of luciferases isolated from different fireflies we have found out what amino acids are located in the positions of Cys residues in *L. mingrelica* luciferase. Instead of Cys, Ser is located in position 86 in luciferases without Cys (Table 1), therefore, it was decided to replace Cys86 for Ser especially since Cys86 in *L. mingrelica* luciferase has a hydrophilic microenvironment. The majority of luciferases (except *L. cruciata* luciferase) contain hydrophobic residues Ile or Val in position 146. However, the analysis of three-dimensional model of the enzyme shows that Cys146 is located on the surface of the molecule in hydrophilic microenvironment. Therefore, it was decided to obtain two mutant forms of *L. mingrelica* luciferase for position 146: Cys146Ser and Cys146Ile.

Table 1. Amino acid residues in the vicinity of Cys86 and Cys146 residues (numeration according to *L. mingrelica* luciferase) for different firefly luciferases

Luciferase	Residue number	
	80 – 88	143 – 149
L. mingrelica	ALCSENCEE	TVTCIKK
P. pyralis	VVCSENSLQ	KLPIIQK
L. lateralis	ALCSENCEE	TVTAIKT
L. cruciata	ALCSENCEE	TVTTIKT
Hotaria unmunsana	ALCSENCEE	TVTCIKT
H. papariensis	ALCSENCEE	TVTCIKT
H. tsushimana	ALCSENCEE	TVTCIKT
H. parvula	ALCSENCEE	TVTCIKT
L. noctiluca	AVCSENSLQ	-LPIIQK
P. rufa	AVCSENSLQ	KLPVIQK
Cylindroiulus distinctus	AVCSENSLQ	SLPVIKK
P. pennsylvanica	SVCSENCVQ	RLPIIKK

METHODS

Site-specific mutagenesis was carried out on the basis of pLR plasmid carrying *L. mingrelica* luciferase gene using polymerase chain reaction method. Mutant forms with Cys86Ser, Cys146Ile, and Cys146Ser replacements were purified by the method[2] to the purity not less than 95% according to SDS electrophoresis data.

RESULTS AND DISCUSSION

Site-specific mutagenesis was used to obtain mutant forms of *L. mingrelica* luciferase with Cys86Ser, Cys146Ile, and Cys146Ser replacements. It was shown that these mutations did not alter catalytic properties of the luciferase ($K_{m,\ LH_2}$, $K_{m,ATP}$, V). Kinetics curves of thermoinactivation of wild-type luciferase and its mutant forms were obtained in 0.05 mmol/L Tris-acetate buffer solution, pH 7.8, containing 10 mmol/L $MgSO_4$, 2 mmol/L EDTA, at 37°C concentration of the enzyme was 10^{-8} mol/L. As can be seen from Fig. 1, the change of Cys86 residue, located in surface cavity, to Ser had no significant effect on thermostability of the luciferase but the substitution of the surface Cys146 residue by Ile and Ser increased thermostability of the enzyme 2 and 3 times, respectively. Hence, introduction of the hydrophilic Ser residue into hydrophilic microenvironment resulted in higher stabilizing effect than the introduction of hydrophobic Ile residue. An important characteristic of firefly luciferase is stability towards organic solvents. In this work we have studied the stability of the enzyme towards dimethylsulfoxide (DMSO) that is used in ATP-metry for the release of intracellular ATP. The dependence of the luciferase activity on DMSO concentration was obtained for the wild-type enzyme and its three mutant forms. The dependencies shown in Fig. 2 indicate that the Cys86Ser and Cys146Ile mutations did not influence the luciferase activity in the presence of DMSO and the activity of the mutant form with Cys146Ser replacement was 2.5 times as high as the activity of the wild-type enzyme.

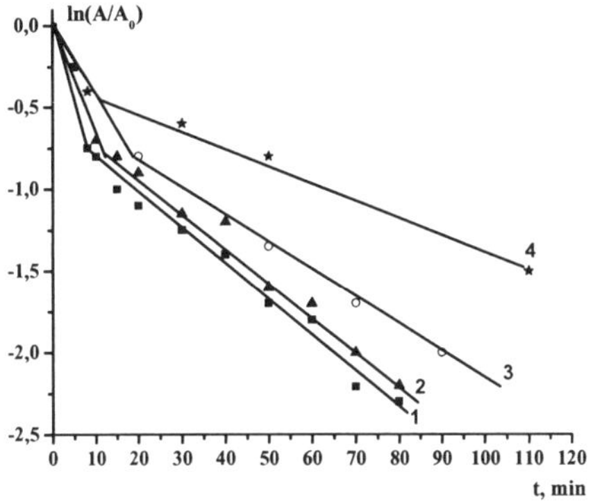

Fig. 1. Time-courses of thermoinactivation of the wild-type (1) and mutant *L. mingrelica* luciferases with the Cys86Ser (2), Cys146Ile (3), Cys146Ser (4) replacements, 37°C

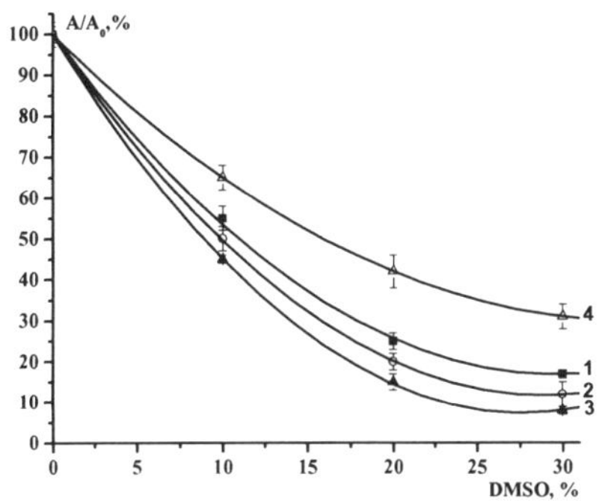

Fig. 2. Activity of the wild-type (1) and mutant *L. mingrelica* luciferases with the Cys86Ser (2), Cys146Ile (3), and Cys146Ser (4) replacements as a function of DMSO concentration

ATP reagents for the assay of microamounts of ATP were obtained on the basis of the wild-type recombinant luciferase and its mutant forms. Calibration curves, i.e., dependencies of bioluminescence intensity on ATP concentration for ATP solution in water and in DMSO (Fig. 3 A, B), are straight lines within ATP concentration range 10^{-9} – 10^{-12} mol/L. The specific activity of the ATP reagents on the basis of the mutant forms was higher than that of the reagent with wild-type luciferase.

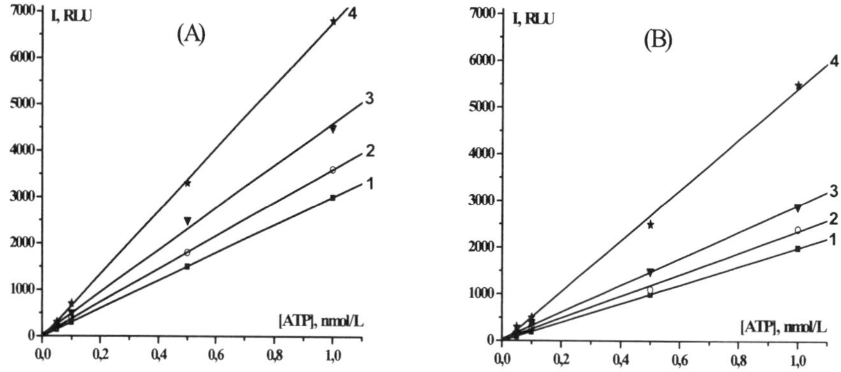

Fig.3. Calibration curves for the assay of ATP in water solution (A) and in DMSO (B) for the ATP-reagents containing the wild-type (1) or mutant *L. mingrelica* luciferases with Cys86Ser (2), Cys146Ile (3), Cys146Ser (4) replacements

For the ATP solution in water, the sensitivities of the ATP-reagents (calibration curve slopes) on the basis of Cys86Ser and Cys146Ile mutant forms were 1.2 and 1.5 times higher than the sensitivity of the standard reagent, respectively. Sensitivity from the reagent prepared from the Cys146Ser mutant form was 2.5 times higher. For the ATP solution in DMSO, the sensitivities for the ATP assay of the ATP-reagents prepared from the wild-type luciferase and Cys86Ser and Cys146Ile mutant forms were close, but the sensitivity of the ATP-reagent on the basis of Cys146Ser mutant form was 2.5 times higher. The last reagent retained 80% of the luciferase activity, whereas the activity of other ATP-reagents was less than 65%.

ACKNOWLEDGEMENTS

This work was supported by a State Contract (No. 02457.11.7002) (Ministry of Science and Education, Russion Federation).

REFERENCES

1. Alter S, DeLuca M. The sulfhydryls of firefly luciferase are not essential for activity. Biochemistry 1986; 25:1599-605.
2. Dementieva E, Zheleznova E, Kutuzova G, Lundovskikh I, Ugarova N. Physicochemical properties of recombinant *Luciola mingrelica* luciferase and its mutant forms. Biochemistry (Moscow) 1996;61:115-9.
3. Ugarova N, Brovko L. Protein structure and bioluminescent spectra for firefly bioluminescence. Luminescence 2002;17:321-30.
4. Conti E, Franks N, Brick P. Crystal structure of firefly luciferase throws light on a superfamily of adenylate-forming enzymes. Structure 1996;4:287-98.
5. Sandalova T, Ugarova N. Model of the active site of firefly luciferase. Biochemistry (Moscow) 1999;64:962-7.

FIREFLY LUCIFERASE AROSE FROM FATTY ACYL-COA SYNTHETASE BY GENE DUPLICATION AND GENE RECRUITMENT

YUICHI OBA,[1] SATOSHI INOUYE[2]

[1] Graduate School of Bioagricultural Sciences, Nagoya University, Nagoya 464-8601, Japan; [2] Yokohama Research Center, Chisso Co., 5-1 Okawa, Kanazawa-ku, Yokohama 236-8605, Japan

INTRODUCTION

In the firefly, light emission is produced by the reaction of D-luciferin, O_2 and firefly luciferase [EC 1.13.12.7] in the presence of ATP and Mg^{2+}. Previously, we found that firefly luciferase has catalytic activity of long-chain fatty acyl-CoA synthetase [EC 6.2.1.3].[1] We also found that a homologous gene of firefly luciferase in *Drosophila melanogaster*, CG6178, is a long-chain fatty acyl-CoA synthetase but not luciferase.[2,3] Recently, we cloned and characterized paralogous genes of firefly luciferase in the Japanese firefly *Luciola cruciata*[4] and orthologous genes of firefly luciferase in non-luminous beetle *Tenebrio molitor*.[5] In this study, we describe an evolutionary process of firefly luciferase based on our recent results.

MATERIALS AND METHODS

Homologous genes of firefly luciferase in *L. cruciata* and *T. molitor* were cloned using degenerate PCR and RACE techniques.[4,5] The histidine-tagged recombinant proteins expressed in *E. coli* were purified by Ni-chelate affinity chromatography.[4,5] Adenylation activity of proteins was determined using [α-^{32}P]ATP.[1-5] After incubation with various carboxylic acids, ^{32}P-AMP produced was separated by TLC and radioactivity was measured using an imaging analyzer. Alignment of amino acid sequences were performed using a MAFFT program (L-INS-i), and a phylogenetic tree was reconstructed using a PAUP program. All sites containing a gap in sequences were excluded and 371 amino acid residues were analysed.

RESULTS

Firefly luciferase as a fatty acyl-CoA synthetase. Substrate specificity for an acyl-CoA synthetase was examined using luciferases from *Photinus pyralis* and *L. cruciata*. These luciferases showed significant activity with medium-chain fatty acids (C10:0 ~ C14:0), palmitolate (C18:1) and EPA (C20:5). For long-chain fatty acids (C16:0 ~ C20:4), acyl-CoA synthetic activity was comparably weak (Table 1). The most suitable substrate was laurate (C12:0) and firefly luciferases have a *kcat*/*Km* ratio of 0.04 $\mu M^{-1}\cdot s^{-1}$ with laurate.[1,3]

CG6178 in *Drosophila melanogaster*. *CG6178* in *D. melanogaster* is the most similar gene to firefly luciferase gene. Recombinant CG6178 showed a broad spectrum of substrate specificity as an acyl-CoA synthetase with various medium- and long-chain fatty acids (Table 1).

Table 1. Luciferase and other adenylate-forming enzymes

Species	Gene (Genbank accession no.)	Number of AA (%)a	PTS1 signal (C-terminus)b	Enzyme activity FACS c	Enzyme activity Lumind
Photinus pyralis	Luciferase (M15077)	550 (68)	+ (-SKL)	+1,3	+
Luciola cruciata	Luciferase (AB220162)	548 (100)	+ (-AKM)	+1,3	+
	LcLL1 (AB196455)	536 (41)	+ (-AKL)	+4	–4
	LcLL2 (AB196456)	545 (46)	+ (-SKL)	–4	–4
Tenebrio molitor	*TmLL-1* (AB098615)	544 (45)	+ (-SKL)	+5	–5
	TmLL-2 (AB098616)	545 (45)	+ (-SKL)	+5	–5
	TmLL-3 (AB098617)	526 (38)	–	+5	–5
Drosophila melanogaster	*CG6178* (NM_142964)	544 (40)	+ (-SKL)	+2,3	–2
Arabidopsis thaliana	*At4CL2* (AF106085)e	556 (33)	–	–6	?
	At4g05160 (AY250839)f	544 (33)	+ (-SKM)	+6	?
	At5g63380 (AY250835)f	562 (34)	+ (-SKL)	+6	?
Pseudomonas sp.	Acyl-CoA synthetase	?	?	+1	–1

[a] Parentheses indicate amino acid identity (%) with *L. cruciata* luciferase. [b] Peroxisome targeting signal 1 was searched with the PTS1predictor (http://mendel.imp.ac.at/mendeljsp/index.jsp). [c] Fatty acyl-CoA synthetic (FACS) activity. [d] Luminescence activity with D-luciferin. [e] 4-coumarate:CoA ligase in A. thaliana. [f] Acyl-CoA synthetase of jasmonic acid precursor in *A. thaliana*.

The most suitable substrate was laurate (C12:0) and CG6178 have a *kcat/Km* ratio of 0.24 $\mu M^{-1} \cdot s^{-1}$ with laurate.[2,3]

LcLL1 and LcLL2 in *Luciola cruciata*. Two paralogous genes of firefly luciferase (*LcLL1* and *LcLL2*) in *L. cruciata* were cloned. The recombinant protein of LcLL1 possessed fatty acyl-CoA synthetic but not luminescence activity. Neither fatty acyl-CoA synthetic activity nor luminescence activity was detected in LcLL2 (Table 1).[4]

TmLL-1, TmLL-2 and TmLL-3 in *Tenebrio molitor*. The mealworm *T. molitor* is a non-luminous beetle. Three homologous genes of firefly luciferase (*TmLL-1*, *TmLL-2* and *TmLL-3*) were isolated. The recombinant proteins showed fatty acyl-CoA synthetic activity with laurate and α-linolenate (C18:3), but luminescence activity with D-luciferin was not observed (Table 1).[5]

DISCUSSION

From phylogenetic analysis of adenylate-forming enzymes, homologous genes of firefly luciferase in insects (*CG6178, LcLL1, LcLL2, TmLL-1, TmLL-2 and TmLL-3*) form a paraphyletic group at the root of the beetle luciferase clade. Together with the results that LcLL1 and TmLL-1~3 possess fatty acyl-CoA synthetic activities, we suggest that beetle luciferase arose from fatty acyl-CoA sythetase localized in peroxisome.[2,3] Recently, Schneider *et al.*[6] reported that peroxisomal acyl-CoA synthetase of jasmonic acid precursors (At4g05160 and At5g63380) have significant activity with medium- and long-chain fatty acids.

They proposed that plant 4-coumarate:CoA ligases also arose from fatty acyl-CoA synthetase (Table 1). We found that a chimeric protein of N-terminal domain of *P. pyralis* luciferase with C-terminal domain of CG6178 retained luminescence activity, suggesting that the N-terminal domain of luciferase was diverged functionally.[7]

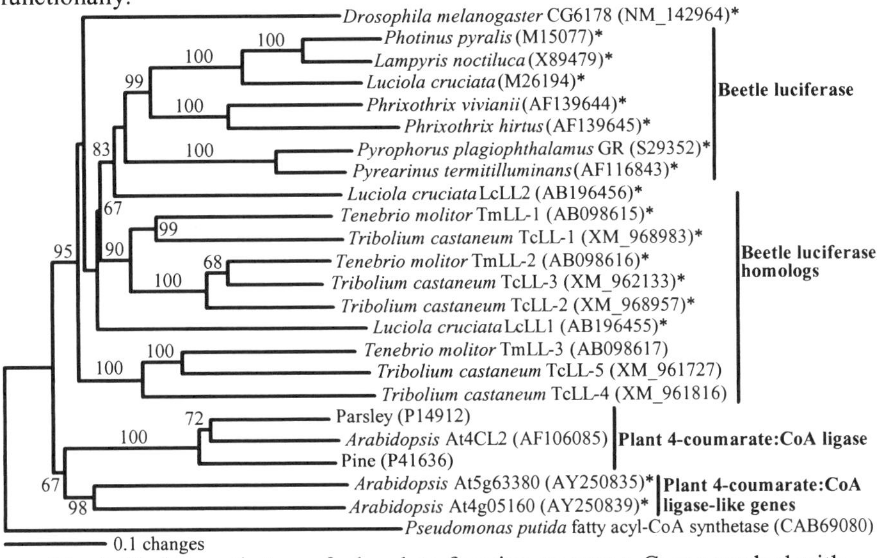

Fig. 1. Phylogenetic tree of adenylate-forming enzymes. Genes marked with asterisks have a peroxisome targeting signal at the C-terminus. Numbers on nodes indicate % bootstrap values from 10,000 replicates. Genbank accessation number is in parentheses. Bootstrap values over 60% are indicated

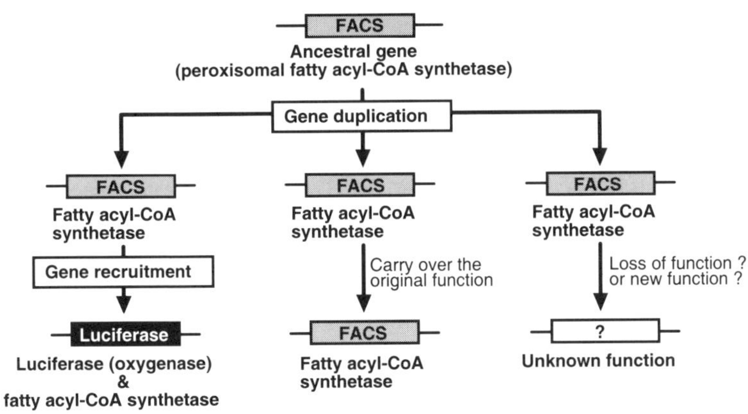

Fig. 2. Schematic representation of evolutionary process of firefly luciferase

Gene duplication and gene recruitment (or gene sharing, in which single protein shares both original and derived functions simultaneously) are major factors for enzyme diversification.[8] Our results of which *T. molitor* and *L. cruciata* have at least 3 homologs indicated the occurrence of several gene duplications on peroxisomal fatty acyl-CoA synthetase during the evolution of beetles. Functional duality of firefly luciferase[1] suggested that gene recruitment occurred in the course of luciferase evolution. In conclusion, we proposed that firefly luciferase originates from a peroxisomal fatty acyl-CoA synthetase in an ancestral insect by enzyme diversification with gene duplication and gene recruitment (Fig. 2).

REFERENCES

1. Oba Y, Ojika M, Inouye S. Firefly luciferase is a bifunctional enzyme: ATP-dependent monooxygenase and a long chain fatty acyl-CoA synthetase. FEBS Lett 2003;540:251-4.
2. Oba Y, Ojika M, Inouye S. Characterization of *CG6178* gene product with high sequence similarity to firefly luciferase in *Drosophila melanogaster*. Gene 2004;329:137-45.
3. Oba Y, Sato M, Ojika M, Inouye S. Enzymatic and genetic characterization of firefly luciferase and *Drosophila* CG6178 as a fatty acyl-CoA synthetase. Biosci Biotechnol Biochem 2005;69:819-28.
4. Oba Y, Sato M, Ohta Y, Inouye S. Identification of paralogous genes of firefly luciferase in the Japanese firefly, *Luciola cruciata*. Gene 2006;368:53-60.
5. Oba Y, Sato M, Inouye S. Cloning and characterization of the homologous genes of firefly luciferase in the mealworm beetle, *Tenebrio molitor*. Insect Mol Biol 2006;15:293-9.
6. Schneider K, Kienow L, Schmelzer E, Colby T, Bartsch M, Miersch O, et al. A new type of peroxisomal acyl-coenzyme A synthetase from *Arabidopsis thaliana* has the catalytic capacity to activate biosynthetic precursors of jasmonic acid. J Biol Chem 2005;280:13962-72.
7. Oba Y, Tanaka K, Inouye S. Catalytic properties of domain-exchanged chimeric proteins between firefly luciferase and *Drosophila* fatty acyl-CoA synthetase CG6178. Biosci Biotech Biochem 2006 (in press).
8. Todd AE, Orengo CA, Thornton JM. Evolution of protein function, from a structural perspective. Curr Opin Chem Biol 1999;3:548-56.

BIOLUMINESCENCE IN CLICK BEETLES (FAMILY ELATERIDAE): MOLECULAR PHYLOGENETIC ANALYSIS, INFERRED FROM 28S RIBOSOMAL DNA

REIKO SAGEGAMI-OBA,[1] YUICHI OBA,[1] HITOO ÔHIRA[2]
[1] *Graduate School of Bioagricultural Sciences, Nagoya University, Chikusa-ku, Nagoya 464-8601, Japan;* [2] *Kitsuneyama, Maigi-cho, Okazaki 444-3511, Japan*

INTRODUCTION

Luminous species have been found in six families of the superfamily Elateroidea: Lampyridae, Phengodidae, Rhagophthalmidae, Omalisidae, Elateridae and Throscidae. D-luciferin is identical between luminous species of Lampyridae, Phengodidae, and Elateridae. Cloning studies revealed that luciferases in lampyrids, phengodids, and elaterids have significant amino acid sequence identity.[1] This suggests that bioluminescence mechanisms of these luminous beetles are essentially identical, and that the common ancestor of the present luminous beetles was hypothesized to be luminous. All known species of Lampyridae, Phengodidae, and Rhagophthalmidae are luminous,[2] but click beetles contain luminous and non-luminous species. Thus, phylogenetic studies should provide insights into the evolution of beetle bioluminescence. We examined the evolutionary relationships between species of Elateridae based on partial sequences of nuclear 28S ribosomal DNA (28S rDNA).

MATERIALS AND METHODS

Taxa and DNA isolation. We analyzed 77 taxa of Elateridae, of which 68 species occur in Japan. Four luminous species were collected in Central and South America. The taxa consist of nine subfamilies and 54 genera.[3] To prevent the degradation of genomic DNA, collected specimens were immediately killed by immersion in 99.5% ethanol and stored at 4°C until use. Total DNA was extracted from the legs of a single specimen using a DNeasy tissue kit (Qiagen) or a QIAamp DNA mini kit (Qiagen).

PCR amplification and DNA sequencing. An approximately 880-bp sequence of the 28S rDNA was amplified by PCR using a GeneAmp 9700 thermal cycler (Applied Biosystems) with following primers: 28S-01, 5'-GAC TAC CCC CTG AAT TTA AGC AT-3'; and 28SR-01, 5'-GAC TCC TTG GTC CGT GTT TCA AG-3'.[4] In some cases, an approximately 810-bp sequence was amplified using the primers 28S-F1 (5'-CGA ACA GGG AAD AGC CCA GC-3') and 28SR-01. The reaction was carried out in a volume of 21 μL, containing 0.38 μM of each primer, 0.19 mM of each dNTP and 0.5 units of Ex Taq polymerase in Ex Taq buffer (Takara). The amplification was: 94°C for 1 min, followed by 35 cycles at 94°C for 1 min, 50°C for 1 min, and 70°C for 2 min, and a final extension step of 1 min at 70°C.

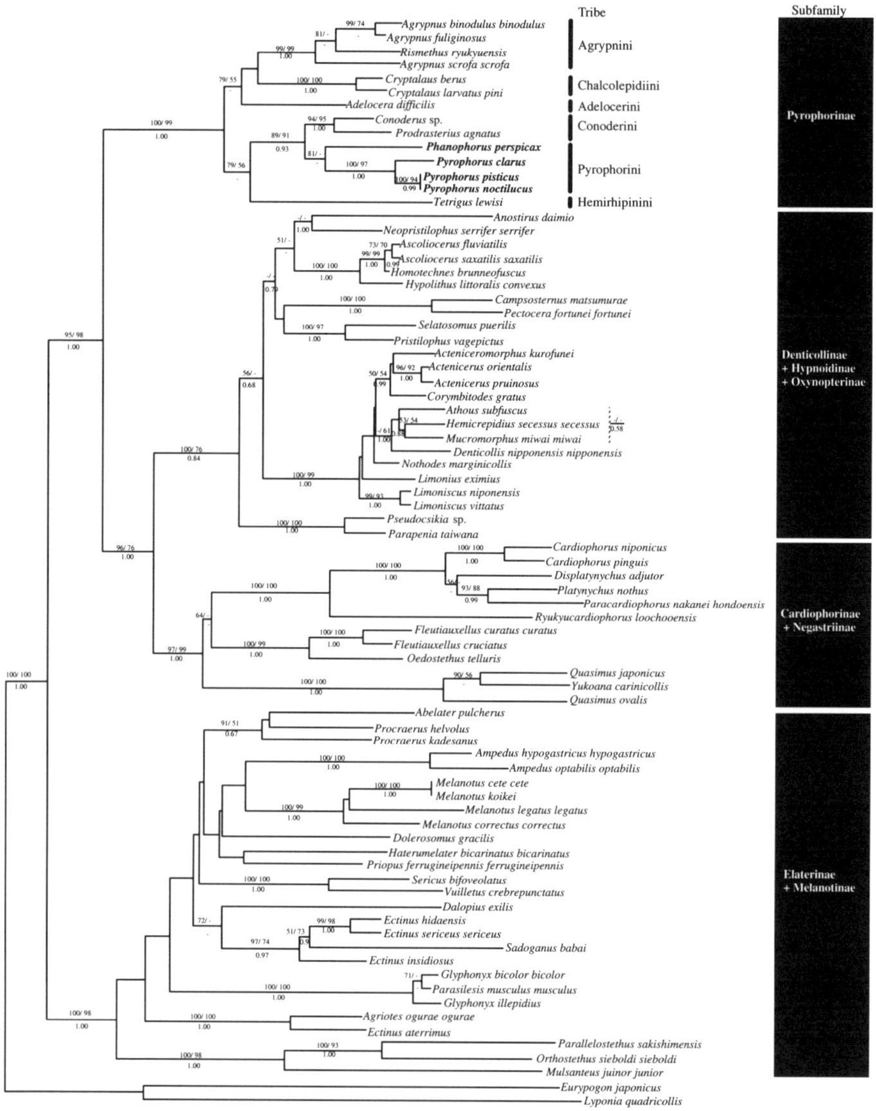

Fig. 1. Neighbor-joining phylogenetic tree of Elateridae resulting from the analysis of the 28S DNA data set. Neighbor-joining (left) and maximum-parsimony (right) bootstrap values >50% are shown above branches. Bayesian posterior probabilities are shown below branches. There were 0.01 substitutions per site

The amplicons were directly sequenced or ligated into the pGEM-T Easy vector (Promega) and sequenced using a BigDye Terminator kit (Applied Biosystems) and an ABI PRISM 3100 genetic analyzer.

Phylogenetic analysis. 28S rDNA sequences were aligned by ClustalX-1.83.1[5] or MAFFT version 5.734.[6] PAUP*4.0 beta 10[7] or MrBayes-3.1.2[8] was used to conduct phylogenetic analyses. For neighbor-joining, maximum-parsimony and maximum-likelihood, trees were searched by PAUP* program as described previously.[9] For Bayesian analysis, each run was performed using default staring parameters and comprised 4,000,000 generations. Bayesian posterior probabilities were calculated from majority-rule consensus of trees sampled every 100 generations once the Markov chain reached stationary (determined by empirical checking of likelihood values) using MrBayes.

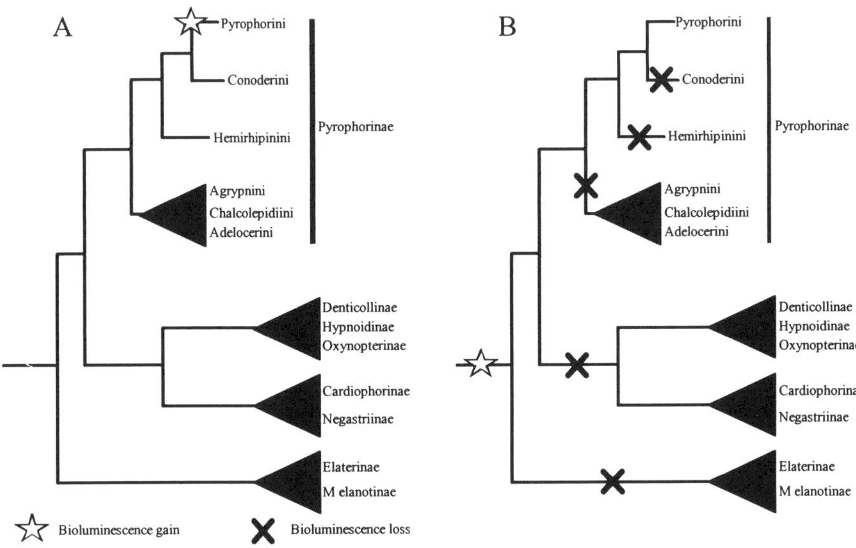

Fig. 2. Character mapping of bioluminescence types on Elateridae phylogeny. Most parsimonious reconstruction of the states "luminous" and "non-luminous" places the ancestral state as non-luminous (A). The luminous state (B) requires five or more steps of bioluminescence loss

RESULTS AND DISCUSSION

In our phylogenetic analyses of Elateridae with 14 outgroup taxa of Elateroidea and one carabid based on 28S rDNA, all 77 elaterids grouped into a single clade with high statistical support,[9] which suggests that Elateridea is monophyletic. All luminous click beetles belong to tribes Pyrophorini and Campyloxenini of subfamily

Pyrophorinae.[10] The four luminous species analyzed in the present study (genera *Pyrophorus* and *Phanophorus*, which belong to the tribe Pyrophorini) form a clade and placed at a derived position of elaterid phylogeny (Fig. 1). The most parsimonious reconstruction of the "luminous" and "non-luminous" states, using the molecular phylogenetic results, places the ancestral state for elaterids as non-luminous (Fig. 2). This suggests that the ancestral condition of the Elateridae was non-luminous and that bioluminescence arose once within a subgroup of the Pyrophorinae. Thus, the bioluminescence of the click beetle should be expected to have evolved independent of that in other luminous beetles, such as Lampyridae and Phengodidae. If this hypothesis is correct, the obvious question is why such a high similarity exists between the luciferases of Elateridae and the other luminous beetles. We suspect that conducting molecular phylogenetic analysis among the six families of luminous beetles will help answer this question.

REFERENCES

1. Wood KV. The chemical mechanism and evolutionary development of beetle bioluminescence. Photochem Photobiol 1995;62:662-73.
2. Branham MA, Wenzel JW. The origin of photic behavior and the evolution of sexual communication in fireflies (Coleoptera: Lampyridae). Cladistics 2003; 19:1–22.
3. Ôhira H.The systematics of subfamilies in Elateridae from Japan. Booklet for a lecture in the 17th annual meeting of the Japanese Society of Coleopterology 1999:4 p.
4. Kim CG, Zhou HZ, Imura Y, Tominaga O, Su ZH, Osawa S. Pattern of morphological diversification in the *Leptocarabus* ground beetles (Coleoptera: Carabidae) as deduced from mitochondrial ND5 gene and nuclear 28S rDNA sequences. Mol Biol Evol 2000;17:137–45.
5. Thompson JD, Gibson TJ, Plewniak F, Jeanmougin F, Higgins DG. The CLUSTAL X windows interface: flexible strategies for multiple sequence alignment aided by quality analysis tools. Nucleic Acids Res 1997;25:4876-82.
6. Katoh K, Misawa K, Kuma K, Miyata T. MAFFT: a novel method for rapid multiple sequence alignment based on fast fourier transform. Nucleic Acids Res 2002;30:3059-66.
7. Ronquist F, Huelsenbeck JP. MRBAYES 3: Bayesian phylogenetic inference under mixed models. Bioinformatics 2003;19:1572-4.
8. Sagegami-Oba R, Oba Y, Ôhira H. Phylogenetic relationships of click beetle (Coleoptera: Elateridae) inferred from 28S ribosomal DNA: Insights into the evolution of bioluminescence in Elateridae. Mol Phylogenet Evol 2006 (in press).
9. Stibick JNL. Classification of the Elateridae (Coleoptera). Relationships and classification of the subfamilies and tribes. Pacific Insects 1979;20:145-86.

EFFECT OF MICROENVIRONMENTAL CHANGES ON KINETIC PARAMETERS OF STEADY-STATE ENZYME-INDUCED BACTERIAL BIOLUMINESCENT REACTION

IE SUKOVATAYA,[1] NA TYULKOVA,[2] EV KAYKOVA[1]

[1]*Dept. of Biophysics Krasnoyarsk State University, 79 Svobodny pr., Krasnoyarsk, 660041, Russia Email: irinas@lan.krasu.ru;*

[2] *Institute of Biophysics, Russian Academy of Science, Siberian Branch, Akademgorodok, Krasnoyarsk, 660036, Russia, Email: biotech@ibp.ru*

INTRODUCTION

Bacterial luciferase is a flavin monooxygenase that catalyzes the oxidation of reduced flavin mononucleotide ($FMNH_2$) and a long chain aldehyde by molecular oxygen to give the corresponding acid, FMN, and blue-green light.[1] These luciferases are members of a large family of $(a/ß)_8$ barrel proteins.[2,3] Its recently determined 3D structure suggests active site resides within a large internal cavity of the a-subunit. It is sufficiently large enough to accommodate $FMNH_2$, O_2, and a aldehyde.[2,4] Amino acid sequence alignment between the two subunits reveals that they share 32% sequence identity. It has also been found that some aliphatic long chain compounds such as ketones, azides, alcohols,[5] as well as different organic solvents[6,7] are inhibitors of luciferase from *Vibrio harveyi*. It is believed that the affinity of aldehydes to luciferases is stipulated by hydrophobic connections.[8] Nevertheless, there is little information available on the nature of bacterial luciferases binding sites. To get more information about the catalytic properties of luciferases isolated from *Photobacterium leiognathi*, we compared the effect of organic solvents on maximum light intensity. Moreover, an important advantage of carrying out the bioluminescence reaction in organic media is the increased solubility of hydrophobic substrates. Besides, understanding the structure-activity relationships of the luciferase under different conditions is of fundamental importance for both theoretical and applicative purposes. In this work, we have studied the effects of increasing the concentrations of several water-miscible organic solvents on kinetic parameters of steady-state enzyme-induced bacterial bioluminescent reaction of luciferase from *P. leiognathi in vitro* with the chemical reduced $FMNH_2$.

METHODS

The luciferase from *P. leiognathi* (strain 208) was highly purified by ion-exchange chromatography.[9] To measure the control, a reaction was carried out in a mixture of the following composition: 10 μL of $(0.07–0.13)\times10^{-5}$ mol/L luciferase, 50 μL of 9.4×10^{-4} mol/L aqueous solution of tetradecanal - A_{14} (Merck, Germany), 840 μL of 2×10^{-2} mol/L phosphate buffer, 0.5 mL of 7.2×10^{-4} mol/L aqueous solution of $FMNH_2$ (Sigma, USA), 0.5 mL of 0.1 mol/L NADH, pH 7. The reaction was initiated by the injection of $FMNH_2$. In this case the light is emitted as a glow with a slow decay. Measurements were carried out with a bioluminometer designed at the Institute of Biophysics (Russian Academy of Sciences, Siberian Branch) at a temperature of 25°C. Reaction parameters were recorded with a 2210 ("LKB-Wallace", Finland) recorder In experiments, the phosphate buffer was substituted for a water-organic mixture. Ethanol, dimethyl sulfoxide (DMSO) and formamide (Sigma, USA) were used as organic solvents. Concentrations of solvents were expressed in volume %. The apparent value of the Michaelis constant (K_m) at various concentrations of organic solvents was determined by traditional initial velocity (I_{max}) methods: Lineweaver-Burk, Hanes and Eadie-Hofstee.[10,11] Concentrations of A_{14} in the reaction mixture were varied (A_{14} = (0,47–47)·10^{-7}M) for each chosen concentration of organic solvents. The experimental results obtained have been statistically processed by Excel for Windows 98.

RESULTS

The effect of ethanol, DMSO and formamide on the kinetic parameters of steady-state enzyme-induced bacterial bioluminescent reaction were studied. Addition of organic solvents into the reaction medium of the bioluminescent reaction changes the kinetic parameters of the light flash. The activity of luciferase dropped sharply with increasing concentration of solvent. The degree of reaction rate inhibition depends on the solvent kinds and their concentrations. In particular, the 50% and complete loss of luciferases activity follows the order: formamide > DMSO > ethanol.

The next step was to define the types of enzyme inhibition of luciferases by different concentrations of these solvents. The inhibition of the luciferase by all solvents tested was not competitive with respect to the substrate A_{14}. It was found that formamide and ethanol behaved like a non-competitive inhibitor. In other words, the substrate and the inhibitor bind with the enzyme active site independently from each other and this increases the reaction rate.[10,11] The inhibition effect of DMSO on the luciferase appears to be a non-competitive type.

Secondary plots from standard graphs (1/[S], 1/V) or (V, V/[S]) or Dixon plots (1/V, [I]) ([S], [I] substrate and inhibitor concentrations, V – catalytic rate) are usually used to investigate reaction kinetics. This graph is linear only if the enzyme has one binding site for the inhibitor.[11] The analysis of the data obtained in these studies shows that formamide partial by non-competitively inhibits the enzyme activity of the luciferase from *P. leiognathi* (Fig. 1). It seems that, in this case, the partial inhibition effect of formamide can be represented by a parabolic function, hence two molecules of the formamide are involved in the luciferase inhibition.

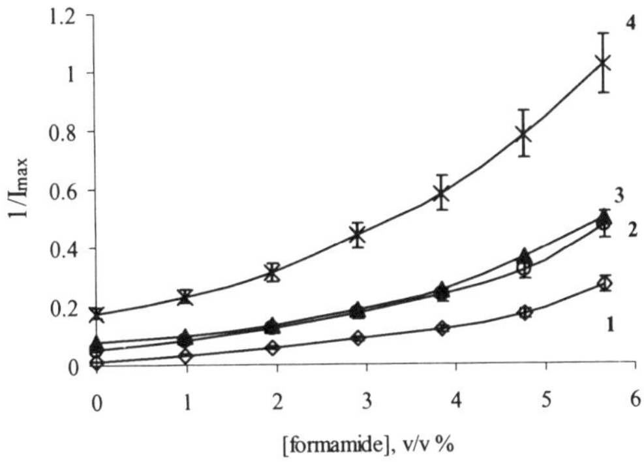

Fig. 1. Dixon plot of for partial non-competitive inhibition of the bioluminescence reaction, catalyzed by the luciferase from *P. leiognathi* for respective to tetradecanal in water- methanol mixtures: 1 – $4.7 \cdot 10^{-6}$ M; 2 – $0.47 \cdot 10^{-6}$ M, 3 – $23.5 \cdot 10^{-7}$ M; 4 – $11.8 \cdot 10^{-7}$ M tetradecanal

The addition of the solvents reduces hydrophobic interactions and increases electrostatic intermolecular interactions. The modifications to K_m reflect the nature of binding of an enzyme with substrates, from the represented data it is possible to draw a conclusion about the primary contribution of these (hydrophobic) and other (electrostatic) forces to the formation of an enzyme-substrate complex.[12-14] Obviously, for the specific binding A_{14} with luciferase where K_m increases with increase of DMSO or ethanol concentration hydrophobic interactions predominate, but for the specific binding in the presence of formamide with luciferase, electrostatic interactions predominate (Fig. 1).

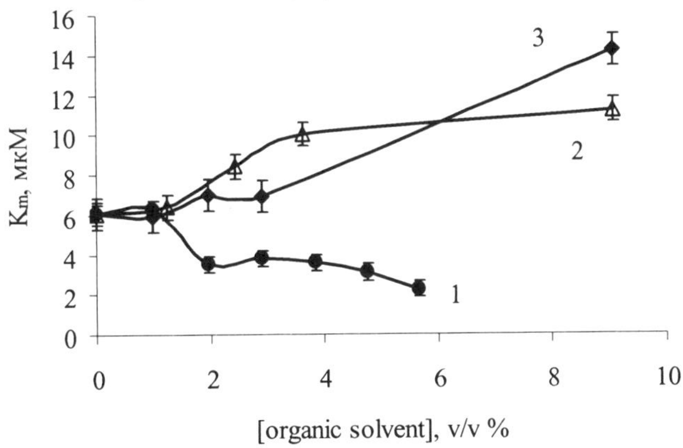

Fig. 2. Effect of aqueous-organic mixtures on the apparent value of Michaelis constant (K_m) luciferase *P. leiognathi* for tetradecanal

ACKNOWLEDGEMENTS

Supported in part by Award No. RUX0-002-KR-06 of the U.S. Civilian Research & Development Foundation (CRDF) and RF Ministry of Education and Science, BRHE Program, grants No. 2006-РИ-19.0/001/454 and 2006-РИ-111.0/002/065 of the RF Ministry of Education and Science.

REFERENCES

1. Baldwin T, Nicoli M, Becvar J, Hastings J. Bacterial luciferase: binding of oxidized flavin mononucleotide. J Biol Chem 1975;250:2763-8.
2. Fisher A, Thompson T, Thoden J, et al. The 1.5 Å resolution crystal structure of bacterial luciferase in low salt conditions. J Biol Chem 1996;271:21956-58.
3. Moore S, James M, O'Kane D, Lee J. Crystal structure of a flavoprotein related to the subunits of bacterial luciferase. EMBO J 1993,12:1767-74.
4. Huang S, Tu S. Identification of a catalytic base in bacterial luciferase by chemical rescue of a dark mutant. Biochemistry 1997;36:14609-15.
5. Holzman T, Baldwin T. Isolation of bacterial luciferases by affinity chromatography on 2,2-diphenylpropylamine-sepharose: phosphate-mediated binding to an immobilized substrate analogue. Biochemistry 1982;21:6194-201.
6. Francisco W, Abu-Soud H, Baldwin T., Raushel F. Interaction of bacterial luciferases with aldehyde substrates and inhibitors. J Biol Chem 1993;268: 24734-41.
7. Curry S, Lieb W Franks N Effects of general anesthetics on bacterial luciferases enzyme from *Vibrio harveyi*: An anesthetic target site with differential sensitive. Biochemistry 1990;29:46441-52.
8. Viswanathan T, Campling M, Cushley R. Interaction of long-chain aldehydes with lusiferase. A carbon-13 nuclear magnetic resonance study. Biochemistry 1979;18:2504-8.
9. Tyulkova N. Purification of bacterial luciferase from *Photobacterium leiognathi* with use FPLS-system. In: Jezowska-Trzebiatowska B. ed. Biological Luminescence. Singapore: World Scientific, 1989:369-74.
10. Cornish-Bowden A. Principles of enzyme kinetics. Moscow: Mir, 1979.
11. Keleti T. Basic enzyme kinetics. Moscow: Mir, 1990:183-244.
12. Gladilin A, Levashov A. Enzyme stability in systems with organic solvents. Biochemistry (Moscow) 1998;63:408-21
13. Khmelnitsky Y, Mozhaev V, Belova A, Sergeeva M, Martinek K. Denaturation capacity: a new quntitative criterion for selection of organic sovents as reaction media in biocatalysis. Eur J Biochem 1991;198:31-8.
14. Gupta M. Enzyme functions in organic solvents. Eur J Biochem 1992;203:25-32.

THERMAL STABILITY AND ABSORPTION SPECTRA OF BACTERIAL LUCIFERASES FROM *PHOTOBACTERIUM LEIOGNATHI* IN WATER-ORGANIC SOLVENTS

IE SUKOVATAYA,[1] NA TYULKOVA,[2] EV KAYKOVA[1]

[1] *Dept of Biophysics Krasnoyarsk State University, 79 Svobodny pr., Krasnoyarsk, 660041, Russia, Email: irinas@lan.krasu.ru;*

[2] *Institute of Biophysics, Russian Academy of Science, Siberian Branch, Akademgorodok, Krasnoyarsk, 660036, Russia Email: biotech@ibp.ru*

INTRODUCTION

Luciferases of luminous bacteria - flavin-dependent monooxygenases catalyze the oxidation reaction of the long-chain aliphatic aldehyde (RCHO) and reduced flavinmononucleotide ($FMNH_2$) involving molecular oxygen to the respective fatty acid with the emission of light quanta in the visible spectrum.[1] The interaction between the protein and the conditions of its functionality plays a major role in understanding of the structure-activity relationships.[2,3] Behaviour of the bacterial luciferases under different conditions is of fundamental importance for both theoretical and applicative purposes. It was found that denaturation by heat and guanidine-HC1 or urea, notably those which cause protein aggregation, could also be reversed.[4] Maximal luminescence intensity was observed at 20-30°C for *Photobacterium leiognathi (P. leiognathi).*[5] We found that, by adding organic solvents in the reaction medium, the quantum yield of bacterial bioluminescence reaction *in vitro* markedly increases.[6,8] Moreover, an important advantage of carrying out the bioluminescence reaction in organic media is the increased solubility of hydrophobic substrates. To get more information about the structure properties of luciferase isolated from *P. leiognathi* we studied the effect of organic solvents on the thermal stability and absorption spectra. If the spectral properties of the of luciferase are influenced by not only the structure of substrates and luciferase, but also properties of reaction mixture, addition of organic solvents might affect the spectrum.

METHODS

The luciferase from *P. leiognathi* (strain 208) was highly purified by ion-exchange chromatography.[9] As a control, a reaction was carried out in a mixture of the following composition: 10 μL of $(0.07–0.13)\times10^{-5}$ mol/L luciferase, 50 μL of 9.4×10^{-4} mol/L aqueous solution of tetradecanal - C_{14} (Merck, Germany), 840 μL of 2×10^{-2} mol/L phosphate buffer, 0.5 mL of 7.2×10^{-4} mol/L aqueous solution of $FMNH_2$ (Sigma, USA), pH 7. The reaction was initiated by the injection of the photorecovered $FMNH_2$. In this case the bioluminescent reaction is a short flash of light with pronounced maximum (I_{max}) and fast decay of bioluminescence (i.e., the enzyme turns over once). Measurements were carried out with a bioluminometer designed at the Institute of Biophysics (Russian Academy of Sciences, Siberian

Branch) at a temperature of 38°C. Different solutions of luciferase were incubated in a thermostated water bath. Aliquots (10 μL) were taken at regular time intervals and the residual enzyme activity was determined as described earlier.

Reaction parameters were recorded with a 2210 ("LKB-Wallace", Finland) recorder. In the experiments, the phosphate buffer was substituted for a water-organic mixture. Methanol (Sigma, USA), acetone, dimethyl sulfoxide (DMSO) (Serva, Germany) and formamide (Serva, Germany) were used as organic solvents. They were used at the highest purity grade available. Concentrations of solvents were expressed in volume %. Absorption spectra were recorded on a UVIKON-943 (Kontron Instruments, Italy) spectrophotometer with a 2 nm bandwidth and a spectral range of 200 - 350 nm. Bioluminescence intensity values are normalized. The experimental results obtained were statistically processed by Excel for Windows 98.

RESULTS

The effects of organic solvents with concentration which led to higher activity or quantum yield of the bacterial bioluminescent reaction *in vitro* on the thermal stability and absorption spectra of luciferases from *P. leiognathi* was studied. The light intensity of the luciferase shows a bell-shaped relationship between the intensity of emitted light and the temperature in phosphate buffer and organic solvents (pH 7).

The thermal stability of luciferase is remarkably different on addition of organic solvent. A comparison of the thermal stabilities of the luciferase was carried out by addition of different concentration of methanol, DMSO, acetone and formamide at a temperature of 38°C. The bioluminescence activity was measured as a function of the time and the $\ln I_{max}$ values were plotted. It was found that luciferase was resistant to heat at low concentrations of methanol (1 - 2.5 % v/v) and 2.5 % v/v DMSO, and its activity was little affected at 38°C (Fig. 1).

Luciferase was also resistant to heat at low concentrations of acetone. Luciferase was unstable in solutions of methanol, acetone and DMSO (e.g., the presence of 2.5, 5 and 10 % v/v DMSO 7, 23 and 27 % of activity, respectively). Luciferase activity was lost after 10 min incubation at 38°C. Under the same conditions the inhibition rate constants were determined as 0.010, 0.087 and 0.099, respectively. Luciferase was unstable in all solutions of formamide at a temperature of 38°C. Therefore, in the presence of methanol, acetone and DMSO in concentrations at which enzyme activity or quantum yield of the bioluminescence reaction is enhanced, the luciferase can be a thermostable enzyme.

Absorption spectra of luciferase were investigated after addition of organic solvents. Qualitatively non-similar spectral changes were observed for organic solvents used. The absorption spectra of luciferase produced in buffer were almost identical to those in methanol (Fig. 2).

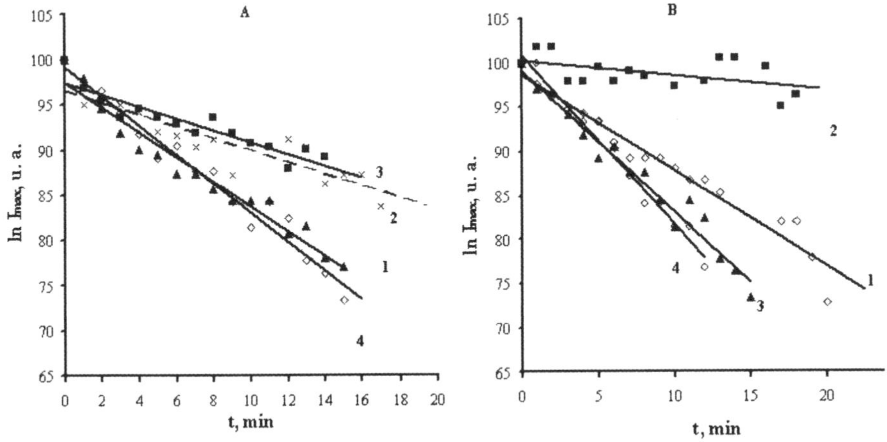

Fig. 1. The relationship between the intensity of emitted light of bioluminescent reaction catalyzed by luciferase from *P. leiognathi* at a temperature of 38°C in different concentrations of organic solvents: (A) – 0 (1), 1 (2), 2.5 (3) and 5 % v/v (4) methanol; (B) – 0 (1), 2.5 (2), 5 (3) and 10 % v/v (4) DMSO

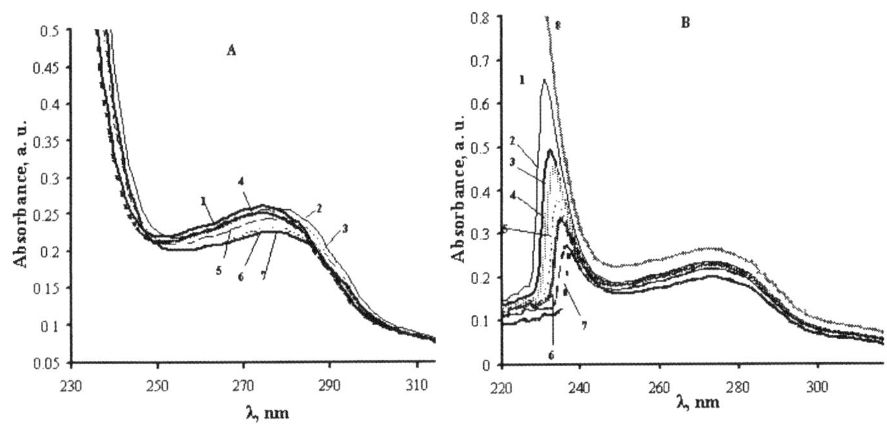

Fig. 2. Absorption spectra of luciferase from *P. leiognathi* in the presence of increasing organic solvents concentrations: (A) 0 (1), 0.5 (2), 1 (3), 2 (4), 7.5 (5), 12 (6) and 22 % v/v (7) methanol; (B) 0.5 (1), 1 (2), 2 (3), 3 (4), 4 (5), 5.6 (6), 7.5 (7) and 22 % v/v (8) DMSO

At both low and higher concentrations of methanol, there was a small inhibitory effect without spectral change. At all concentrations of DMSO and methanol the shape of the absorption spectra was not changed. In contrast, acetone and formamide addition induced a change of spectral shape. The absorption spectrum of luciferase in DMSO and 3.3 % v/v acetone shows two maxima. In the presence of increasing DMSO concentrations this first maximum of absorption (230 nm) is slowly red shifted to 240 nm and its intensity decreases. The DMSO-induced change of absorption spectra was bimodal with two emission maxima at 232 and 274 nm. The acetone-induced change of absorption spectra was bimodal with two emission maxima at 240 and 287 nm.

ACKNOWLEDGEMENTS

The research described in this publication was made possible in part by Award No. RUX0-002-KR-06 of the U.S. Civilian Research & Development Foundation (CRDF) and RF Ministry of Education and Science, BRHE Program, grants No. 2006-РИ-19.0/001/454 and 2006-РИ-111.0/002/065 of the RF Ministry of Education and Science.

REFERENCES

1. Baldwin T, Nicoli M, Becvar J, Hastings J. Bacterial luciferase: binding of oxidized flavin mononucleotide. J Biol Chem 1975;250:2763-8.
2. Friedland J, Hastings J. The reversibility of the denaturation of bacterial luciferase. Biochemistry 1967;6:2893-900.
3. Tyulkova NA, Sandalova TP. The comparative study of the temperature effect on the bacterial luciferase. Biochimiya (Moscow) 1986;62:275-87.
4. Khmelnitsky Y, Mozhaev V, Belova A, Sergeeva M, Martinek K. Denaturation capacity: a new quntitative criterion for selection of organic sovents as reaction media in biocatalysis. Eur J Biochem 1991;198:31-41.
5. Gupta M. Enzyme functions in organic solvents. Eur J Biochem 1992;203:25-32.
6. Sukovataya I, Tyulkova N. Effect of dielectric properties of media on kinetic parameters of bioluminescent reaction. Vestnik Moskovskogo Universiteta: Khim 2000;41:8-11.
7. Sukovataya I, Tyulkova N. Kinetic analysis of bacterial bioluminescence in water-organic media. Luminescence 2001;16:271-3.
8. Sukovataya I, Tyulkova N. Effect of organic solvents on bacterial bioluminescence reaction. Vestnik Moskovskogo Universiteta: Khim, 2003; 44:9-12.
9. Tyulkova N. Purification of bacterial luciferase from *Photobacterium leiognathi* with use FPLS-system. In: Jezowska-Trzebiatowska B. ed. Biological luminescence. Singapore: World Scientific, 1989:369-74.

CHARACTERIZATION OF A FISH SYMBIONT OF *PHOTOBACTERIUM PHOSPHOREUM* WITH ALTERED SPECTRAL PROPERTIES

S WHITAKER,[1] PJ HERRING,[2] AK CAMPBELL,[3] PJ HILL,[1] CED REES[1]

[1]School of Biosciences, University of Nottingham, Sutton Bonington Campus, Loughborough, Leics, LE12 5RD, UK; [2]George Deacon Division (room 494/02), National Oceanography Centre, Empress Dock, Southampton SO14 3ZH, UK; [3]Department of Medical Biochemistry and Immunology, Cardiff University, Heath Park, Cardiff, CF14 4XX, UK

INTRODUCTION

Photobacterium phosphoreum is found widely distributed in a variety of habitats in the world's oceans both as free living organisms and as symbionts associated with a wide range of deep-sea fish species. On isolation the different members of the *Photobacterium* group can be separated on the basis of biochemical tests. However distinct clades or sub-species have been identified on the basis of sequence analysis of specific regions of the 16S rDNA and *lux* sequences[1] and it has been proposed that *Photobacterium kishitanii* is the specific bioluminescent symbiont of deep-sea fishes.[2] Here we describe the characterisation of a light organ (LO) symbiont that has closest sequence homology to *P. phosphoreum* but possesses distinct bioluminescent properties and therefore may represent a sub-species of *P. phosphoreum* that can be found in symbiotic relationship with deep sea fish.

MATERIALS AND METHODS

Photobacterium strains were routinely grown using artificial sea water broth (ASW) at 22°C. To islolate bacteria the light organ of the fish was swabbed and then spread directly onto NSW agar (Nutrient broth containing 60:40 (v/v) sea water/distilled water) and incubated at 20°C. Colonies on the plate were screened for bioluminescence after 18 h growth. During growth experiments cell mass was monitored at 600 nm and bioluminescence measured using a Turner luminometer. Fluorescent emission spectra were determined using a Photex Ebc-01 camera with a monochromotor at room temperature. DNA extraction was carried out using the GES method[4] and PCR was carried out using *Taq* DNA polymerase and buffers (Amersham, Little Chalfont, UK) using 30 cycles of amplification.

RESULTS

During a research cruise, a fish of the genus *Physiculus* possessing a light organ was caught in a night trawl at a depth of 400-505 m at 19 11.1N, 58 59.8E using a RMT8 net. Colonies with a small circular translucent colony morphology and intense blue bioluminescence were cultured from the light organ of the fish. Biochemical testing was carried out and the results indicated that the bacterium belonged to the genus *Photobacterium* and had characteristics that matched those described for

P. phosphoreum in Bergey's manual[3] (data not shown) except for the colour of the bioluminescence produced.

The whole of the 16S rDNA and a region of *luxA* (bases 143-693) were amplified by PCR and sequenced. BLAST analysis of *Physiculus* LO symbiont 16S sequences indicated best match to the type strain *P. phosphoreum* ATCC 11040 rather than any of the sequenced *P. kishitanii* 16S genes, particularly in the region containing most ambiguity between strains[2] (Fig. 1). Three single base differences were also identified in other regions of the sequence (data not shown). Alignment of *luxA* sequences gave closest match to *P. phosphoreum* AK-3[2] (92% identity) confirming the identification of the *Physiculus* LO symbiont as a member of *P. phosphoreum* species rather than the *P. kishitanii* clade (data not shown).

```
Pp          70 ...TAACAG AWWGAWAGC TT GCTWTCWWT GCTGACGAGCG
Psym.       70 ...TAACAG ATCGATAGC TT GCTATCAAT GCTGACGAGCG
Pk          70 ...TAACAG AAARRAAGC TT GCTTYYTTT GCTGACGAGCG
```

Fig. 1. Alignment of 16S sequence corresponding to *E. coli* bases 70-104. Pp, *P. phosphoreum;* Psym, *Physiculus* LO symbiont; Pk, *P. kishitanii.* Underlining indicates the region participating in stem-loop formation. W = A or T, R= A or G, and Y= C or T

However, the *Physiculus* LO symbiont strain displayed some very obvious differences in spectral properties to *P. phosphoreum* ATCC 11040 but had similar spectral properties to *P. phosphoreum* NCIMB 844 which is known to contain a lumazine (LumP) protein (Table 1). PCR primers homologous to conserved regions of *lumP* (bases 1-23 and 404-425 of sequence M37625) were used to amplify a 450 bp product from *Physiculus* LO symbiont genomic DNA. Sequence analysis gave 52.8% similarity to the published *lumP* sequence and this would account for the spectral shift seen.

Table 1. Spectral analysis of *Photobacterium* isolates. λmax, peak fluorescence emission; ½ bandwidth, the spectral distribution of the fluorescence at half the peak light intensity

Isolate	Emission spectra maxima (λ_{max})	½ band width (nm)
P. fischeri MJ-1	485	84 (452-536)
P. phosphoreum NCIMB 844	478	60 (452-512)
Physiculus LO symbiont	472	62 (454-516)

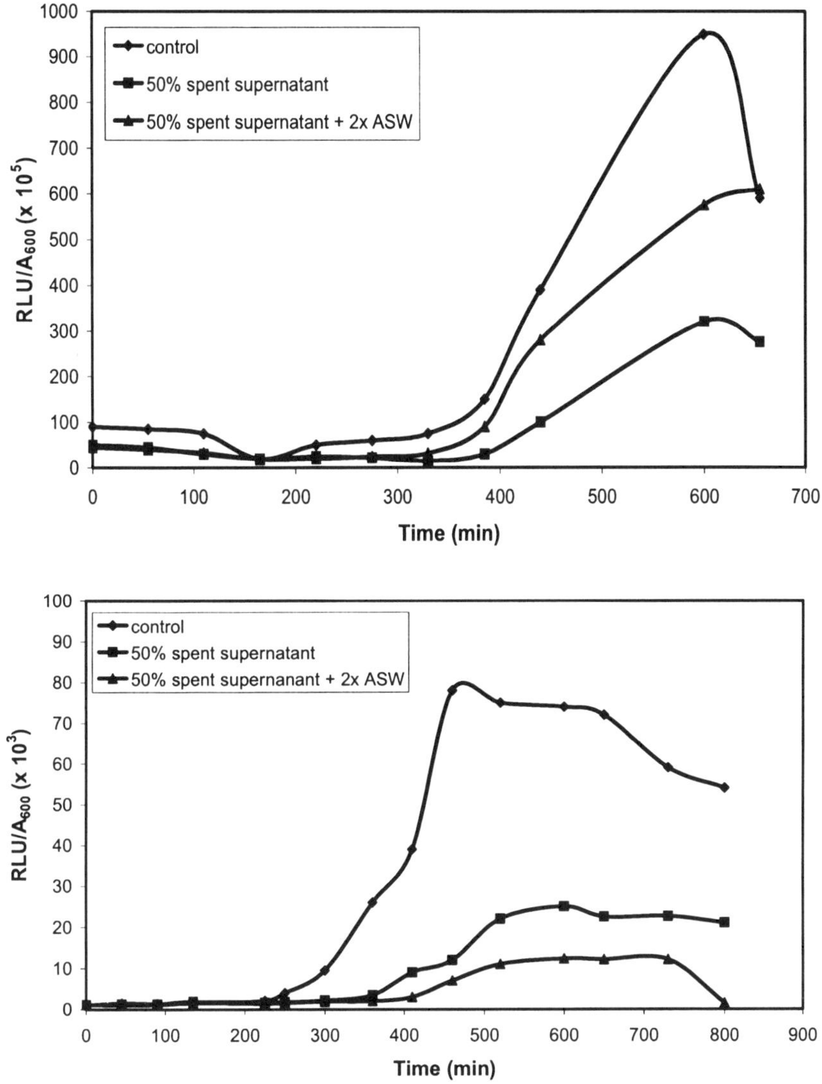

Fig. 2. Cultures of the *Physiculus* LO symbiont (top panel) or *P. phosphoreum* ATCC 11040 (bottom panel) were grown in ASW at 22°C for 18h, centrifuged and the culture supernatant sterilised using a 0.45 μm pore size filter. Bacteria were inoculated into ASW broth alone (♦), ASW mixed 1:1 with filtered supernatant (■) or double strength ASW mixed 1:1 with filtered supernatant (▲). Data are presented as RLU/A600nm to correct for differences in cell mass. NB: scale in top panel multiplied 100X more than bottom panel

Another distinguishing feature was the brightness of the bioluminescence emitted. To determine if the *lux* genes were regulated by an autoinduction mechanism, fresh bacterial cultures were supplemented with filtered supernatant of overnight cultures of the same organism. An additional control of using double strength ASW mixed 1:1 with filtered supernatant was also used to rule out effects on *lux* gene expression caused by nutrient depletion For *P. phosphoreum*, as expected, the time of induction during growth was brought forward by adding culture supernatant known to contain autoinducer molecules. For the *Physiculus* LO symbiont, no such shift was seen and light levels were 100-fold higher than those gained for *P. phosphoreum* ATCC 11040 (Fig 2). When culture supernatants from these two organisms were separated using thin layer chromatography and assayed for the presence of inducer molecules,[4] three short chain inducer molecules were detected in culture supernatants of *P. phosphoreum* but only one in samples from the *Physiculus* LO symbiont which corresponded in Rf value to either *N*-(3-oxo)-hexanoyl-L-homoserine lactone (OOHL) or *N*-hexanoyl-L-homoserine lactone HHL (data not shown).

The analysis of two genetic loci confirms the identification of this bacterium as *P. phosphoreum* species and therefore members of this group can also be found as symbionts in the light organs of deep sea fish. However differences in the 16S and *luxA* sequences, the novel lumazine protein and the different pattern of bioluminescence control suggest that this organism represents an undescribed sub-species which may be able to adapt to this life style better than other members of the group.

REFERENCES

1. Dunlap PV, Ast JC. Genomic and phylogenetic characterization of luminous bacteria symbiotic with the deep-sea fish *Chlorophthalmus albatrossis*. Appl Environ Microbiol 2005;71:930-9.
2. Ast JC, Dunlap PV. Phylogenetic resolution and habitat specificity of members of the *Photobacterium phosphoreum* species group. Environ Microbiol 2005;7:1641-54.
3. Garrity G. Bergey's Manual of Systematic Bacteriology. New York: Springer, 2001.
4. McClean KH, Winson MK, Fish L, Taylor A, Chhabra SR, Camara M, et al. Quorum sensing and *Chromobacterium violaceum*: exploitation of violacein production and inhibition for the detection of N-acylhomoserine lactones. Microbiology 1997;143:3703-11.

PART 2

INSTRUMENTATION

LUMINOMETERS FOR INDUSTRIAL APPLICATIONS OF RAPID MICROBIOLOGY

FRITZ BERTHOLD,[1] KLAUS HAFNER,[2]
MARTIN SCHREIBERS,[3] VEIKKO TARKKANEN[4]
[1] Berthold Technologies GmbH & Co. KG, Calmbacher Straße 22, 75323 Bad Wildbad, Germany, Email: fritz.berthold@Berthold.com; [2] 3D Engineering, Coppistraße 55, 04157 Leipzig, Germany, Email: hafner@3dengineering.de;
[3] Celsis International B.V., Amperestraat 13, 6372 BB Landgraaf, The Netherlands, Email: mschreibers@celsis.nl; [4] 503, Chemin Pied Marin 2, 84380 Mazan, France, Email: vt@lumi1.com

HISTORICAL PERSPECTIVE

In late '70s, releasing reagents for nucleotides from cells facilitated a fast and nearly quantitative extraction of nucleotides from microbial cells and allowed low-toxic systems for rapid and reproducible measurement of endogenic microbial ATP.[1,2] A parallel progress in instruments resulted in the use of photon counters and programmable injection/mixing of reagents into a cuvette positioned in front of a photomultiplier detector.[3] Due to a rapid decay of the luminescence signal this was the most versatile approach at that time.

In the course of the following decade, the original concept was extensively automated, comparable with the development of sample handling used in scintillation counters for radioactivity measurement.[4,5] Reliability of bioluminescence depends on protocols , and reagents, but also on instruments. New methods alone do not function without these. After appropriate instruments became available, a multitude of new applications can be seen in the scientific literature, like rapid sterility control as a routine procedure.

Introduction of the 96 well microplates offered more automation possibilities during the '90s. This allowed processing increasing sample numbers with a reduced cost. This development occurred in parallel with the introduction of new luminescent labels, and sparked new major uses of luminescence as replacement of radioactivity.

MICROBIAL CONTAMINATION SCREENING IN THE FOOD INDUSTRY

In a typical screening situation in the dairy industry the abundant, non-microbial ATP must be hydrolyzed prior to the release of the microbial one. In a subsequent step the released microbial ATP is quantified.

Reagents for this measurement sequence have undergone improvements, like rapid quantitative hydrolysis of non-microbial and a quantitative extraction of microbial ATP. New luminescence reagents with their glow-type characteristics do not require injection in front of the light detector any more.

Normally, three reagents are supplied. The first in sequence is an enzyme system removing all non-microbial ATP like such from non-microbial cells. This typically takes 5-6 minutes. The second releases ATP from microbial cells during one to seven minutes while retarding the influence of the first reagent. The third reagent (luciferase) addition follows immediately and produces photons which are typically measured for 1 to 3 seconds. The whole sequence requires a reproducible, intense mixing of the reaction components.

PROBLEMS IN INDUSTRIAL HIGH THROUGHPUT INSTRUMENTS

While inconsistencies of results in bioluminescence are normally ascribed to poor methods, a connection between instrument and sample consistence has not been sufficiently analyzed. If we consider sequential addition and mixing of three reagents into a non-homogeneous sample and expect reproducible results, we are faced with new problems. Manual luminometers were based on the use of cuvettes and built so that the reagents were injected with sufficient speed and proper angle in order to ensure immediate mixing and a reproducible reaction. Automatic cuvette based instruments allowed to increase sample throughput, but are difficult to interface with robotic systems, and handling large numbers of cuvettes is cumbersome.

Microplates have now become the format of choice for high sample throughput and automation. However, their use in industrial luminometry was associated with a number of problems. Insufficient reagent mixing, carryover of sample and reagent material, congestion of reagent injection lines, and sample splashing led to unreliable results. Moreover, the need for frequent decontamination of instrument parts was not welcomed in routine laboratories.

It became apparent that, for samples with high and changing viscosities, a precise reagent supply must be followed by intense shaking/mixing, at the same time avoiding any contamination of injection tips or splashing of liquids. Present instrument configurations, including shaking/mixing functions, do not fully meet the requirements for industrial routine.

LATEST DEVELOPMENTS

We have designed a novel type microplate luminometer, called XS4, addressing the issues with presently available instruments, which normally have been developed for research or clinical use, but not for an industrial environment.

Experience has shown that on-board reagent addition is a major source of luminometer malfunction, therefore requiring substantial improvements. Before beginning the first measurement, the reagent injection system must be primed, meaning to fill the reagent ducts and the injector pumps with reagents, replacing any other liquids like wash solutions, or air. To save on reagent cost, priming or dead volume is minimized by closest proximity of reagent container, pump to the injection point. The XS4 can be equipped with up to 4 injectors, sufficient for all presently known or anticipated applications. Occasional splashing of reagents is almost unavoidable, e. g. in the priming phase where air and reagent mixing may lead to droplet formation, but operator errors are even more frequently the cause.

During priming, a trough with a large enough opening is positioned below the injection points, from where the liquids flow into an internal waste container which is continuously emptied into an outside container.

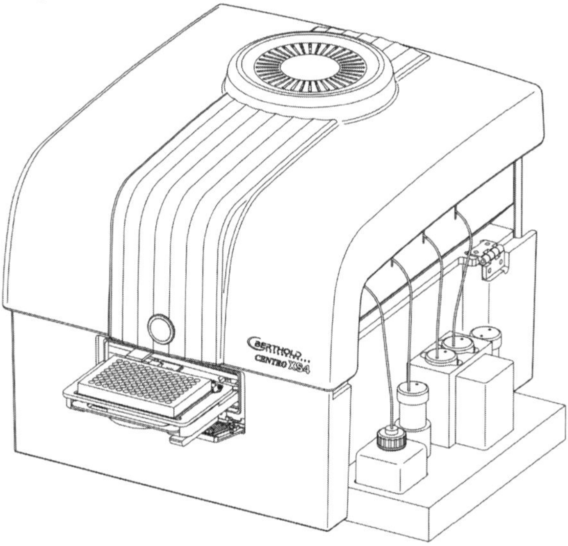

Fig. 1. XS4 Industrial Microplate Luminometer

The instruments cover is easily opened for access to the working platform This allows a quick inspection of injection tips for any clogging or other damage and, if necessary, a practically fool-proof exchange of parts. Most reagent spillings occur below the working platform. Therefore, this can also be opened easily, without any need to disconnect liquid or electric connections, for inspection and cleaning of the compartment containing the xy-table and plate holder. This is recommended as a routine procedure. The entire electronics are contained in yet another separate compartment, out of reach of any liquids.

An active on-board vortexer (patents applied for) with its own motor and oscillating plate holder allows fast mixing even of high-viscosity samples. This makes it suitable for samples with varying viscosities and homogeneities, to be mixed with extractant and light emission reagents for the detection of bacterial contaminations.

The standard procedure is now that any reagent dispensing is immediately followed by at least a short (typically 1s) vortexing step. Longer time vortexing is applied during incubation periods. In a typical assay for *E. coli*, a standard deviation of 7 % with injection alone went down to 2 % for injection plus vortexing. A built-in light standard assures proper calibration of the photon counter each time the system is started.

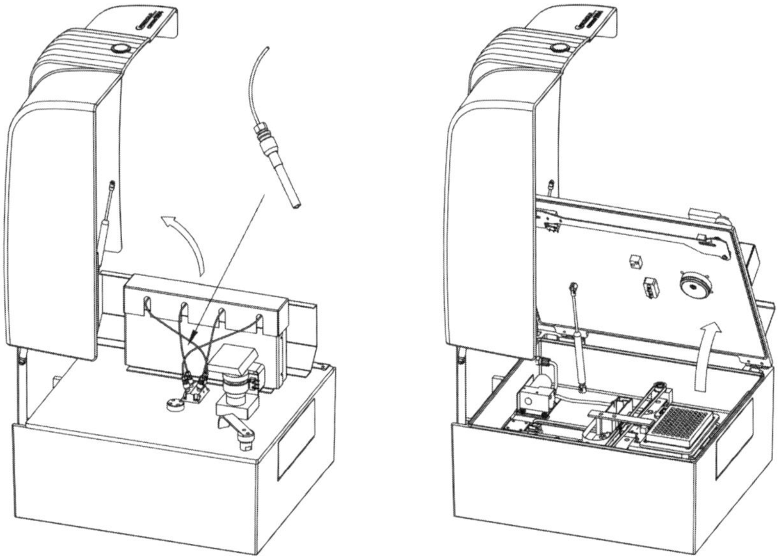

Fig. 2. XS4 Luminometer. Left: cover opened. Right: working platform opened

The XS4 is designed for routine use, also by unskilled operators, in a rugged industrial environment, and allows maintenance and even most repair functions to be performed by the user. Nevertheless, its sensitivity is equal to that of the best research luminometers so that it would also be a good choice for research use requiring highest reliability.

ACKNOWLEDGEMENTS

We thank Michael Peukert for supporting us with his valuable expertise.

REFERENCES

1. Kolehmainen SE, Tarkkanen V. Selective determination of nucleotides in viable somatic and microbial cells. United States Patent 1981:4,303,752.
2. Berthold F, Kolehmainen S, Tarkkanen V. Use of a method, particularly an automatic method, for measuring light as well as improvements and apparatuses for carrying out the same. United States Patent 1983:4,390,274.
3. Stanley PE. Commercially Available luminometers and low-level light imaging devices. Methods Enzymol 2000;305:96-103.
4. Berthold F, Herick K, Siewe R M. Luminometer design and low light detection. Methods Enzymol 2000;305:62-87.
5. Stanley PE, Smithers R, Simpson WJ. eds. In: A Practical Guide to Industrial Uses of ATP-Luminescence in Rapid Microbiology. Lingfield, UK: Cara Technology Limited, 1997.

APPLICATION OF A LUMINESCENCE MICROSCOPE WITH NOVEL OPTICAL SYSTEM FOR DETECTION OF THE GENE EXPRESSION PATTERN OF INDIVIDUAL CELLS

Y HATTA-OHASHI,[1] N HAYASAKA,[2] T TAKAHASHI,[1]
Y SHIGEYOSHI,[2] H SUZUKI[1]

[1] Research & Development Division, Olympus Corporation, Hachioji, Tokyo 192-8512, Japan; [2] Dept. of Anatomy and Neurobiology, Kinki University School of Medicine, Osaka-Sayama, Osaka 589-8511, Japan; Email: yoko1_ohashi@ot.olympus.co.jp

INTRODUCTION

Luciferase is particularly useful as a reporter to monitor gene expression in living cells and organisms because of its simplicity, good S/N ratio and high quantitativity.[1,2] The short half-life of the luciferase reporter compared with other markers, chloramphenicol acetyl transferase and green fluorescent protein, allows accurate and continuous monitoring of transcription in living cells.[3] In luciferase promoter assays, a luminometer has been generally used for detection of luciferase activity, with capability of measuring only total luminescence from the cell population. We therefore developed a luminescence microscope with a novel optical system, which makes it possible not only to measure the light intensity but also to obtain detailed information on the morphogenesis and gene expression pattern of individual cells. To evaluate this imaging system, we applied it for detection of (1) the heat shock protein (HSP) 70B promoter activity in HeLa cells and (2) the circadian rhythmicity of a clock gene (PER2) expression in the suprachiasmatic nucleus slices (SCNs).

MATERIALS AND METHODS

HSP70B promoter assay. HeLa cells were plated on a 35 mm culture dish with Dulbecco's modified Eagle's medium containing (DMEM) 10% fetal calf serum (FCS) and were transfected transiently with the pGL3 basic promoter vector (Promega) that contains the HSP70B promoter region.[4,5] At 24 hours after transfection, the medium was replaced with DMEM containing 10% FCS supplemented with 10 mM HEPES (pH 7.2). For heat stimulus, cells were treated for 1 h at 43°C, and then 200 μM D-luciferin (Promega) was added before the imaging experiment.

Animals and slice culture. Per2::luc knock-in mice were originally generated by Joseph S. Takahashi at Northwestern University as described previously.[6] Briefly, mPer2::luc knock-in mice were generated by homologous recombination in which a luciferase gene (from pGL3, Promega) was fused in-frame to the 3'-end of the endogenous mouse Per2 gene. The mice were kept under 12 h light: 12 h dark conditions (L: D = 12: 12) until they were used for experiments. Coronal brain slices

(300 μm thick) were prepared from the Per2::luc knock-in mice using a Microslicer (Dosaka, Japan) and slices were cut with scalpel blades in about 2-mm square pieces including bilateral SCN slices and optic chiasm. The SCN slices were then put on a membrane (MILLICELL-CM, Millipore) and cultured at 37°C in a 35 mm culture dish with 1.2-mL of culture medium including 500 μM D-luciferin.[7]

Optical imaging of luminescence. Images were acquired using the luminescence microscope LUMINOVIEW (LV100, Olympus) attached a cooled charge-coupled device (CCD) camera ORCA-AG (Hamamatsu Photonics) and DP-30 (Olympus) (light was completely shielded with a black box). The operating temperature of the CCD camera was set to -30°C. In this system, the optical parameters such as numerical aperture of the objective and tube lens, and total magnification were optimized for luminescence imaging of a single cell. The dish was kept at 37°C in the humidified chamber of the microscope during observation. Detailed observation conditions are described in each figure legend. Luminescence intensity from single cells was measured as an average value in a region of interest (ROI) encircled for each cell using MetaMorph software (Universal Imaging).

RESULTS AND DISCUSSION

We examined the HSP70B promoter activity on HeLa cells using the luminescence microscope. Increase in HSP70B promoter activity in HeLa cells was observed by heat stimulation. In Fig. 1B, each set of symbols represents luminescence intensity for an individual cell. We found that chronological response of promoter activity induced by heat stimulation was not homologous within individual cells. It implies that cellular processes during physiological responses can be detected at the cellular level.

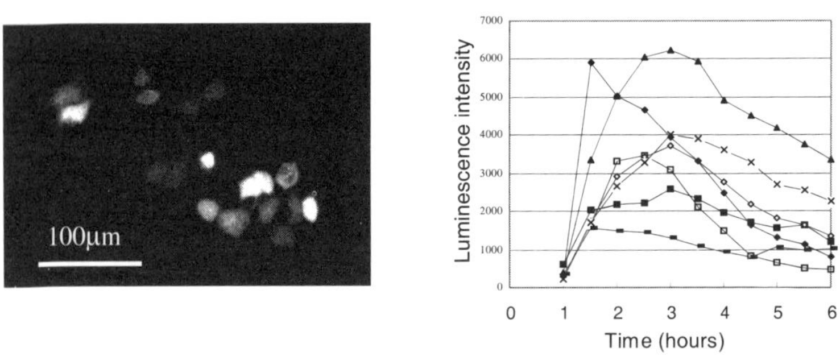

Fig. 1 Luminescence imaging of the HSP70B promoter activity in HeLa cells. A: Luminescence image of HeLa cells expressing luciferase gene under the HSP70 promoter. Images were taken by 5 min exposure at 30 min intervals and binning with 1x1 of 1344x1024 pixels using 20x objective lens (numerical aperture (NA) 0.75). B: Time course analysis of HSP70B promoter activity after 1 hour heat stimulation. Each set of symbols represents luminescence intensity for an individual cell.

We next studied circadian rhythmicity of gene expression in the SCN slices. Images were captured every hour for six days. The time course of luminescence from individual cells within the SCN is shown in Fig. 2A and 2B. Various patterns of circadian rhythmicity were observed within the SCN compared with the principal phase peak of the whole SCN as shown in Fig. 2B (solid line). This result revealed that time-lapse bioluminescence imaging enabled detection of circadian rhythmicity of PER2 expression in each cell within the cultured SCN. The circadian rhythmicity of the individual SCN cells has mainly been analyzed electrophysiologically.[8]

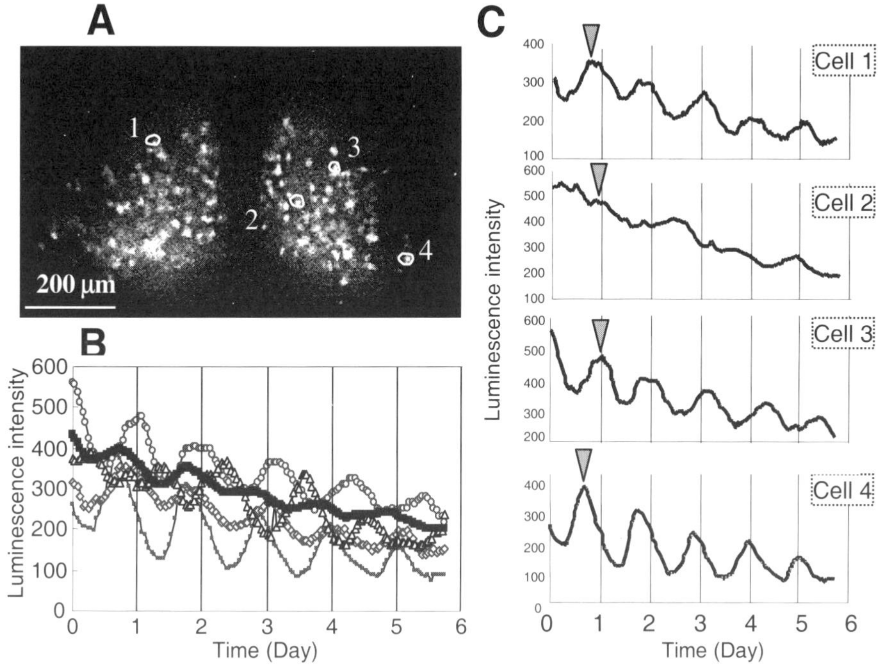

Fig. 2 Luminescence imaging of circadian rhythmicity of individual SCN cells
A: Per2::Luc luminescence image from cultured SCN. Images were taken by 58 min exposure with binning 2x2 of 3144x1024 pixels using 40x objective lens (NA 0.6).
B: The PER2::Luc rhythm of the whole SCN (solid line) and four representative SCN cells (open symbols). C: Different circadian profiles of individual SCN cells.
The arrow indicates the peak of the PER2 expression for each cell.

Only recently, the single-cell circadian rhythmicity of clock gene expression with optical bioluminescence imaging was reported in several studies,[1,9] using highly sensitive cooled CCD camera (e.g., cryogenic CCD camera).

Here, we showed high-resolution luminescence imaging of individual cells using our newly developed luminescence microscope, with a conventional cooled CCD camera. Another study using our luminescence imaging system also demonstrated that introduction of certain mutant clock genes into cells resulted in arrhythmicity in the individual cells.[9] In conclusion, this luminescence imaging system can be applied for studies on the dynamic monitoring of gene expression in individual living cells.

REFERENCES

1. Welsh DK, Yoo SH, Liu AC, Takahashi JS, Kay SA. Bioluminescence imaging of individual fibroblasts reveals persistent, independently phased circadian rhythms of clock gene expression. Curr Biol 2004;29;14:2289-95.
2. Welsh DK, Kay SA. Bioluminescence imaging in living organisms. Curr Opin Biotechnol 2005;16:73-8.
3. Norris AJ, Stirland JA, McFerran DW, Seymour ZC, Spiller DG, Loudon AS, et al. Dynamic patterns of growth hormone gene transcription in individual living pituitary cells. Mol Endocrinol 2003;17:193-202.
4. Wu B, Hunt C, Morimoto R. Structure and expression of the human gene encoding major heat shock protein HSP70. Mol Cell Biol 1985;5:330-41.
5. Hunt C, Morimoto RI. Conserved features of eukaryotic hsp70 genes revealed by comparison with the nucleotide sequence of human hsp70. Proc Natl Acad Sci USA 1985;82:6455-9.
6. Yoo SH, Yamazaki S, Lowrey PL, Shimomura K, Ko CH, Buhr ED, et al. PERIOD2::LUCIFERASE real-time reporting of circadian dynamics reveals persistent circadian oscillations in mouse peripheral tissues. Proc Natl Acad Sci USA 2004;101:5339-46.
7. Nakamura W, Honma S, Shirakawa T, Honma K. Clock mutation lengthens the circadian period without damping rhythms in individual SCN neurons. Nat Neurosci 2002;5:399-400.
8. Yamaguchi S, Isejima H, Matsuo T, Okura R, Yagita K, Kobayashi M, et al. Synchronization of cellular clocks in the suprachiasmatic nucleus. Science 2003;302:1408-12.
9. Sato TK, Yamada RG, Ukai H, Baggs JE, Miraglia LJ, Kobayashi TJ, et al. Feedback repression is required for mammalian circadian clock function. Nat Genet 2006;38:312-9.

NOVEL TOOLS FOR QUANTIFICATION OF LUMINESCENCE

PC HICKEY
LUX Biotechnology Ltd. King's Buildings, Edinburgh, EH9 3JF, UK
Email: ph@luxbiotech.com

INTRODUCTION

A novel range of tools have been developed for the calibration of luminescence detection equipment such as plate readers and bio-imaging systems. The range of devices, known as Glowells™, have the unique feature that they fit into micro plates and may be measured alongside biological samples. Glowells™ are available in a range of formats, colours and intensities. Luminescence is emitted from gaseous tritium light sources (GTLS) which have an extremely stable output and predictable decay over time. Different intensities are achieved by attenuation of the light source using high quality glass neutral density filters. The light output of every Glowell™ unit is calibrated and traceable to national standards. This allows instruments to be calibrated during experiments. In this chapter we provide experimental data on the characteristics of Glowells™ including measurement of the spectral output and decay over time. Using a Low Light Imaging Standard™ (LLIS™) we visualised the light scatter angle, investigate different CCD camera exposure times, and provide a comparison with a living organism in a Petri dish.

MATERIALS AND METHODS

Emission spectra. The emission spectra of Glowells™ were measured using a Fluoromax™ spectrofluorometer. The excitation source was turned off and spectra were measured at 1 nm intervals. **Luminescence decay.** Glowells™ were measured at regular intervals using a calibrated silicon photodiode detector (manufactured by Hamamatsu) traceable to National Physics Laboratory (NPL) standards. **CCD Imaging.** Samples were imaged using a thermoelectrically cooled camera containing the Sony ICX085 CCD chip. Light scatter was measured by placing the LLIS™ adjacent to a block of 2% agarose. ***In vivo* imaging.** A green LLIS™ was placed alongside a culture of the bioluminescent fungus *Armillaria mellea,* grown on 2% malt extract agar a Petri dish. The intensity plot was achieved using "Image-J" software.

RESULTS AND DISCUSSION

Emission spectra. Data was obtained for 4 colours of Glowells (Fig. 1). All showed Gaussian curves, except for the RED unit. The scattered spectrum of the RED Glowell™ is likely to be due to components in the GTLS phosphor (e.g. Europium). The different coloured standards may be chosen for specific assays e.g. BLUE (~450 nm) for aequorin and YELLOW (~560 nm) for firefly luciferase.

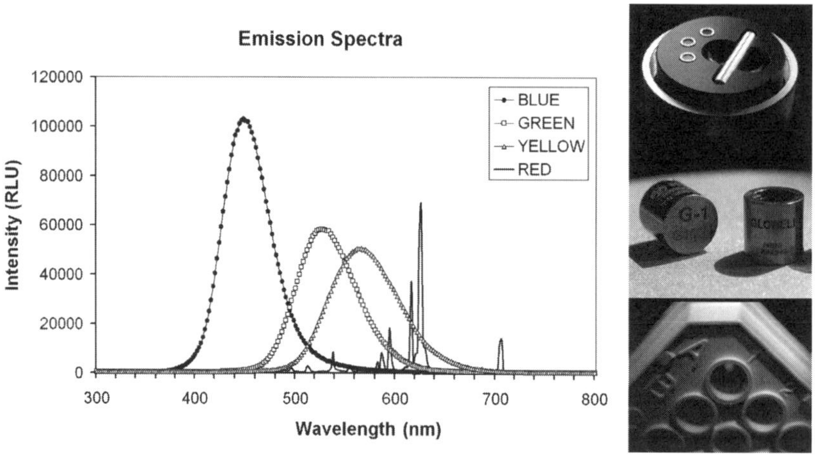

Fig. 1. *Left:* Emission spectra from 4 colours of Glowell™; *Right:* Images showing Glowell™ standard kit containing 3 units of different intensity; close up of individual 96-well inserts showing laser etched serial number; Photograph of Glowell™ emitting luminescence in a 96-well plate

Luminescence decay. The reduction in luminescence is a function of the half-life of the gaseous tritium and stability of the phosphor scintillator. For calibration purposes this decay must be limited and predictable. Green Glowells™ were measured repeatedly by an Independent Calibration Laboratory (ICL) over a period of 18 months. The results obtained are traceable to NPL standards.

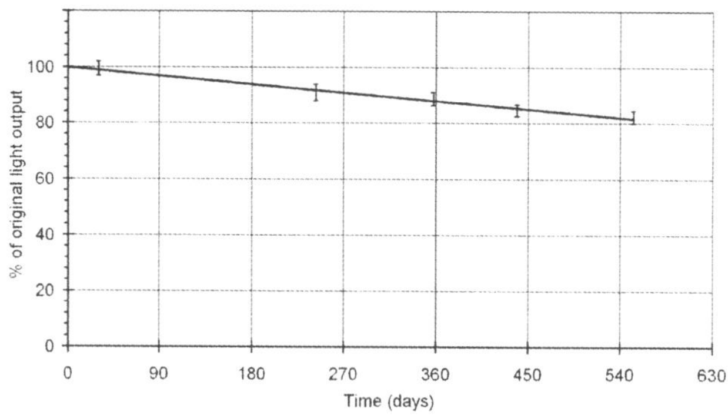

Fig. 2. Reduction in luminescence output over time measured using G1 green Glowells™ (R^2 =0.9917)

The analysis of a set of green Glowells™ showed very stable reduction in light output over time, measured over 18 months (Fig. 2). This is equivalent to approximately 1% per month, or 12% over a year. From the GTLS manufacturer's data, the steepest decay in light output is seen in the first month after the sources are manufactured. Therefore, in production of Glowell™ standards, all light sources are "aged" for a minimum of 6 weeks before being assembled into finished units.

CCD imaging. The use of Charge Coupled Device (CCD) and low light cameras is becoming increasingly popular in biological imaging applications e.g. microbial cultures, chemiluminescent gels, whole organisms etc. Table 1 shows the data from a calibrated LLIS™ and calculation of luminous flux.

Table 1. Photometric calibration of LLIS™. Measurements have been corrected for background and converted into luminous flux (Lumens) at 525 nm

LLIS™ OUTPUT	RELATIVE INTENSITY (%)	MEASUREMENT pW	LUMINOUS FLUX (LUMENS)
APERTURE 1	50	1768.5	1.07^{-06}
APERTURE 2	10	415.8	2.55^{-07}
APERTURE 3	1	33.1	2.07^{-08}
APERTURE 4	0.1	3.7	2.15^{-09}
APERTURE 5	0.01	BACKGROUND	BACKGROUND

By taking into consideration the photometric data supplied with every unit, it is possible to estimate a camera's sensitivity and dynamic range. This can be used to make a decision on the purchase and optimise routine set up of the camera for individual requirements. We have used the LLIS™ to measure light scatter and evaluate the sensitivity and dynamic range of a CCD camera (Fig. 3).

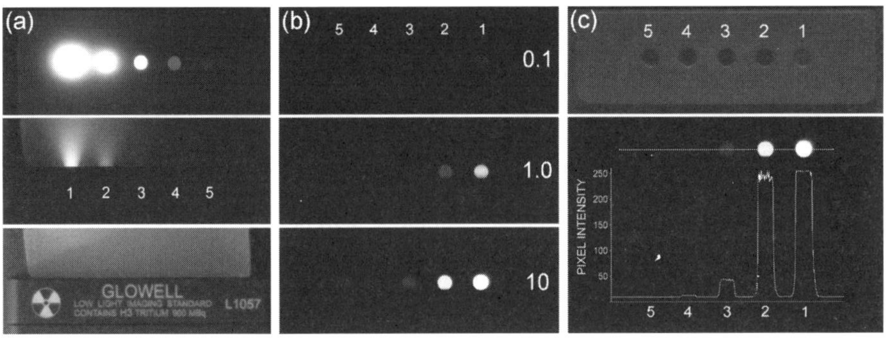

Fig. 3. (a) Light scatter from the LLIS™; (b) Different exposures times; (c) Intensity plot of light from CCD image

The angle of light scatter corresponds to an angle of approx 25°. Figure 3(b) indicates the dynamic range is in the region of 4 orders of magnitude. Aperture 4 requires a 10 s exposure to be detected, therefore it is estimated that a 100 s exposure would be required to image the 5th aperture. The intensity plot reveals that aperture 1 is saturated, and the camera's limit of detection is aperture 4.

***In-vivo* imaging.** *In-vivo* luminescence imaging is a powerful tool for studying a variety of organisms from plants to whole animals. To demonstrate the effectiveness of the LLIS™ for this application, we have shown a comparison with a bioluminescent fungus (Figure 4). By directly comparing the LLIS™ to biological specimens, any variation in samples can be quickly identified. The LLIS™ also allows researchers to standardise data e.g. using different equipment and adjusting exposure times accordingly, both saving time and maintaining consistency.

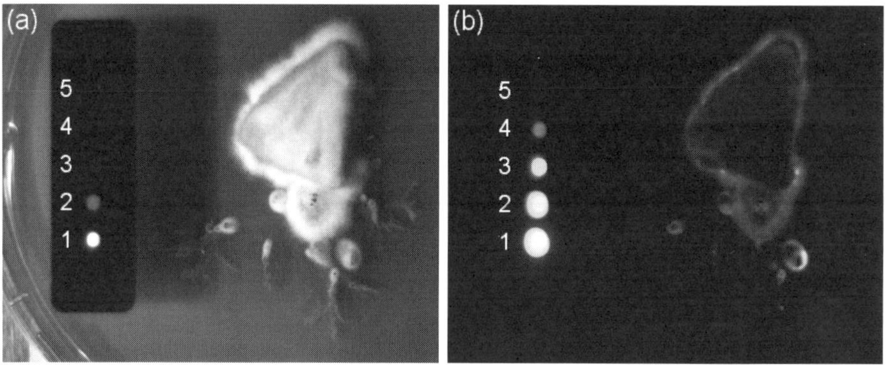

Fig. 4. Comparison of LLIS™ with a bioluminescent fungus.
(a) Image taken in ambient light. (b) Long exposure image acquired in the dark.
Note the luminescence is brightest in actively growing areas of the colony

CONCLUSIONS

Glowells™ are precision devices available in various formats for virtually any instrument that measures light. The decay of light output in Glowells™ is < 1% per month in the first two years after manufacture. Glowells™ help resolve issues arising from day-to-day variation in biological samples, enabling direct standardisation of data and evaluation of instrument sensitivity. Glowells™ are ideal experimental controls that can be used simultaneously in experiments, e.g. within 96 well plates and alongside living organisms. Fluorescent versions are in development.

ACKNOWLEDGEMENTS

Thanks to Catherine Rowley-Williams, Emma Perfect and Steven Magennis. Further information on Glowell™ products is available at www.luxbiotech.com.

REFERENCES

1. Pizzonia J. A CCD sensor optimized for chemiluminescence, fluorescence, and brightfield applications. Amer Biotechnol Lab 2006; Oct: 14-5.
2. Jenks PJ. Glowell™ - The First True Biotechnology CRM. RM report 2004; Vol. 3 No.2.

LUMINESCENCE MICROSCOPE FOR REPORTER ASSAY OF SINGLE LIVE CELLS

H SUZUKI,[1] S DOSAKA,[2] Y OHASHI-HATTA,[1] T SUGIYAMA[1]
[1] *R & D Division, Olympus Corporation, Hachioji, Tokyo 192-8512, Japan;*
[2] *Olympus Engineering Corporation, Hachioji, Tokyo 192-0045, Japan*
Email: hirobumi2_suzuki@ot.olympus.co.jp

INTRODUCTION

The bioluminescent system (luciferase reporter assay system) is widely used for the study of gene expression, signal transduction and other cellular activities. The luciferase assay is conventionally performed by the photon-counting luminometer method. In this system, light emitting from cells is measured as integrated value through all cells. Recently, we needed to monitor expression activity of genes of interest in each cell spatially and temporally. To obtain a luminescence image of cells, ultra low-light imaging technology, such as photon-counting CCD and liquid nitrogen cooled CCD cameras, has been required.[1-4] Furthermore, the imaging system is also applied for tissue and embryo luminescence imaging.[5,6] However the high sensitive CCD camera is expensive, and resolution power of the photon-counting CCD camera is insufficient for image analysis.

We have successfully developed a new optical system for bioluminescence imaging of single live cells using a conventional cooled CCD camera and microscope. Using this microscope, the promoter activity of c-*fos* gene and tetracycline operator activity were assayed at single cell level.

METHODS

Reporter gene construction. Complete region of c-*fos* gene promoter[7] is inserted into the pGL3 basic promoter vector (Promega). For tetracycline induction assay, the GL3 luciferase gene is into multicloning site of the Tet on vector (Invitrogen) under CMV promoter. HeLa cells are transfected by the lipofectamine (Invitrogen) with the c-*fos* reporter constructs, and HeLa stable cell line expressing tetracycline repressor is transfected by Polyfect (Qiagen) with the Tet-luciferase construct.

Luminescence imaging of single cells. Fig. 1 shows a diagram of luminescence microscope developed for single live cell imaging (Luminoview LV-100, Olympus). Light emitting from cells (a) is collected by UAPO20X (Olympus) objective lens (c) with NA (numerical aperture) 0.75, and the light passed through the tube lens (d) creates an image on the cooled CCD camera (e). The CCD sensor chip is cooled to 5°C (DP-30, Olympus) and –60°C (ORCA-II ERG, Hamamatsu Photonics).

RESULTS AND DISCUSSION

The brightness of the image is related to the NA squared of the objective lens and to magnification squared. We optimized these optical parameters for luminescence imaging of single live cells as follows: $(NA/\beta)^2 > 0.01$, β is total magnification of the image. In this system, the value of $(NA/\beta)^2$ is equivalent to 0.35.

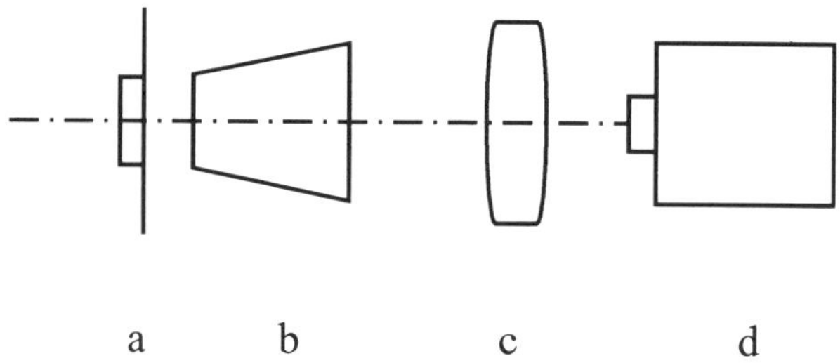

Fig. 1. Diagram of luminescence microscope developed for single live cell imaging. Luminoview LV-100 model microscope (Olympus), consisting of: a) sample stage, b) objective lens, c) tube lens and d) cooled CCD camera

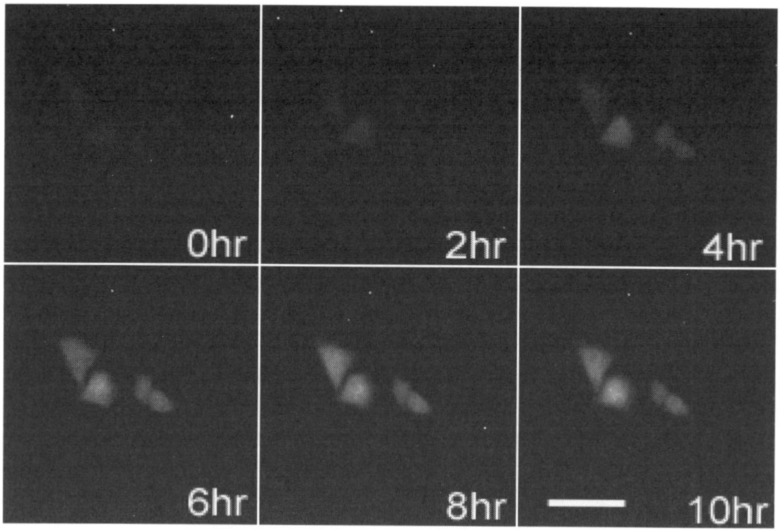

Fig. 2. Luminescence images of HeLa cells transfected Tet-luciferase vector. Luciferase gene expression is induced by tetracycline addition (2 μg/mL), and luminescence images are captured by 1 min exposure with 2 h interval over a period of 10 h (DP-30 CCD camera). 1 mM luciferin is required in culture medium (D-MEM). Scale bar indicates 100 μm

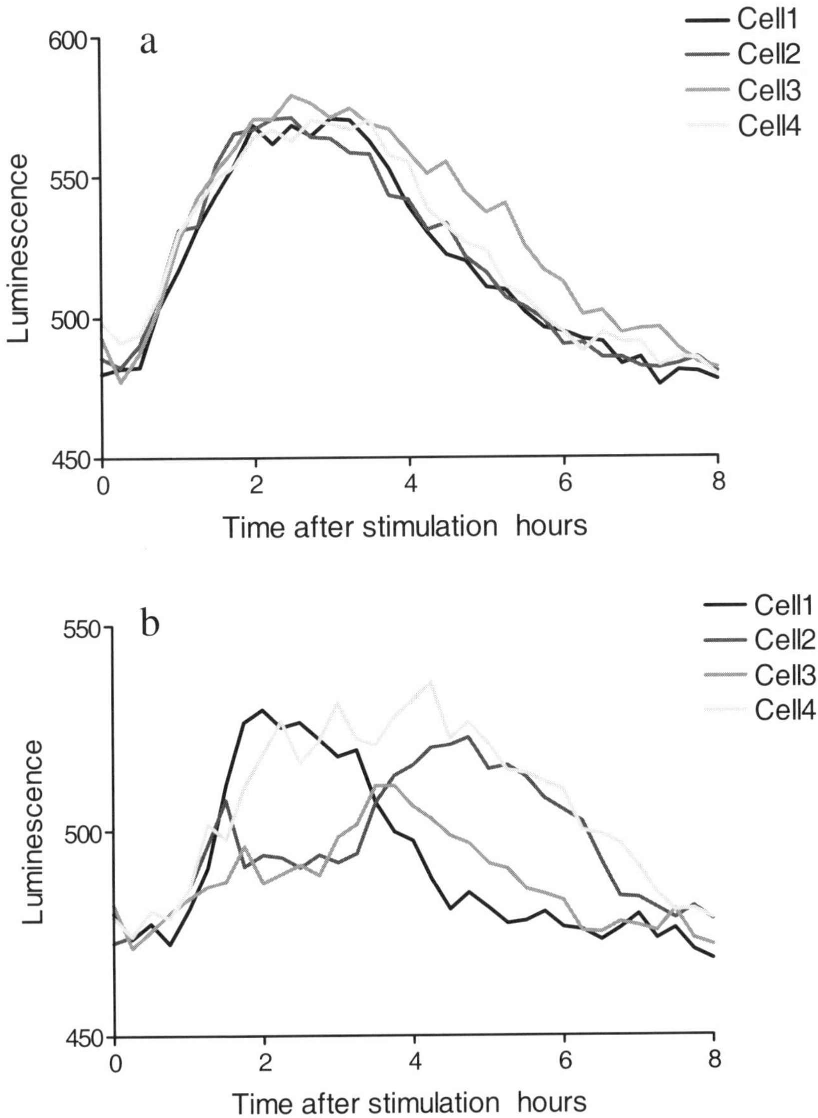

Fig. 3. Time course analysis of c-*fos* promoter activity in each cell. HeLa stable cell line transfected c-*fos* promoter construct is stimulated by 10 % FBS (a) and 100 μM ATP (b) in culture medium containing with 0.5 mM luciferin. Luminescence image is captured by 14 min exposure with 15 min interval for 8 h (ORCA-II ERG CCD camera). Binning 2x2 is required for 1344x1024 pixels CCD sensor chip

Fig. 2 shows HeLa cells transfected Tet-luciferase vector. Luciferase gene expression is induced by tetracycline addition, and its activity reaches plateau in 6 hours. The luminescence microscope constructed here proved that light emitted from single live cells can be detected by a conventional cooled CCD camera within 1 min exposure. Fig. 3 shows time course analysis of c-*fos* promoter activity in each cell. On serum (FBS) stimulation, c-*fos* promoter activity behaves homogeneously in each cell (Fig. 3a). This is equivalent to results obtained by luminometer. Nevertheless, ATP stimulation leads to heterogeneous response in promoter activity among cells temporally (Fig. 3b). These results are reproducible using both transiently transfected cells and stably transfected clonal cell lines. The heterogeneity and temporal variability in the degree of gene expression among cells was studied on prolactin promoter of lactotrophic cells by Takakusa *et al*, who pointed out that it poses the nature of stochastic phenomena controlling the rate of gene transcription in each cell.[4] The data presented here suggest this imaging promoter assay system will provide answers to the outstanding problems in luminometric promoter assays, such as transfection efficiency, physiological condition of each cell (e.g., cell cycle), heterogeneous promoter activity among cells and spatial distribution of heterogeneous cell populations in tissue.

REFERENCES

1. Rutter GA, White MR, Tavare JM. Involvement of MAP kinase in insulin signalling revealed by non-invasive imaging of luciferase gene expression in single living cells. Current Biol 1995;5:890-9.
2. White MRH, Mauko M, Amet L, Elliot G, Braddock M, Kingsman AJ, et al. Real-time analysis of the transcriptional regulation of HIV and hCMV promoters in single mammalian cells. J Cell Sci 1995;108:441-55.
3. Kennedy HJ, Violler B, Rafiq I, Kahn A, Rutter GA. Upstream stimulatory factor-2 (USF2) activity is required for glucose stimulation of L-pyruvate kinase promoter activity in single living islet β cells. J Biol Chem 1997;272: 20636-40.
4. Takasuka N, White MR, Wood CD, Robertson WR, David JR. Dynamic changes in prolactin promoter activation in individual living lactotrophic cells. Endocrinol 1998;139:1361-8.
5. Nakamura W, Yamazaka S, Takusa NN, Mishima K, Block GD. Differential response of *Period 1* expression within the suprachiasmatic nucleus. J Neurosci 2005;25:5481-7.
6. Masamizu Y, Ohtsuka T, Takashima Y, Nagahara H, Takenaka Y, Yoshikawa K, et al. Real-time imaging of the somite segmentation clock: Revelation of unstable oscillators in the individual presomitic mesoderm cells. Proc Nat Acad Sci 2006;103:1313-8.
7. Chen C, Clarkson RW, Xie Y, Hume DA, Waters MJ. Growth hormone and colony-stimulating factor 1 share multiple response elements in the c-*fos* promoter. Endocrinol 1995;136:4505-16.

PART 3

ASSAYS

ELEVATION OF CHOLINE CONCENTRATION IN CARDIAC TROPONIN-I POSITIVE HUMAN PLASMA

MACIEJ ADAMCZYK, R. JEFFREY BRASHEAR, PHILLIP G. MATTINGLY
Abbott Laboratories, Diagnostics Division, Dept. 09MJ, Bldg. Ap20, 100 Abbott Park Road, Abbott Park, Il, 60064, USA Email: maciej.adamczyk@abbott.com

INTRODUCTION

In humans, myocardial ischemia can lead to tissue necrosis with an accompanying increase in the plasma concentration of cardiac-specific troponins (cTnI and cTnT) that is related to the severity of the injury. Thus these biomarkers are useful in the diagnosis of acute myocardial infarction. Plasma cardiac troponin levels are also elevated in other etiologies such as myocarditis[1] or myocardial ischemia-reperfusion injury[2,3] as a result of coronary bypass surgery. Recently it was reported that the efflux of choline paralleled that of cardiac troponin during experimentally induced global ischemia/reperfusion in the isolated rat heart model[4] under conditions expected to produce tissue necrosis.[5] Reperfusion after a period of induced global ischemia (30 min) resulted in a biphasic release of choline in the model system, while ischemic preconditioning (four, 5 min periods of induced ischemia) blocked the second phase of choline efflux attributed to the degradation of membrane phospholipids mediated by cytostolic phospholipase A2. Choline has thus gained recent attention as a cardiac marker, particularly in patients with suspected acute coronary syndrome (ACS).[6] In this study we compared the plasma choline concentration in neat human plasma in a cardiac troponin-I positive sample population with that of a normal blood donor population. The method employed was based on the enzymatic generation of hydrogen peroxide from choline using choline oxidase and followed by detection using a chemiluminescent acridinium-9-carboxamide.[7, 8]

MATERIALS AND METHODS

Reagents. 9-[[(3-Carboxypropyl)[(4-methylphenyl)sulfonyl]amino]-carbonyl]-10-(3-sulfopropyl)acridinium inner salt was prepared as previously described.[9] Choline bitartrate (Sigma-Aldrich) was dried to constant weight over phosphorus pentoxide under vacuum in a Chem-Dry® drying apparatus (Laboratory Devices, Inc., Holliston, MA, USA) at 102°C. All aqueous solutions were prepared with HPLC grade water (EM Science) with the exception of 0.25 M sodium hydroxide, which was obtained as a premixed solution (VWR International, West Chester, PA, USA). All other reagents were obtained commercially and used as received. All prepared aqueous reagents were stored in narrow mouth, amber HDPE bottles (Nalgene-Nunc) at 2–8°C unless otherwise indicated.

Instrumentation. Chemiluminescent measurements were made on a Berthold Mithras LB940 multimode microplate reader (Berthold Technologies, Oak Ridge, TN, USA) equipped with three reagent injectors, PMT cooling and temperature control. Precision pipetting used EDP3-Plus® electronic pipettes (Rainin Instrument, LLC, Woburn, MA, USA). Assays used 96-well, non-treated black polystyrene, round-bottom trays (Costar, Corning Inc., Corning, NY, USA).

Sample preparation. Frozen (-20°C) lithium-heparin plasma samples that had tested positive for cardiac troponin-I (n = 161) and plasma samples from normal blood donors (n = 161) were obtained from the Abbott Specimen Collection Bank and thawed on ice immediately prior to analysis. All the samples were IRB approved for research use.

Assay protocol. Choline standards (150, 75, 50, 30, 20, 10, 5, 0 μM; 0.2 M phosphate buffer, pH 8) and samples (1 μL) were pipetted in replicates of 6 into the microplate, which was then allowed to equilibrate to 28°C in the microplate reader for 15 min. After that time each well was sequentially assayed. Choline oxidase (*Alcaligenes* sp.) solution (10 μL; 10 U/mL, 0.2 M phosphate buffer, pH 8, 0.1% Na cholate) was dispensed from injector 1, then after a 2 s delay the chemiluminescent indicator 3 (10 μL; 4 μM, aq 0.1 % Na cholate) and aq sodium hydroxide (30 μL; 0.25 N) were sequentially dispensed from injectors 3 and 2, respectively. The signal was recorded in 0.1 s intervals for a total of 2 s (Fig. 1).

Data Handling. Data generated was reduced using MikroWin 2000. A point-to-point calibration curve (Fig. 2) was constructed by plotting the average maximum signal (RLUmax) achieved at each standard concentration. Sample values were then calculated using the software FIT function. Sample populations were compared using MedCalc® v 0.1.01.

RESULTS AND DISCUSSION

Normal blood donor samples, by definition, meet the minimum criteria for blood donations established by the FDA: age 17–70, weight > 50kg, general healthy (no infection, high blood pressure, or anemia, etc.) and passing the medical history screening. Other factors affecting the plasma choline concentration, i.e., dietary intake and time since last meal, were not considerations. Cardiac troponin-I positive samples were greater than 0.2 ng/mL, exceeding the upper reference limit of 0.16 ng/mL for the assay used (Bayer ACS:Centaur).

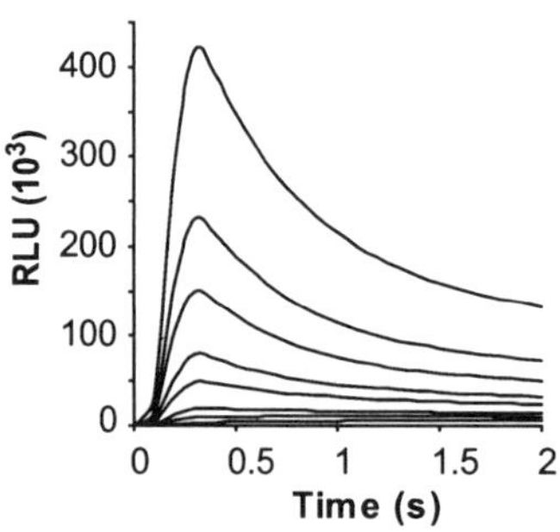

Fig. 1. Chemiluminescence profiles

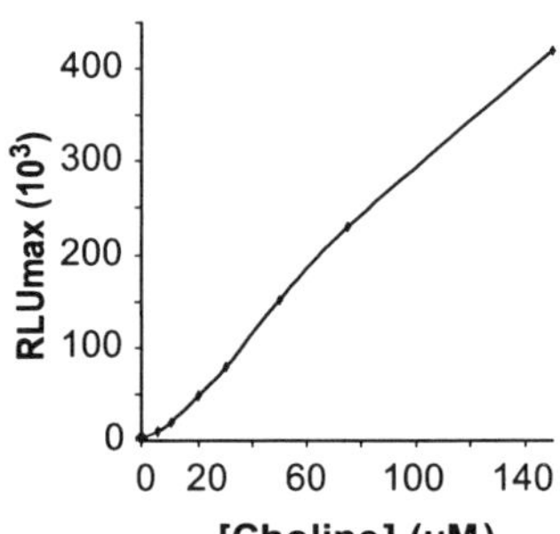

Fig. 2. Standard curve

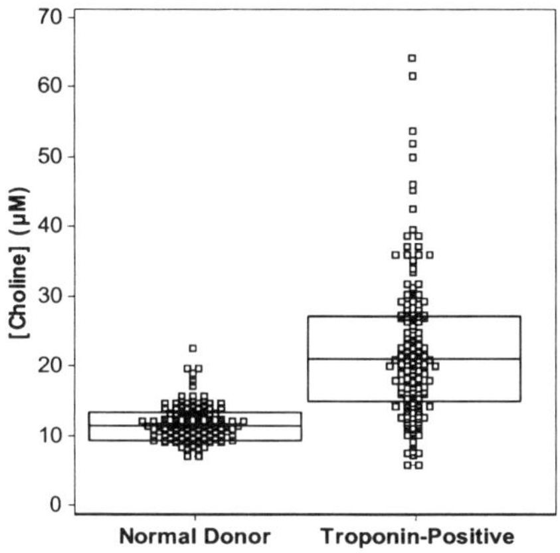

Fig. 3. Plasma choline in normal-donor and cTnI positive populations

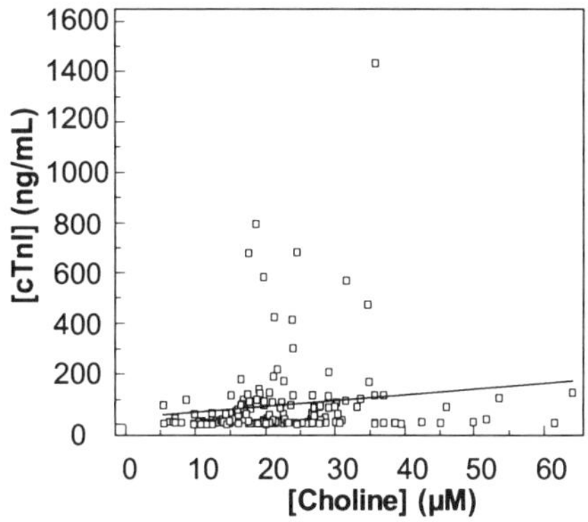

Fig. 4. Correlation between plasma cTnI and choline

The box and whisker data comparison graph in Fig. 3 shows the choline levels in the two populations. The median level of choline in the cTnI-positive population (20.60 μM, 25th–75th percentiles, 14.60–26.80 μM) was significantly higher than in the normal donor population (11.28 μM, 25th–75th percentiles, 9.73–13.13) (Mann-Whitney, p<0.0001). There was only a weakly positive correlation between the reported level of cardiac troponin-I in the samples and the measured choline concentration (Fig. 4, Spearman's coefficient of rank correlation (*rho*) 0.248, significance level P=0.0017, 95% CI for *rho* 0.097–0.388). Receiver operator curve (ROC) analysis is shown in Fig. 5. At a cutoff level of 14.6 μM, the present plasma choline assay showed sensitivity of 74.5%, specificity of 91.3% and AUC of 0.861.

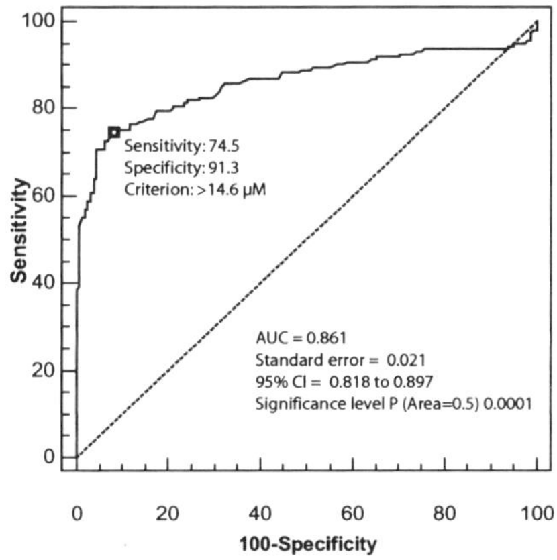

Fig. 5. ROC curve for plasma choline in normal donor blood and cTnI positive populations

By comparison, the ARCHITECT STAT Troponin-I Assay has sensitivity of 60%, specificity of 95.4% and AUC of 0.918 at a cTnI concentration of 0.3 ng/mL, 0-6 h post-admission.

CONCLUSIONS

There was significant overlap between the plasma choline level measured in normal donor and cardiac troponin-positive populations. We speculate that this could be due to the relatively fast equilibration of free plasma choline with the phospholipid pool (hours) and the persistence of circulating cardiac troponins (days).

REFERENCES

1. Alpert JS, Thygesen K, Antman E, Bassand JP. Myocardial infarction redefined—a consensus document of the Joint European Society of Cardiology/American College of Cardiology Committee for the redefinition of myocardial infarction. J Am Coll Cardiol 2000;36:959–69.
2. Thielmann M, Marggraf G, Neuhauser M, Forkel J, Herold U, Kamler M, et al. Administration of C1-esterase inhibitor during emergency coronary artery bypass surgery in acute ST-elevation myocardial infarction. Eur J Cardiothorac Surg 2006;30:285–93.
3. Enc Y, Karaca P, Ayoglu U, Camur G, Kurc E, Cicek S. The acute cardioprotective effect of glucocorticoid in myocardial ischemia-reperfusion injury occurring during cardiopulmonary bypass. Heart Vessels 2006;21:152-6.
4. Bruhl A, Hafner G, Loffelholz K. Release of choline in the isolated heart, an indicator of ischemic phospholipid degradation and its protection by ischemic preconditioning: no evidence for a role of phospholipase D. Life Sci 2004;75: 1609–20.
5. Hearse DJ. Myocardial protection during ischemia and reperfusion. Mol Cell Biochemistry 1998;186:177–84.
6. Apple FS, Wu AH, Mair J, Ravkilde J, Panteghini M, Tate J, et al. Future biomarkers for detection of ischemia and risk stratification in acute coronary syndrome. Clin Chem 2005;51:810–24.
7. Adamczyk M, Brashear RJ, Mattingly PG. Rapid high-throughput detection of peroxide with an acridinium-9-carboxamide: A homogeneous chemiluminescent assay for plasma choline. Bioorg Med Chem Lett 2006;16: 2407–10.
8. Adamczyk M, Brashear RJ, Mattingly PG, Tsatsos PH. Homogeneous chemiluminescent assays for free choline in human plasma and whole blood. Anal Chim Acta 2006;579:61–7.
9. Adamczyk M, Chen Y-Y, Mattingly PG, Pan Y. Neopentyl 3-triflyloxypropanesulfonate. A reactive sulfopropylation reagent for the preparation of chemiluminescent labels. J Org Chem 1998;63:5636–9.

HIGH THROUGHPUT DETECTION OF HYDROGEN PEROXIDE: VALIDATION OF HOMOGENEOUS CHEMILUMINESCENT ASSAYS FOR CHOLINE IN HUMAN PLASMA AND WHOLE BLOOD

MACIEJ ADAMCZYK, R. JEFFREY BRASHEAR,
PHILLIP G. MATTINGLY, PANAGIOTA H. TSATSOS
Abbott Laboratories, Diagnostics Division, Dept. 09MJ, Bldg. Ap20,
100 Abbott Park Rd., Abbott Park, Il, 60064, USA
Email: maciej.adamczyk@abbott.com

INTRODUCTION

Choline is a fascinating molecule, whose metabolism/function is intertwined with that of phospholipid synthesis and intercellular signaling, cholinergic neurotransmission, and homocysteine biosynthesis. The concentration of free choline is highly regulated in humans and can be perturbed by diet, pregnancy and pathological conditions such as ischemia and renal failure. Additionally, a number of enzymes of current interest as biomarkers/drug targets lead directly (lysophospholipase D/autotaxin[1]) or indirectly (phospholipase A2[2]) to choline. Thus, a rapid high throughput method for choline analysis is highly desirable. We have developed a method of quantitation of choline in undiluted human plasma and ultrafiltered whole blood based on the enzymatic generation of hydrogen peroxide from choline using choline oxidase and acridinium-9-carboxamide chemiluminescent detection.[3, 4]

MATERIALS AND METHODS

Reagents. 9-[[(3-Carboxypropyl)[(4-methylphenyl)sulfonyl]amino]-carbonyl]-10-(3-sulfopropyl)acridinium inner salt was prepared as previously described.[5] Choline bitartrate (Sigma-Aldrich) was dried to constant weight over phosphorus pentoxide under vacuum in a Chem-Dry® drying apparatus (Laboratory Devices, Inc., Holliston, MA, USA) at 102°C. All aqueous solutions were prepared with HPLC grade water (EM Science) with the exception of 0.25 M sodium hydroxide, which was obtained as a premixed solution (VWR International, West Chester, PA, USA). All other reagents were obtained commercially and used as received. All prepared aqueous reagents were stored in narrow mouth, amber HDPE bottles (Nalgene-Nunc) at 2–8°C unless otherwise indicated.

Instrumentation. All chemiluminescent measurements were performed on a Berthold Mithras LB940 multimode microplate reader (Berthold Technologies U.S.A. LLC, Oak Ridge, TN, USA) equipped with three reagent injectors, PMT cooling and temperature control. Precision pipetting was done using EDP3-Plus® electronic pipettes (Rainin Instrument, LLC, Woburn, MA, USA). All assays were performed in 96-well, non-treated black polystyrene, round-bottom trays (Costar, Corning Inc., Corning, NY, USA). Ultrafiltration of samples was carried out in Nanosep® 10K Omega™ centrifugal devices (Pall Corp., East Hills, NY, USA) on a

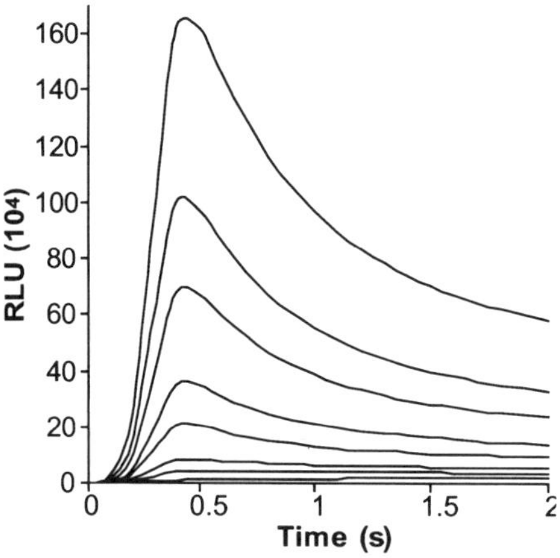

Fig. 1. Chemiluminescence profiles

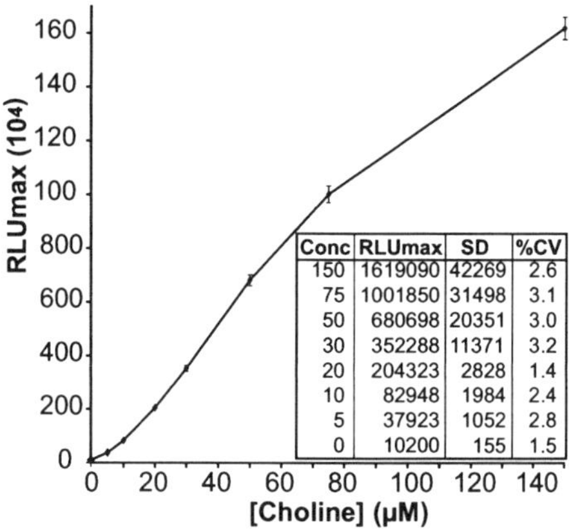

Conc	RLUmax	SD	%CV
150	1619090	42269	2.6
75	1001850	31498	3.1
50	680698	20351	3.0
30	352288	11371	3.2
20	204323	2828	1.4
10	82948	1984	2.4
5	37923	1052	2.8
0	10200	155	1.5

Fig. 2. Standard curve

Hermle Z230MA microcentrifuge (B.Hermle AG, Gosheim, Germany) at 15000 rpm for 15 min. LC-MS/MS measurements were performed using an Agilent HPLC system (Agilent Technologies) equipped with a cyanopropyl bonded phase column (Supelcosil LC-CN, 3μ. 7.5 cm × 4.6 mm) and temperature-controlled autosampler coupled to an API 4000 LC/MS/MS equipped with a TurboIonSpray probe (Applied Biosystems).

Sample preparation. Frozen human plasma collected on lithium heparin obtained from the Abbott Specimen Collection Bank was thawed on ice, and then centrifuged for 5 min (15000 rpm) before directly assaying. Frozen human blood collected on EDTA obtained from the Abbott Specimen Collection Bank was thawed on ice and then an aliquot (100 μL) was ultrafiltered before analysis.

Chemiluminescent assay protocol. Choline standards (150, 75, 50, 30, 20, 10, 5, 0 μM; 0.2 M phosphate buffer, pH 8) and samples (4 μL) were pipetted in replicates of 4 into the microplate, which was allowed to equilibrate to 28°C in the microplate reader for 15 min. After that time each well was sequentially assayed. Choline oxidase (*Alcaligenes* sp.) solution (40 μL; 10 U/mL, 0.2 M phosphate buffer, pH 8, 0.1% Na cholate) was dispensed from injector 1, then after a 2 s delay the chemiluminescent indicator 3 (40 μL; 4 μM, aq 0.1 % Na cholate) and aq sodium hydroxide (100 μL; 0.25 N) were sequentially dispensed from injectors 3 and 2, respectively. The signal was recorded in 0.1 s intervals for a total of 2 s (Fig. 1). The data generated was reduced using MikroWin 2000 supplied with the Berthold Mithras instrument. A point-to-point calibration curve (Fig. 2) was constructed by plotting the average maximum signal (RLUmax) achieved at each standard concentration. Sample values were then calculated using the software FIT function.

LC-MS/MS assay protocol. Choline standards, plasma or whole blood samples (20 μL) and a solution of d_9-choline in acetonitrile (10 μM, 100 μL) were vortexed for 5 s, then centrifuged (10 min, 10000 rpm). Afterwards, the supernatant (20 μL) was transferred to an autosampler vial containing aq. ammonium formate (15 mM, pH 3.5)/methanol/acetonitrile [1 mL, 10:6.5:3.5 v/v/v]. Standards and samples were loaded into the autosampler equilibrated to 4°C. The HPLC column was equilibrated with mobile phase [aq. ammonium formate (15 mM, pH 3.5)/methanol/acetonitrile (10:6.5:3.5 v/v/v)] and the samples (15 μL) were eluted (600 μL/min). The signal for the choline peak and d_9-choline internal standard (retention time, 2.72 min) was monitored through multiple reaction monitoring (MRM) [declustering potential: 33 V; collision energy: 30 V; collision cell exit potential: 3 V; curtain gas: 30 psi; nebulizer gas: 40 psi; turbo gas: 25 psi; ion spray voltage: 5100 V; Turbo Ion Spray temperature: 550 °C] using the following transitions: choline, m/z 104 → 60, d9-choline m/z 113 → 69. Data for duplicate replicates were elaborated using the supplied Analyst software.

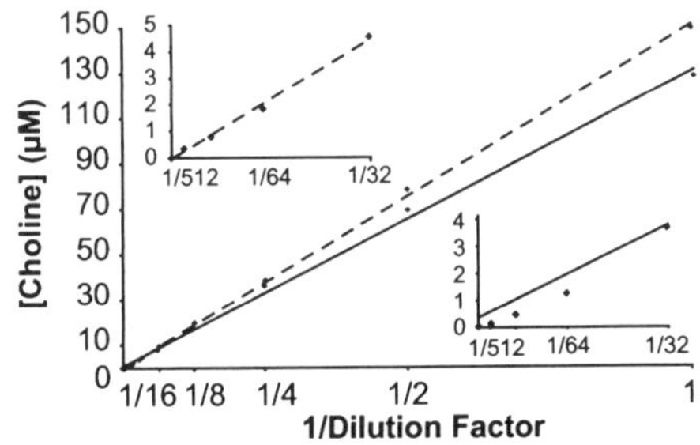

Fig. 3. Dilution linearity: ultrafiltered lysed whole blood (---); neat plasma (–)

RESULTS AND DISCUSSION

Precision. A pooled plasma sample with a low choline concentration (15 μM) was prepared from 39 individual samples that had been previously analyzed. A portion was aliquoted and stored frozen at –70°C while the remainder was spiked with the choline stock solution to give high choline concentration pools (32, 42 μM), which was likewise aliquoted and frozen. Pooled samples of whole blood were similarly prepared (40, 55 μM). Each of the sample pools was analyzed in four runs per day for two consecutive days, constituting 48 results per sample pool. The within run %CV ranged from 1.4–2.1 %, while the total %CV was 1.4–4.3 %.

Recovery. Either of two concentrated solutions of choline in phosphate buffer (52.6 μL, 200 μM or 400 μM) was spiked into aliquots (1000 μL) of each plasma (15.1, 30.3 μM) or whole blood pool (33.6, 46.3 μM) to increase the initial concentration of the aliquot by 10 μM and 20 μM, respectively. Recovery was calculated using the following formula: ([Measured]/ [Expected]) × 100. Recovery in the neat plasma samples ranged from 90–95 %, while recovery in ultrafiltered lysed whole blood was 99–106 %.

Dilution Linearity. A plasma and whole blood sample with a high choline concentration were chosen from previously analyzed samples. The neat plasma sample was serially two-fold diluted in the zero calibrator. The ultrafiltrate prepared from the whole blood sample was similarly diluted. Both samples diluted linearly (plasma: y=128.91x + 0.3846 R^2 = 0.9976; whole blood: y=150.77x + 0.0869 R^2 = 0.9994.)

Interferences. Specificity (< 15% interference) was demonstrated by spiking plasma and whole blood samples with a series of endogenous substances (bilirubin, triglycerides, protein, hemoglobin) and common drugs (ascorbic acid, aspirin, heparin, etc.). Hemoglobin (550 mg/dL) showed a positive interference of 84.5% in the neat plasma assay; the interference dropped to 12% at 137 mg/dL where the sample was still visibly colored. Nifedipine, an oxidizable dihydropyridine drug, caused 19.9% negative interference in the neat plasma assay.

LC-MS/MS Correlation. Plasma (n = 56) and whole blood samples (n = 27) were analyzed the chemiluminescent and LC-MS/MS methods. The chemiluminescent method correlated well with LC-MS/MS in the analysis of both sample matrices (plasma: slope, 0.9122 R = 0.9890; whole blood: slope, 0.9391 R = 0.9708.)

CONCLUSIONS

We developed a simple, semi-automated, high-throughput microplate assay for quantifying choline in unextracted human plasma and lysed whole blood. The specificity of assay resides in the use of the enzyme choline oxidase, which converts choline to betaine with the concomitant production of hydrogen peroxide. The peroxide is detected using a chemiluminescent acridinium-9-carboxamide. The use of chemiluminescence allows assaying of choline in the range of 0–150 μM with a limit of detection of 0.1 μM.

REFERENCES

1. Ferguson CG, Bigman CS, Richardson RD, vanMeeteren LA, Moolenaar WH, Prestwich GD. Fluorogenic phospholipid substrate to detect lysophospholipase D/autotaxin activity. Org Lett 2006;8:2023-6.
2. Koenig W, Twardella D, Brenner H, Rothenbacher D. Lipoprotein-associated phospholipase A2 predicts future cardiovascular events in patients with coronary heart disease independently of traditional risk factors, markers of inflammation, renal function, and hemodynamic stress. Arterioscler Thromb Vasc Biol 2006;26:1586-93.
3. Adamczyk M, Brashear RJ, Mattingly PG. Rapid high-throughput detection of peroxide with an acridinium-9-carboxamide: A homogeneous chemiluminescent assay for plasma choline. Bioorg Med Chem Lett 2006;16:2407-10.
4. Adamczyk M, Brashear RJ, Mattingly PG, Tsatsos PH. Homogeneous chemiluminescent assays for free choline in human plasma and whole blood. Anal Chim Acta 2006;579:61-7.
5. Adamczyk M, Chen Y-Y, Mattingly PG, Pan Y. Neopentyl 3-triflyloxypropanesulfonate. A reactive sulfopropylation reagent for the preparation of chemiluminescent Labels. J Org Chem 1998;63:5636-9.

DEVELOPMENT OF BIOLUMINESCENT PYROPHOSPHATE ASSAY USING PYRUVATE PHOSPHATE DIKINASE AND ITS APPLICATION TO SNPS ANALYSIS

H ARAKAWA,[1] E MUNAKATA,[1] R OBINATA,[1] M MAEDA,[1]
S SUZUKI,[2] M KAMAHORI,[3] H KAMBARA[3]
[1] *School of Pharmaceutical Sciences, Showa University, Tokyo 142-8555, Japan;*
[2] *Research & Development Division, Kikkoman Corporation, Chiba 278-0037, Japan;* [3] *Central Research Laboratory, Hitachi Ltd, Tokyo 185-8601, Japan*
Email: arakawa@pharm.showa-u.ac.jp

INTRODUCTION

DNA analysis is an important technology for diagnosising infectious disease and tailored medication. In this study, we developed a novel bioluminescent assay for pyrophosphate and it was applied to SNPs (Single Nucleotide Polymorphism) analysis using single base extension reaction. The principle of this method is as follows: A specific primer within each aliquot possessing a short 3' end of the base of interest was hybridized to the single-stranded DNA template, subsequently, one of either dATPa-S, dTTP, dGTP and dCTP, and (exo-) Klenow DNA polymerase were added and incubated for 1 min. Pyrophosphate released by DNA polymerase is converted to ATP by pyruvate phosphate dikinase (PPDK) in presence of pyruvate phosphate and AMP, and the concentration of ATP is determined using the firefly luciferase reaction. When this technique was applied to the K-*ras* gene, it was possible to determine clearly one point mutation in codon 12. This method, which does not require expensive equipment, can be utilized to rapidly monitor one point mutation per gene.

MATERIALS AND METHODS

Chemicals. *PPDK* from *Microbispora rosea* subsp. *aerata* (EC2.7.9.1) and thermostable *Luciola cruciata* firefly luciferase (EC 1.13.12.7) were obtained from Kikkoman Co. (Chiba, Japan). Pyrophosphate, dNTP Mixture and (exo-) Klenow DNA Polymerase were acquired from Takara Shuzo Co., Ltd (Osaka, Japan). a-thio-dATP (dATPa-S) was from Amersham Biosciences (USA). Primers were synthesized by Takara Shuzo Co., Ltd (Osaka, Japan). Luciferin was obtained from Sigma Chemical Co (St. Louis, MO, USA). Other reagents were of analytical grade.

Methods: Bioluminescence SNPs analysis with the novel PPi assay. A specific primer within each aliquot possessing a short 3' end of the base of interest was hybridized to the single-stranded DNA template (p53 gene or K-*ras* gene (1~50 pmol)). Subsequently, one of either dATPa-S, dTTP, dGTP and dCTP, and (exo-) Klenow DNA polymerase were added and incubated for 1 min. PPi produced during the one base extension assay was measured with the PPDK-bioluminescence assay. That is, determination of the single base extension was assessed by comparison of the emission intensity upon addition of the four dNTP types.

Bioluminescent detection of pyrophosphate. Pyrophosphate solution or extension assay solution (10 μL) was introduced to a test tube; subsequently, 10 μL of PPDK luciferin-luciferase solution (2.34 U/mL PPDK, 0.2 mM luciferin, 5.5 U/mL luciferase 0.0125 mM AMP, 0.04 mM PEP, 0.005 U/mL apyrase, 0.05 mM DTT, 5% treharose, 1 mM EDTA, 7.5 mM $MgSO_4$, 30 mM BES, pH 8.0) was injected. Emission intensity was monitored for 30 seconds using the subminiuture luminescence reader (Hitachi Ltd., Japan).

RESULTS AND DISCUSSION

The principle of bioluminescence SNPs analysis is based on the luciferin /luciferase reaction following production of PPi in extension reaction, which is converted into ATP in the presence of PPDK (Fig. 1).

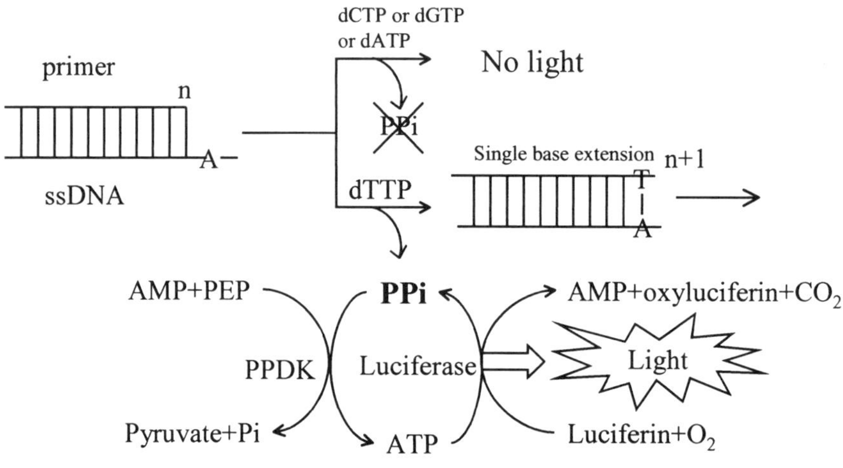

Fig. 1. Principle of bioluminescence SNPs analysis

PPi assay using PPDK/luciferase method[2] developed previously for PCR product was improved slightly for this analysis. In short, the reaction conditions were set in order to monitor sequentially the extension assay in the event that light emission may disappear quickly as shown in Fig. 2-a. Detection of 5.0×10^{-16} mol/assay of PPi was possible by this method as shown in Fig. 2-b. High sensitivity exceeding that afforded by the ATP-sulfurylase / luciferase detection system[1] and PPDK/luciferase for the PPi assay[2] was obtained. The reproducibility of the standard curve was 3.2 -

6.3% (CV, n=6). Subsequently, this novel bioluminescence assay was applied to the SNPs analysis in p 53 gene and K-*ras* gene.

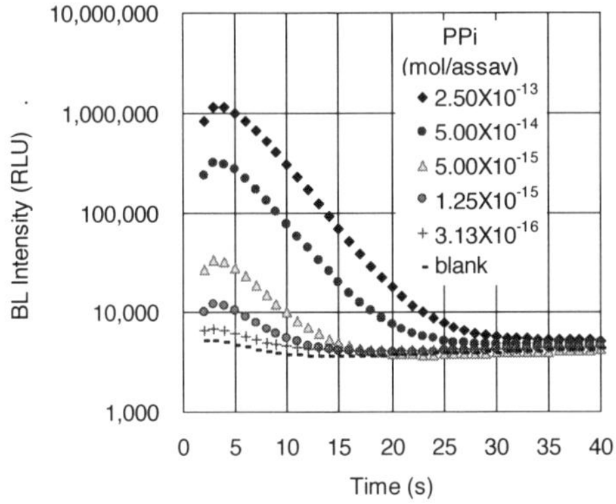

Fig. 2-a. Time course of luminescent intensity

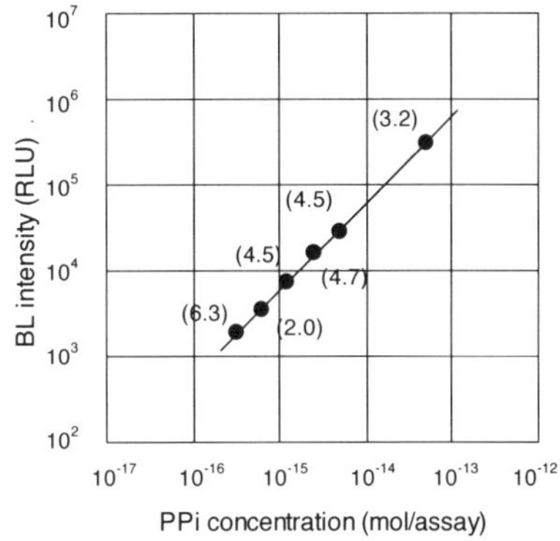

Fig. 2-b. Standard curve

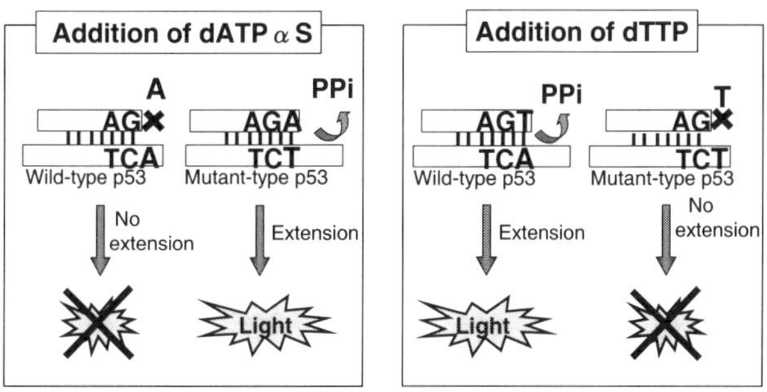

Fig. 3. Principle of bioluminescent detection on SNPs analysis for p 53 gene

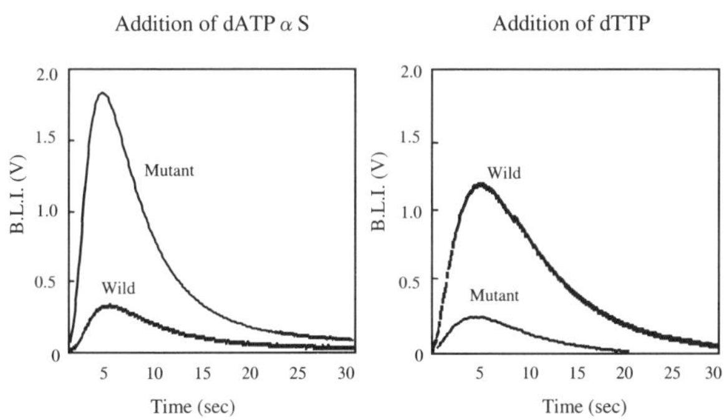

Fig. 4. Emission patterns with mutation and wild-type p53 gene

Table 1. SNPs analysis

	S/N ratio*			
K-ras DNA	dATPα S	dGTP	dCTP	dTTP
Ala (5'-GC-3')	0.7	134.7	1.3	4.0
Gly (5'-GG-3')	1.1	1.3	258.9	2.5
Cys (5'-TG-3')	0.7	1.2	139.4	2.4
Ser (5'-AG-3')	0.9	1.0	139.0	2.6
Asp (5'-GA-3')	1.0	1.9	1.3	144.0
Val (5'-GT-3')	104.3	1.7	1.0	2.4
Arg (5'-CG-3')	0.8	1.8	131.3	3.1

$$^{*}\text{S/N ratio} = \frac{\text{Bioluminescent intensity of extension product with DNA}}{\text{Bioluminescent intensity of extension product without DNA}}$$

This technique was applied to the p53 gene (Fig. 3). The results showed that it was possible to determine clearly the wild type containing A and mutant type containing T at the identical position of the mutation site in one minute (Fig. 4). Next this technique was applied to the K-*ras* gene which have seven SNPs in codon12. Wild type is GGT for Gly, and mutation types are CGT for Arg, GCT for Ala, TGT for Cys, AGT for Ser, GAT for Asn and GTT for Val. In case of Ala mutation, a base of mutation site is C. When complementary dGTP is added in extention assay, single base extension reaction is generated and light is emitted by PPDK/luciferase reaction. Light emission was not seen with other nucleotides. S/N ratios of emission intensity obtained with reactions of dATPa-S, dTTP, dGTP and dCTP for each K-*ras* SNPs were shown in Table 1.

As a result, S/N ratio in adding the complementary nucleotide for each SNPs showed values of 104.3 - 258.9. At K-*ras* of Gly, since there are two G at the mutation site, continued two base extension was generated with dCTP, and the emission intensity showed about double the value in comparison with single base extention. This method allowed determination of one point mutation for K-*ras* gene. Application of this method to SNP analysis for tailored medications appears possible.

REFERENCES

1. Imamura O, Arakawa H, Maeda M. Simple and rapid bioluminescent detection of two verotoxin genes using allele specific PCR of *E. coli* O157:H7. Luminescence 2003;18:107-12.
2. Arakawa H, Karasawa K, Igarashi T, Suzuki S, Goto N, Maeda M. Detection of cariogenic bacteria genes by a combination of alle-specific polymerase Chain reactions and a novel bioluminescent pyrophosphate assay. Anal Biochem 2004;333:296-306.

BIOLUMINESCENT METHOD FOR DETECTION OF PROTEINASE ACTIVITY IN UHT MILK

LY BROVKO,[1] C HAJDOK,[1] C TAPP,[1] C RAMSAY,[2] M DRISCOLL,[2] MW GRIFFITHS[1]

[1] *Canadian Research Institute for Food Safety, University of Guelph, ON N1G 2W1, Canada, E-mail: lbrovko@uoguelph.ca;* [2] *Biotrace International Plc, The Science Park, Bridgend, Wales CF31 3NA, United Kingdom*

INTRODUCTION

Thermostable proteinases from *Pseudomonas* spp. are an important source of spoilage of Ultra-High Temperature (UHT) treated dairy products. These enzymes destroy casein micelles resulting in milk gelation. Thus, the level of protease activity in UHT milk is a good indicator of product shelf-life. Several methods to assay protease in milk have been described, but they are not sufficiently sensitive, rapid or easy to perform to meet the requirements of the dairy industry. Recently, a homogeneous one-step bioluminescent method was proposed for the detection of the therapeutically-important protease – Caspase-3.[1] This method is based on the cleavage of a synthetic peptide conjugated with luciferin and subsequent release and detection of the latter by firefly luciferase. Though very sensitive (LOD [limit of detection] ~2-3 mU/mL) and fast (1 h), this method is probably too specific for detection of proteases in milk. The peptide conjugated with luciferin has a highly conservative recognition site for Caspase-3. To apply this approach for detection of proteases in milk several different substrates should be synthesized that are specific to each group of milk proteases. This may significantly increase the cost of the analysis. Another problem with development of protease assays is identifying the necessary detection limit. Conditions for protease activity assays are not standardized and this has led to difficulties in identifying a suitable reference method and, consequently, determining levels of protease activity that limit the shelf-life of UHT products.

Firefly luciferase widely used in analysis due to the very high sensitivity of its detection by bioluminescence. Another advantage of bioluminescent methods is their ability to detect low levels of luciferase activity directly in turbid samples, such as bacterial suspensions or milk and even from inside the body of small animals without any pre-treatment.[2] One distinctive property of firefly luciferase is its pronounced susceptibility to proteolysis.[3] We investigated the possibilities of using firefly luciferase as a substrate of milk proteinase thus providing a method for assessing the level of proteinase activity in UHT milk.

MATERIALS AND METHODS

Reagents and instumentation. Luciferin-luciferase reagents for protease assays were supplied by Biotrace International (Bridgend, UK). ATP (Sigma; 0.1 mM) was prepared daily from a frozen 1 mM stock solution and kept on ice during the time of the experiment. The following proteases from Sigma-Aldrich were used – Trypsin (EC 3.4.21.4) 11,700 BAEE U/mg; Pronase E (EC 3.4.24.31) 5.3 U/mg based on casein proteolysis; Subtilysin Carlsberg (EC 3.4.21.62) 8.8 U/mg based on casein hydrolysis. Reconstituted skim milk (RSM) was prepared from skim milk powder (10% w/v) and autoclaved for 15 min at 121°C to eliminate bacteria and inactivate all proteinases. Protease fluorescent detection kit (Sigma) was used according to manufacturer's instructions with 24 h incubation time. Intensity of bioluminescence was measured using a BG-1 tube luminometer from GEM Biomedical (Sparks, NV, USA). Fluorescence was measured using the Victor® Multilabel Plate Reader from Perkin Elmer Wallac, Canada.

Two-step bioluminescent method for proteinase assay in milk. The experimental protocol was as follows: 100 μL aliquots of LLR were mixed with 100 μL of RSM milk (diluted 1:4 in buffer) that contained different concentrations of specific proteinase in the range 0 – 10 U/mL. The mixture was incubated in a water bath at 30 °C. Bioluminescence was measured at time 0 and after 20 min of incubation by addition of 20 μL of 0.1 mmol/L ATP.

One-step bioluminescent method for proteinase assay in milk. The experimental protocol was as follows: 100 μL of LLR were mixed with 20 μL of RSM milk (diluted 1:2 in buffer) containing different concentrations of specific proteinase in the range 0 – 0.2 U/mL. The bioluminescent reaction was initiated by adding 50 μL of 0.1 mmol/L ATP and bioluminescence was monitored during 2 h.

RESULTS AND DISCUSSION

Optimization of bioluminescent Luciferin-Luciferase Reagent (LLR) composition. Firefly luciferase is highly susceptible to proteolysis, thus monitoring the change of its activity over time provides a way to assess proteinase activity. However, to detect low levels of proteolytic activity, minor changes in luciferase activity must be detected, which can be difficult considering the accuracy of measurements ranging on average from 5 to 10%. Another problem with the detection of proteolytic activity in milk is that milk itself has very high protein content and the added protease substrate (luciferase in our case) competes with milk proteins for active sites of enzymes. The strategy that has been used for reagent optimization was based on the following assumptions: 1) concentration of luciferase should be small enough so even low protease activity will result in more than 15% loss of luciferase activity; 2) concentration of luciferase should be high enough to be able to compete with milk proteins for proteolysis; 3) luciferase should be stable at test conditions; 4) bioluminescent signal should be constant in time (glow-type) to provide the possibility of one-step analysis. The optimized LLR composition showed high thermostability (90-95% remaining activity after 1 h at 30°C) and a very stable signal at 30 μM of ATP. These conditions were used in the following experiments.

Two-step bioluminescent method for proteinase assay in milk. The ratio of bioluminescent readings at 20 and 0 min was calculated and plotted against concentration of proteinase added (Fig. 1). The percent of residual activity was strongly dependent on the amount of proteinase added in the range of protease activity used (53-2650 mU/mL). A limit of detection (LOD, 85% of residual activity) of approximately 50 mU of Pronase E per mL in diluted milk was obtained and this is equivalent to 200 mU/mL of non-diluted milk.

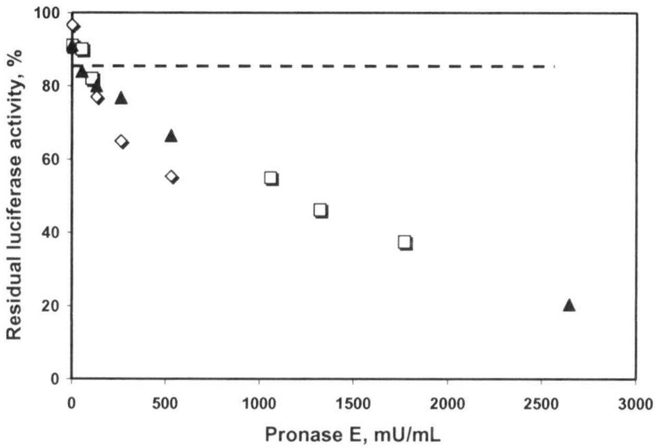

Fig. 1. Residual luciferase activity after 20 min incubation at 30°C in the presence of Pronase E. Different symbols represent data obtained in 3 different experiments. Dashed line represents LOD – 85% of residual activity

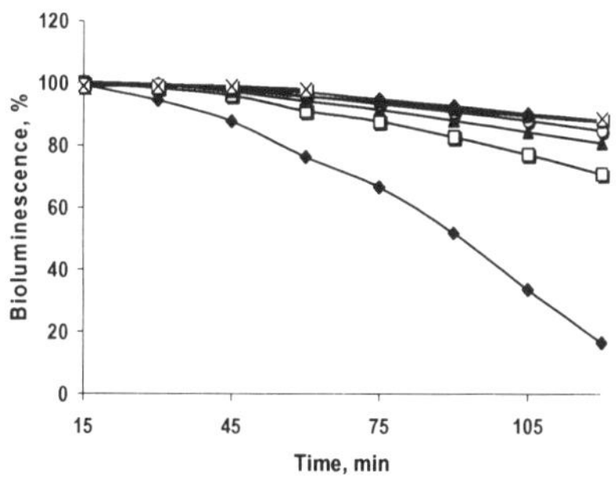

Fig. 2. Time course of bioluminescence intensity for milk samples containing Pronase E (mU/mL): 166 – (♦), 83 –(□); 41 (▲); 20 (●); 10 (○); 5 (Δ); 0 (×)

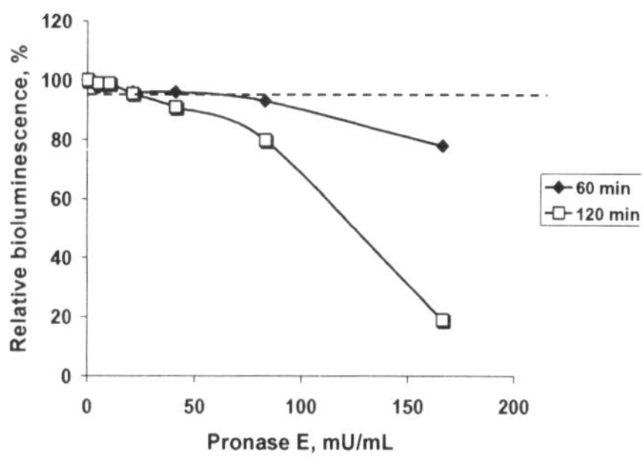

Fig. 3. Relative bioluminescence intensity at 1 and 2 h for milk samples containing Pronase E. Dashed line represents LOD (95% bioluminescence for the sample with 0 proteinase)

One-step bioluminescent method for proteinase assay in milk. The results obtained using the one-step assay are shown in Fig. 2 and 3. The decline in bioluminescence was dependent on the incubation time with the milk containing proteinase activity > 20 mU/mL. The accuracy of time-course bioluminescence measurements is high (CV<5%); thus providing a way to detect proteinase activity in milk with a LOD around 25 mU/mL. This is equivalent to 50 mU/mL for undiluted milk which slightly better than the value obtained with the two-step procedure. The results generated by the one-step bioluminescent method for detection of proteinase activity in milk spiked with different Trypsin concentrations (0 - 6.2 ng/mL) were compared with those obtained using the commercial fluorescent method based on proteolysis of FITC labelled casein (Fig. 4, a-c)

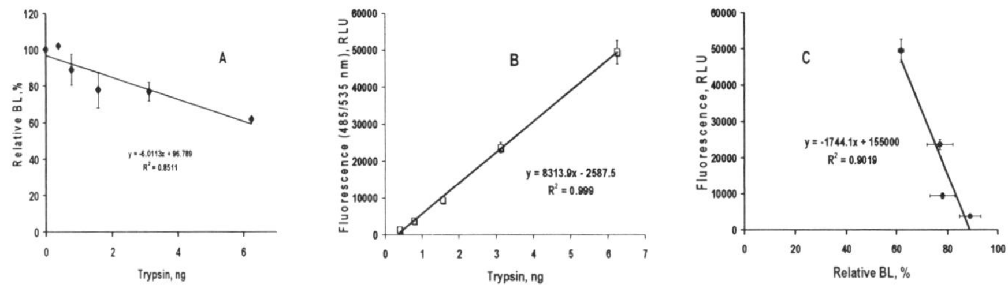

Fig. 4. Comparison of bioluminescent and fluorescent methods of detection of proteinases in milk. Averaged results of 2 different experiments. A – Relative decline in bioluminescence for milk samples containing 0-6.25 ng/mL of Trypsin after 1 h of incubation at room temperature; B – Fluorescence intensity of milk containing Trypsin (0-6.25 ng/mL) and FITC-casein after incubation of 24 h at 37°C; C – Regression analysis

In preliminary experiments, it was shown that the fluorescent method is not sufficiently sensitive to detect Trypsin concentrations (0-6.25 ng/mL) in milk after 1 h incubation. A 1 h incubation time at room temperature was used for BL method and 24 h incubation at 37°C was used for the FL assay. There was high correlation between the methods (R^2= 0.90) at these conditions. The one-step bioluminescent assay format is faster than the fluorescent method. It has a high sensitivity and can be easily adapted for high throughput analysis.

REFERENCES

1. O'Brien MA, Daily WJ, Hesselberth PE, Moravec RA, Scurria MA, Klaubert DH, et al. Homogeneous, bioluminescent protease sssays: Caspase-3 as a model. J Biomol Scr 2005;10:137-48.
2. Greer III LF, Szalay AA. Imaging light emission from the expression of luciferases in living cells and organisms: a review. Luminescence 2002;17:43-74.
3. Thompson JF, Geoghegan KF, Lloyd DB, Lanzetti AJ, Magyar RA, Anderson SM, et al. Mutation of protease-sensitive region in firefly luciferase alters light emission properties. J Biol Chem 1997;272:18766-71.

DETECTION AND CONTROL OF *SALMONELLA* TYPHIMURIUM GROWTH USING SPECIFIC BACTERIOPHAGE

LY BROVKO,[1] NA ROMANOVA,[1] B ALLAIN,[2]
R MANDEVILLE,[2] MW GRIFFITHS[1]
[1] *Canadian Research Institute for Food Safety, University of Guelph, ON N1G 2W1, Canada, E-mail: lbrovko@uoguelph.ca;*
[2] *BiophagePharma Inc., 6100 Royalmount, Montreal QC H4P 2R2, Canada*

INTRODUCTION

Arguably the best way to prevent food-borne infections is by adhering to strict hygiene standards. However, some of the pathogens that cause such infections have a high level of resistance to disinfectant treatments. It was shown recently that adaptation of microorganisms to sanitizers commonly used in the food industry can occur and disinfectants may also result in enhanced resistance to antibiotics.[1] Accordingly, there is an urgent need for novel methods to detect and control the growth of pathogens in food-processing environments that can reveal and eliminate both sensitive and resistant strains of bacterial pathogens. A promising approach to eliminate a target pathogen is to use specific lytic bacteriophages. It was previously shown that certain bacteriophages can lyse both antibiotic sensitive and resistant bacterial strains effectively,[2] thus providing the means for both detection and elimination of pathogens, including antibiotic-resistant strains. The number of bacteriophages present in the environment is almost endless. However, very few bacteriophages have been characterized well enough in terms of their specificity and lytic activities for them to be considered as potential therapeutic agents. This is partly due to the time- and labour-consuming methods used for phage assays. We propose to use ATP bioluminescence to monitor the effect of phages on bacterial cell growth and lysis and thus provide a means for high throughput phage screening. In this work, the ability of two different bacteriophages to detect and control growth of *Salmonella* Typhimurium was investigated using ATP-bioluminescence.

MATERIALS AND METHODS

Bacteriophages and Bacterial Cultures. Two lytic bacteriophages BP1 and BP6 specific for *Salmonella* Typhimurium were isolated and purified by Biophage Pharma Inc. (Montreal, Quebec, Canada). Bacteriophages were propagated using the agar overlay method.[3] Enumeration of phage particles was performed by plaque assay and expressed in Plaque Forming Units (pfu). Bacterial strains used in this work were obtained from the CRIFS (Canadian Research Institute for Food Safety) culture collection and cultured in Tryptose Soya Broth or Agar (TSB and TSA, respectively) at 37°C. Enumeration of bacteria was performed by the plate count technique and expressed in Colony Forming Units (cfu). For ATP measurements, bioluminescent luciferin-luciferase Reagent (BLR) Aqua Trace® from Biotrace International (Bridgend, UK) was used. Extracellular and total (intracellular plus extracellular) ATP were measured and calculated using an internally standardized method, as described previously.[4] For measurements of extracellular ATP, 50 μL cultures of bacteria or bacteria and phages were mixed with 100 μL of BLR and the intensity of bioluminescence (I_{smp}) was recorded using a GEM Biomedical tube luminometer (Sparks, NV, USA). This was followed by the addition of 10 μL of a standard ATP solution to the reaction mixture and the resulting bioluminescence was recorded ($I_{smp} + I_{std}$). For measurements of total (intracellular plus extracellular) ATP in a sample, 50 μL of cell suspension were treated with 150 μL of an ATP releasing agent (Biotrace, Bridgend, UK) and mixed with 100 μL of BLR. Bioluminescence intensity of the sample was then measured followed by the addition of 10 μL of ATP standard and a repeat measurement of bioluminescence intensity. ATP standards were prepared daily from stocks of 0.1 mM solution of ATP (Sigma) in water that were stored at -20°C. Freshly prepared ATP standards were kept on ice during experimentation.

RESULTS AND DISCUSSION

Correlation between plaque formation and extracellular ATP release. Specificity of BP1 and BP6 phages was tested with both the plaque assay (Table 1) and the extracellular ATP measurements (Fig. 1). Efficient phage-mediated cell lysis was observed only at a multiplicity of infection (MOI) ≥ 100 for both phages. Under these conditions, an increase of extracellular ATP was observed 15-20 min post infection and it reached a maximum after 100 min of infection. The concentration of ATP released by phage BP1 was less then for BP6 (Fig. 1A). In all cases plaque formation was accompanied by release of intracellular ATP which indicates a correlation between ATP and phage infectivity (plaque formation).

Table 1. Specificity of BP1 and BP6 phages as determined by plaque assay

Bacterial strain	Plaque formation		Bacterial strain	Plaque formation	
	BP1	BP6		BP1	BP6
Salmonella agona SA 941132	++	++	*S. saintpaus* SA940970	-	-
S. agona S-107	-	-	*S. schwansenburg* S—1303	-	-
S. brandenburg SA 941266	+	++	*S.* typhimurium DT108	+++	+++
S. brandenburg S-474	+	+	*S.* typhimurium DT104	++	++
S. enteritidis SA 932451	+	+	*S.* typhimurium SA 941256	++	++
S. enteritidis EN-2588	+	++	*S. thompson* SA 941182	ND	-
S. infantis SA941223	ND	-	*S. thompson* TH-862	ND	-
S. hadar SA 941256	ND	-	*E. coli* ATCC1129	-	-
S. hadar HR-1015	ND	-	*E. coli* K-12	-	-
S. heidelburg SA 941270	++	++	*Pseudomonas aureus*	-	-
S. heidelburg H1900	+				

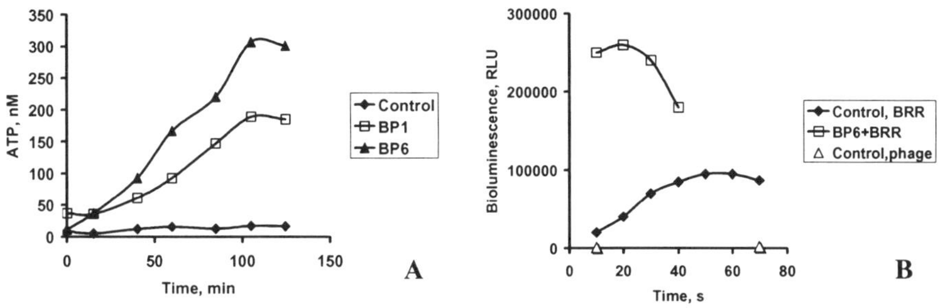

Fig. 1. Time courses of ATP release during the lysis of *Salmonella* Typhimurium cells (10^6 cfu/mL) by both phages (A, MOI=1000, 37°C) and by ATP-releasing reagent (BRR) (B)

Moreover, the ATP measurement is a much faster technique: results are obtained within 1-2 h instead of 8-16 h for the plaque assay. Furthermore, phage-infected *Salmonella* cells were significantly more susceptible to quaternary ammonium compounds – positively charged surfactants used as intracellular ATP releasing agents (Fig. 1B). This difference was detectable 8 min after infection and did not depend on cell concentration or time post-infection. Adsorption of phage on the bacterial cell wall possibly resulted in structural changes that allowed the surfactant to disrupt the cell wall more rapidly.

Phage-based detection of *S.* Typhimurium. The selectivity of bacteriophage-mediated lysis of *S.* Typhimurium was used to develop a method for specific bacterial detection. Extracellular ATP was measured 1 h post-infection in a mixture of *S.* Typhimurium with *S.* hadar or *S.* infantis, in the presence of BP6 phage (Fig. 2). Data indicated that *S.* Typhimurium can be detected in the presence of a 5-fold excess of other *Salmonella* serovars without compromising sensitivity of the assay.

ATP-Bioluminescence for investigating mechanism of action and efficiency of phage as a control agent for *S.* Typhimurium growth. As was mentioned above, the extensive release of ATP in the medium following phage infection was observed only at high MOI. The ability of the phages to control and possibly eliminate *S.* Typhimurium growth was investigated at MOI 1000 by measuring both extracellular and total ATP. It was observed that after a fast increase in extracellular ATP within 10-15 min post infection (Fig. 1A) there was a substantial 30- 40-fold increase in total and intracellular ATP 1-2 h post-infection (Fig. 3), which was not accompanied by an increase in cell number.

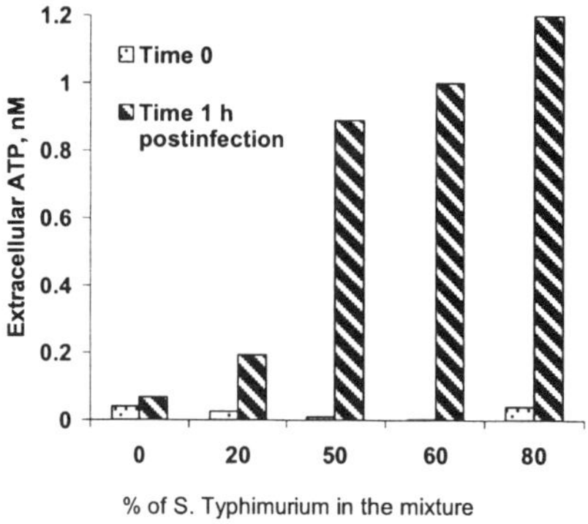

Fig. 2. Dependence of extracellular ATP on percent of *S.* Typhimurium DT108 present in a mixture with S. Hadar immediately and 1 h after infection by BP6 phage: 10^6 cfu/mL, MOI=100

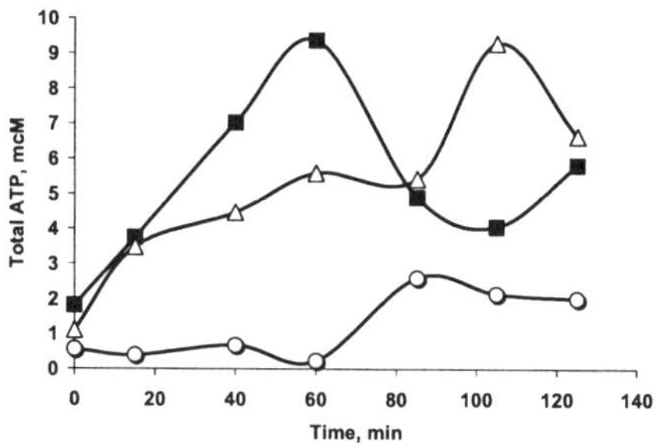

Fig. 3. Time course of total ATP concentration in the samples of *S.* Typhimurium DT108 non-infected (○) or infected by BP6 phage (■) or BP1phage (Δ)

Two hours post-infection there was >2 log reduction in bacterial numbers and a 1 log increase in pfu. This unexpected increase in intracellular ATP in cells infected by phage can be explained by extensive phage mediated synthesis of ATP that is needed for phage particle assembly in bacterial cells prior to their lysis.[5] However, further investigation is necessary to elucidate the mechanism of bacteriophage infection. Use of bioluminescent ATP-metry allowed us to characterize the lytic ability of bacteriophages and assess their potential for specific bacterial detection and growth control in a faster and easier way than conventional plating methods.

REFERENCES

1. Romanova N, Favrin S, Griffiths M. Sensitivity of *Listeria monocytogenes* to sanitizers used in meat industry. Appl Environ Microbiol 2002;68:6405-9.
2. Whichard JM, Sriranganathan N, Pierson FW. Suppression of *Salmonella* growth by wild-type and large-plaque variants of bacteriophage Felix O1 in liquid culture and on chicken frankfurters. J Food Protect 2003;66:220-5.
3. Sambrook J, Russel DW. Molecular cloning. A laboratory manual. Spring Harbor, NY: Cold Spring Harbor Laboratory Press, Vol. 1, 2001.
4. Romanova NA, Brovko LY, Moore L, Pometun E, Savitsky AP, Ugarova NN, et al. Assessment of photodynamic destruction of *Escherichia coli* O157:H7 and *Listeria monocytogenes* by using ATP bioluminescence. Appl Environ Microbiol 2003;69:6393-8.
5. Arisaka F. Assembly and infection process of bacteriophage T4. Chaos 2005; 15:Art. No. 047502.

IMPROVED SENSITIVITY METHOD FOR RAPID HYGIENE MONITORING USING ATP BIOLUMINESCENCE

M DRISCOLL, C RAMSAY, J WATKIN
Biotrace Limited, The Science Park, Bridgend, CF31 3NA, UK
E-mail: markd@biotrace.co.uk

INTRODUCTION

ATP bioluminescence is a widely accepted tool for hygiene monitoring in the food and drink industry, providing rapid results. The level of ATP on the surface of food production facilities is a good indicator of bacterial contamination as well as residual food residues. Poorly cleaned / sterilised surfaces will provide excellent growth conditions for e.g. airborne micro-organisms. The simple and quick assay procedure (< 1 min) promotes routine use, and has several advantages over conventional microbiology techniques. For example, traditional techniques produce CFU results after 2 – 5 days and are unable to detect non-culturable micro-organisms.

As well as food manufacturers, other markets for surface hygiene are important, such as hospitals, clinics, and pharmaceutical companies. Whilst ATP bioluminescence is sufficiently sensitive for most applications at around 10^3 cfu/mL,[1] clients requiring sterility-level sensitivity have had to continue with more traditional methods.

To serve these markets, we propose to use an enhanced sensitivity method for rapid hygiene monitoring using adenylate kinase (AK) as a cell marker.[2] This method makes use of the fact that AK is ubiquitous in virtually all living cells. Supplementing the ATP bioluminescence reagents with an excess of highly purified ADP and magnesium ions, allows the sensitive assay of AK, in a coupled reaction linked to the detection of ATP by luciferase/luciferin:

$$2\text{ADP} \underset{\text{AK}}{\overset{Mg^{2+}}{\longleftrightarrow}} \text{AMP} + \text{ATP}$$

Using this method, linear amplification of ATP over time can be achieved. Since low-AK residual organic residues can also indicate poor hygiene, it is required that the improved rapid hygiene test should be able to detect both ATP and AK sensitively. The main problem with developing an AK rapid hygiene assay until now has been the high ATP backgrounds of the purified ADP reagent. This meant that although low levels of bacteria could be detected, this was at the expense of ATP detection, which was not equivalent to current ATP bioluminescence. For example, 1-10 femtomoles of ATP can typically be detected using ATP-bioluminescence, but with AK bioluminescence, around 100 – 200 femtomoles of ATP can be detected due to the high ADP backgrounds.

Further reduction of ADP backgrounds, already purified by column chromatography, has been achieved in this study using apyrase with a high ratio of ATPase:ADPase activity, stabilised by ultrafiltration to remove the apyrase.

MATERIALS AND METHODS

Instrumentation. A Uni-Lite® luminometer (Biotrace Limited, Bridgend, UK) was used for the measurement of photons of light from the bioluminescence reaction.

Reagents. Reagents for measurement of ATP contained highly purified AK-free luciferase (*Photinus pyralis* EC 1.13.12.7), D-luciferin, buffer salts, metal co-factors, ethylene diaminetetraacetic acid, dithiothreitol and a range of proprietary stabilizers and kinetic enhancers. Reagents for the measurement of cellular adenylate kinase contained twice purified ADP, a proprietary extractant containing cationic and non-ionic surfactants, and a magnesium buffer (Biotrace Ltd., Bridgend, UK). *Escherichia coli* (NCIMB 10243) was from the NCIMB, Aberdeen, UK.

Cells of *E. coli* were grown for 24 hours at 35°C in Tryptone Soya Broth with shaking at 150 rpm. Stock cells were obtained by centrifuging the cells during late stationary phase, washing twice with ATP-free water, and storing in 25 mmol/L HEPES, pH 7.75. Cells were stored at 4°C for 2 h before use to allow them to complete their growth cycle. Cells were serially diluted (10-fold) in sterile water immediately prior to assay.

Purification of Firefly Luciferase for AK Assays. Several methods have been described previously for the purification of firefly luciferase.[3-5] Commercially available luciferases contain high amounts of contaminating adenylate kinase, and this must be removed.

Purification of ADP for AK Assays. ADP (A5285) obtained from Sigma, Poole, UK was purified chromatographically using a HPLC method with anion-exchange column material to remove contaminant ATP. The ADP was further treated by incubating with apyrase (Sigma, A-6410) to further lower the backgrounds to less than 50 RLU (Relative Light Units).

AK Hygiene Assays. Adenylate kinase was assayed in a hygiene swab-test format, moistening a swab bud by dipping it into an ATP-free disposable cuvette containing AK assay reagent (highly purified ADP and proprietary surfactant solution, 200 μL). A sample of micro-organism (10 μL) was then added to the swab bud. A mixture of sterile water (200 μL), magnesium buffer (100 μL) and purified ADP/surfactant solution (100 μL) was mixed in a separate disposable cuvette, and the swab was added to this, then spun briefly on a vortex mixer. After a 30 min incubation at room temperature, the swab was removed and excess liquid squeezed off against the side of the cuvette. A sample (200 μL) was removed into another disposable cuvette, and AK-free luciferase (100 μL) added. Emitted light was measured immediately in a Biotrace Uni-*Lite*®. For controls, *E. coli* was substituted with sterile water.

ATP Hygiene Assays. As for AK hygiene assays, except a 10-fold dilution series of ATP, from 10^{-7} mmol/L to 10^{-14} mmol/L, was assayed instead of *E. coli*. In addition, the 30 min incubation step was removed, as there was no need to incubate ATP samples.

RESULTS AND DISCUSSION

Improved sensitivity. Previous work[2,6] has shown that it is possible to achieve over 10-fold increases in sensitivity for the detection of micro-organisms using bioluminescence methods in a liquid sample when adenylate kinase is used as a cell marker instead of ATP. There has been limited work published on the performance of AK reagents in a hygiene-swab based test.

Typical results obtained for hygiene-swab AK assays are shown in Fig. 1. The results show that the AK hygiene assay could detect 1 CFU per swab, corresponding to 90 CFU/mL (based on background mean + 2sd, or 6 CFU per swab based on background mean x2). This represents a significant improvement in sensitivity compared to ATP hygiene assays.

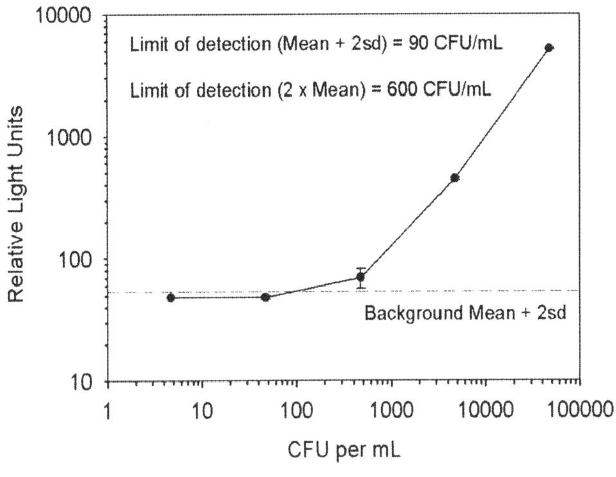

Fig. 1. A dilution series of *E. coli* assayed for AK using highly purified ADP and a hygiene swab method

It was possible to detect 7000 CFU/mL (background mean + 2sd) when assaying with "ATP only" bioluminescence reagents (results not shown), in agreement with previously published data (1). Therefore, an 80-fold improvement in the detection of *E. coli* in a hygiene based swab test has been achieved.

Since low-AK residual organic residues can also indicate poor hygiene, it is required that the improved rapid hygiene test should be able to detect both ATP and AK sensitively. With the improved reduction in ADP backgrounds, we have detected 3.5 femtomoles of ATP in the AK hygiene test (Fig. 2) on a food grade stainless steel surface. This is in the region of that typically achieved with standard ATP bioluminescence tests.

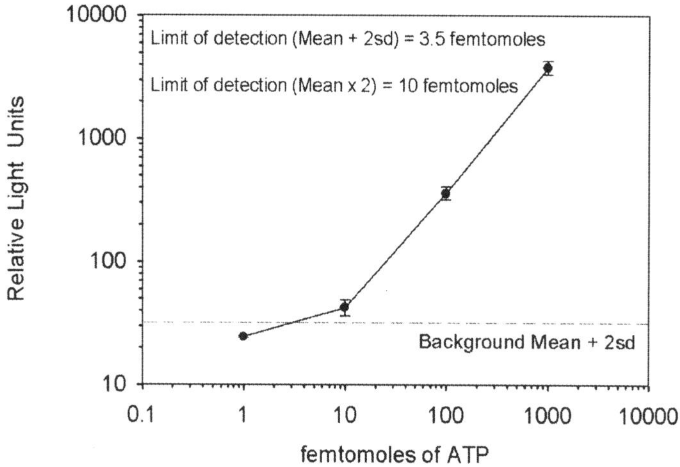

Fig. 2. A dilution series of ATP assayed using the AK hygiene swab method

The significant reagent improvements, specifically the background of the ADP substrate, has allowed the test to sensitively detect micro-organisms, whilst providing an acceptable level of ATP detection for those surface samples which may be high in ATP from food residues, but low in AK from micro-organisms.

REFERENCES

1. Stanley PE. A review of bioluminescent ATP techniques in rapid microbiology. J Biolumin Chemilumin 1989;4:375-80.
2. Squirrell DJ, Murphy MJ. Adenylate kinase as a cell marker in bioluminescent assays. In: Campbell AK, Kricka LJ, Stanley PE. eds. Bioluminescence and Chemiluminescence: Fundamental and Applied Aspects. Chichester: Wiley, 1995:486-9.
3. DeLuca M, Mc Elroy WD. Purification and properties of firefly luciferase. Methods Enzymol 1978;57:3-15.
4. Devine JH, Kutuzova GD, Green VA, Ugarova NN, Baldwin TO. Luciferase from the east European firefly *Luciola mingrelica*: cloning and nucleotide sequence of the cDNA, overexpression in *Escherichia coli* and purification of the enzyme. Biochem Biophys Acta 1993;1173:121-32.
5. Nielsen R, Rasmussen H. Fractionation of extracts of firefly tails by gel filtration. Acta Chem Scand 1968;22:1757-62.
6. Squirrell DJ, Murphy MJ. Rapid detection of very low numbers of microorganisms using adenylate kinase as a cell marker. In: Stanley PE, Simpson WJ, Smither R. eds. A Practical Guide to Industrial Uses of ATP-luminescence in Rapid Microbiology. Lingfield: Cara Technology Ltd., 1997:107-13.

DETERMINATION OF ANTIOXIDANT CAPACITY AND REACTIVITY USING HEMIN-CATALYZED LUMINOL CHEMILUMINESCENCE

CR ECKERT,[1] CH ESTEVES,[1] EL BASTOS,[1] P ROMOFF,[2] WJ BAADER[1]
[1] *Instituto de Química, Universidade de São Paulo - Av. Lineu Prestes, 748 Bloco 12S CEP 05508-900, São Paulo – SP, Brazil*
E-mail: wjbaader@iq.usp.br; [2] *Centro de Ciências e Humanidades, Dept. de Química, Universidade Presbiteriana Mackenzie – São Paulo – SP, Brazil*

INTRODUCTION

The chemiluminescence of luminol can be induced by free radical and/or redox processes and the simplicity and sensitivity of this technique is an advantage in the development of analytical methods for the monitoring of free radical production or consumption.[1] Several of the methods used for measuring total antioxidant capacity of plasma or serum employ azoinitiators, due to their capacity to decompose, producing peroxyl radicals at a constant rate, and luminol chemiluminescence.[2] Previously, our group has described a screening method for antioxidant potential determination based on luminol – hemin – hydrogen peroxide chemiluminescence.[3,4] The emission depletion, caused by an antiradical compound added during the chemiluminescence decay was shown to be proportional to the number of reactive species trapped and the difference between the areas of the emission decay curves, obtained in the absence and the presence of the potential antioxidant, has been utilized as a measure for the antiradical capacity of the sample. The technique can therefore be used for the measurement of the antiradical capacity of pure compounds and complex mixtures.

MATERIAL AND METHODS

The luminol/hydrogen peroxide/hemin antiradical assay was performed as described earlier;[3] measurements were performed on a fluorimeter Varian Eclipse and quantification achieved by using the suppression area (S_{supp}) observed upon delayed addition of the antiradical compound. The values for the antiradical capacity (n) were determined from the linear correlation between suppression area and the sample concentration in comparison with Trolox as reference compound ($n_{Trolox} = 2.0$). The relative antiradical reactivity (R) of the compounds studied was determined by measuring the decrease in emission intensity (I_{supp}) directly after antiradical addition, normalized with the initial emission intensity ($I_{supp} = (I_0 - I) / I_0$). The relative reactivity (R) is obtained from the inclination of the linear correlation between I_{supp} and the additive concentration.

RESULTS AND DISCUSSION

The antiradical capacity of a series of flavonols, flavones, flavanones and catechins was determined by means of the luminol/hydrogen peroxide/hemin CL assay and the values expressed in relation to the antiradical standard Trolox (Table 1).

Table 1. Antiradical capacities (n) and structural properties of flavonoids

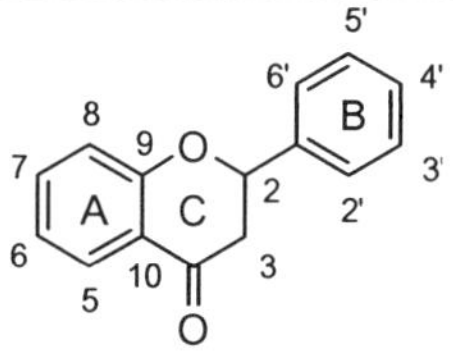

Compound	n	o-di-OH	A5-OH	A6-OH	A7-OH
quercetin	3.60	+	+	-	+
rutin	1.34	+	+	-	+
kaempferol	1.04	-	+	-	+
fisetin	1.50	+	-	-	+
morin	0.87	-	+	-	+
3,7-dihydroxyflavone	1.35	-	-	-	+
3,6-dihydroxyflavone	1.74	-	-	+	-
apigenin	0.90	-	+	-	+
luteolin	3.20	+	+	-	+

For the flavonols, quercetin showed the highest activity, whereas fisetin and kampferol, which have one hydroxyl group less than quercetin, had considerably lower n values. Rutin (3-O-glycosyl-quercetin) has a lower activity as compared to quercetin as it lacks one free hydroxyl group (Table 1). Morin, which differs from quercetin only in the substitution pattern in ring B, shows the lowest n values among the flavonols studied, indicating the importance of *ortho*-dihydroxy substitution in ring B for antiradical activity. However, in contrast to the latter results, 3,7- and 3,6-dihydroxyflavones show considerable activity, a fact which may question the importance of the *ortho*-dihydroxy substitution in ring B and point out an important contribution of hydroxyl substitution in ring A. A comparison of the results obtained with the flavones apigenin and luteolin, however, points out the importance of the *ortho*-dihydroxy substitution in ring B and the high values obtained for luteolin may indicate a minor importance of the C3-hydroxy group for antiradical activity (Table 1).

In order to gain a better insight into the influence of the hydroxyl substitution pattern of flavonoids to their antiradical activity, we have initiated a study on the antiradical behavior of substituted phenols and polyhydroxylated benzene derivatives, using the luminol CL assay.

The results obtained with the derivatives studied shows that, depending on the substitution pattern, the phenolic compounds act as moderate inhibitors of the radical induced luminol CL, showing medium to low and very low n values. Some of the derivatives show competitive inhibition of emission intensity from which the relative radical reactivity (R) can be determined. For one compound, resorcinol, it is possible to obtain n values as well as R values (Table 2).

Table 2. Antiradical capacity (n), relative reactivity (R) and emission enhancer effect of phenolic derivatives upon the luminol/hemin/H_2O_2 chemiluminescence assay

Compound	n	Relative Reactivity (R) (μM^{-1})	Enhancer Effect [AOH] (μM)
phenol	[a]	0.00687	[c]
catechol	0,37	[b]	[c]
resorcinol	1,16	0,414	[c]
hydroquinone	[a]	[b]	1,25 - 40
pyrogallol	[a]	[b]	2,5 - 50
phloroglucinol	[a]	0,00644	[c]
p-fluorophenol	[a]	0,0119	[c]
p-bromophenol	[a]	0,00022	[c]
p-iodophenol	[a]	0,00043	[c]
p-nitrophenol	[a]	0,00845	[c]
p-cyanophenol	0,00034	[b]	[c]
p-methoxyphenol	4,00	[b]	[c]
p-aminophenol	1,60	[b]	[c]

[a] no n values could be obtained; [b] no relative reactivity values R could be obtained; [c] no enhancer effect is observed

Two of the compounds studied show a significant enhancer effect, which is concentration dependent (Table 2). Although this enhancer effect has not yet been quantified, its qualitative observation is an important observation.

Two of the dihydroxybenzene derivatives show moderate antiradical capacity, the 1,2-substituted derivative having a considerably lower value than the 1,3-derivative, a surprising result in view of the importance of *ortho*-disubstitution for flavonoid antioxidant activity. No n-values could be obtained for hydroquinone and phloroglucinol due to its enhancer effect. Also for pyrogallol, the other trihydroxy-derivative studied, no n-value is obtained due to its low capacity, however, the relative reactivity of this compound could be determined as it shows no enhancer effect (Table 2). Antiradical capacity values (n) were obtained for only three of the substituted phenols studied; the electron rich methoxy and amino derivatives showing surprisingly high values. The electron acceptor substituted p-cyanophenol has an extremely low n-value and no capacity values were obtained for the halo-substituted phenols and the nitro-derivative (Table 2). Although no clear-cut trend can be observed from the results obtained up to now, it appears from these preliminary results that 1,2-dihydroxy-substitution is not important for antiradical activity in these model compounds and electron-donor substituted phenols appear to possess high antiradical capacity.

Similar values for the relative reactivity (R) are observed for phenol, phloroglucinol and *p*-nitrophenol, whereas resorcinol possesses an R value two orders of magnitude higher (Table 2). No obvious tendency is observed for the R-values of the *p*-halophenols, the fluoro-derivative being the most reactive and iodophenol showing, in contrast, a higher reactivity than the bromo-derivative. The enhancer effect, observed for hydroquinone and pyrogallol and not yet quantified, can be understood with the action of AOH in the recycling of the active form of the hemin and a participation of the antioxidant radical in the chain propagation.

ACKNOWLEDGEMENTS

Financial support has been provided by Fundação de Amparo à Pesquisa do Estado de São Paulo – FAPESP, Coordenação de Aperfeiçoamento de Pessoal de Nível Superior – CAPES and MackPesquisa, Universidade Persbiteriana Mackenzie.

REFERENCES

1. Lissi E, Salimhanna M, Pascual C, Delcastillo MD. Evaluation of total antioxidant potential (TRAP) and total antioxidant reactivity from luminol-enhanced chemiluminescence measurements. Free Radic Biol Med 1995; 18:153-8.
2. Prior RL, Cao G. In vivo total antioxidant capacity: comparison of different analytical methods. Free Radic Biol Med 1999;27:1173-81.
3. Bastos EL, Romoff P, Eckert CR, Baader WJ. Evaluation of antirradical capacity by H2O2-hemin-Induced luminol chemiluminescence. J Agric Food Chem 2003;51:7481-8.
4. Baader WJ, Bastos EL, Romoff P, Kato MJ. A simplified assay for the evaluation of natural products antioxidant potential. In: Bioluminescence and Chemiluminescence: Perspectives for the 21st Century. Roda A, Pazzagli M, Kricka LJ, Stanley PE. eds. Chichester: Wiley, 1999:303-6.

APPLICATIONS OF ADENYLATE KINASE DETECTION USING BIOLUMINESCENCE

NICHOLAS FOOTE, MARC GREEN AND SARAH GREGORY
Acolyte Biomedica Ltd, Salisbury, SP4 OJQ, UK

INTRODUCTION

Microbial adenylate kinase (AK) can be driven to produce ATP by the reaction 2ADP → AMP + ATP.Measuring the resulting ATP by means of firefly luciferase bioluminescence provides a highly sensitive assay for the presence of micro-organisms (AK rapid® technology). Compared with the direct measurement of ATP, AK technology provides significantly enhanced sensitivity and more reliable detection of micro-organisms since the level of AK per cell for a particular bacterial species remains essentially constant, regardless of physiological state.[1] This makes the technique suitable to be used in a range of applications where the need is to enumerate organisms rapidly or to measure the rate of microbial growth.

The RapiTECT GN™ kit provides the reagents to perform a qualitative and/or quantitative assay for Gram negative organisms in a liquid sample. Typically enumeration results can be obtained in minutes and growth-based results within a few hours.

METHODS

Assay method for Gram negative bacteria. The BacLite® RapiTECT GN™ kit available from Acolyte Biomedica Ltd comprises three main reagents: Bacterial Lysis Reagent, Purified ADP, and a Luciferase / Luciferin Reagent. The former two reagents are mixed 1:1 before use. Samples must contain magnesium ions as a co-factor for both AK and luciferase reactions; if it is not present in sufficient amounts in the samples it must be added separately. A typical assay procedure is:

1. Place 100 µL sample in cuvette.
2. Add 10 µL magnesium acetate solution (if necessary).
3. Add 50 µL mixed Bacterial Lysis reagent / Purified ADP and mix.
4. Leave for precisely 5 min. During this time, adenylate kinase in the sample converts ADP to ATP.
5. Add 50 µL Luciferase / Luciferin Reagent and read for 1 sec in a luminometer.

Enumeration assays. As the relationship between luminometer output (Relative Light Units; RLU) and organism number is linear, a calibration graph obtained on one day can be used to give an accurate quantification of the same organism on a subsequent day.

RESULTS

Typically, examples of AK assays are shown in Fig. 1. On day 0, a calibration graph was generated by preparing a suspension of each of bacterium in ¼-strength Ringers solution. These suspensions were decimally diluted: each dilution was assayed using the RapiTECT GN™ test, and an equivalent volume of sample was plated out on to a nutrient media plate. Standard graphs were generated by plotting log CFU against log RLU for each dilution and adding a line of best fit through the points. The process was repeated on a further 3 days (over a 6 day period), but in these cases the data points have not been fitted.

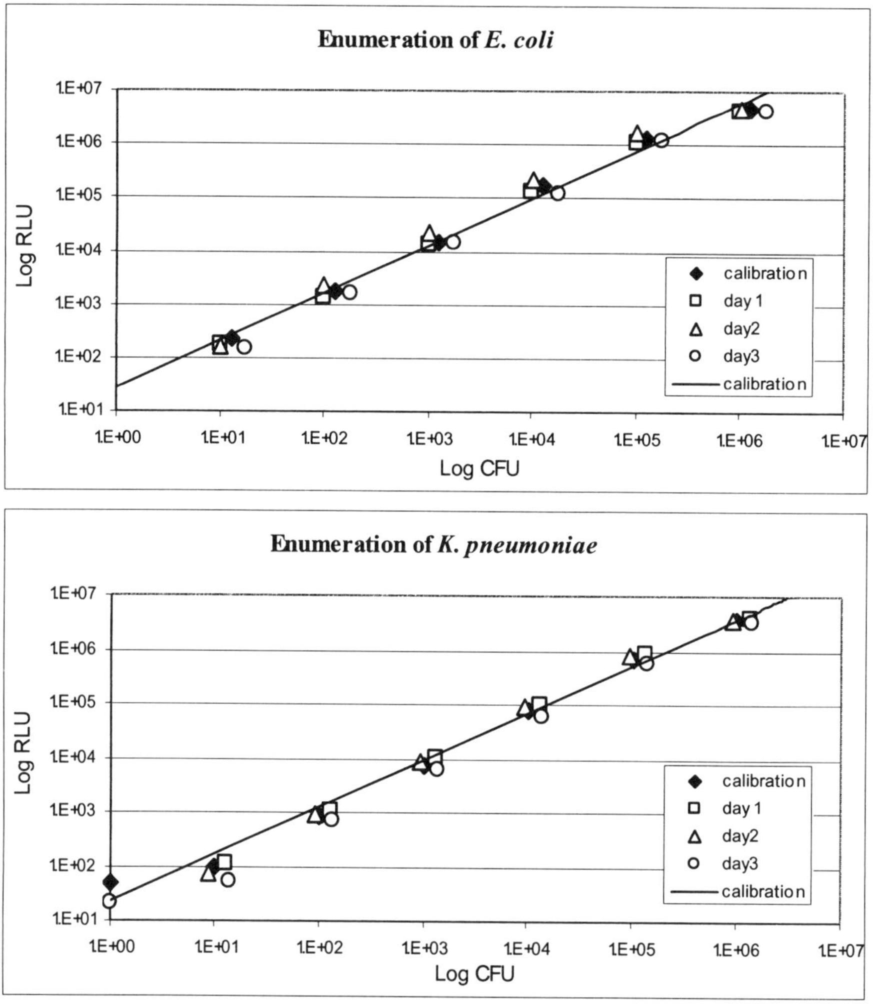

Fig. 1. The use of AK assays with calibration graphs for enumerating Gram negative bacteria

Growth assays. Samples from a liquid culture can be assayed using the AK method to give a rapid indication of growth, and to assess the effect on growth of added substances such as antibiotics. Fig. 2 shows growth curves for two Gram negative bacteria. The organisms were grown in broth at 37°C, with and without the antibiotics ceftazidime and gentamicin at a range of concentrations between 0.5 and 16 mg/L. At hourly intervals aliquots were removed and tested with the BacLite® RapiTECT GN™ kit.

***E. coli* /Ceftazidime**

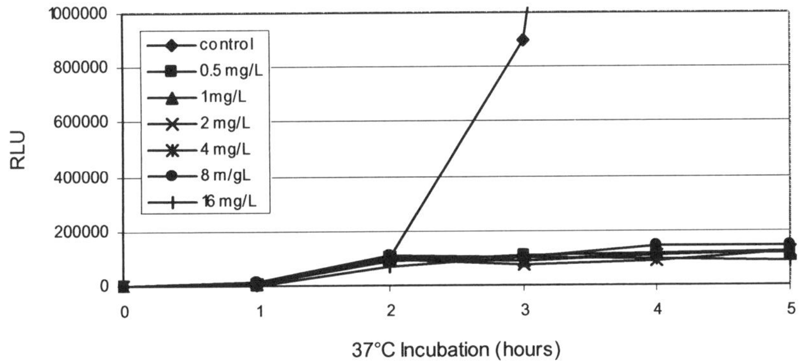

***P. aeruginosa* /Gentamicin**

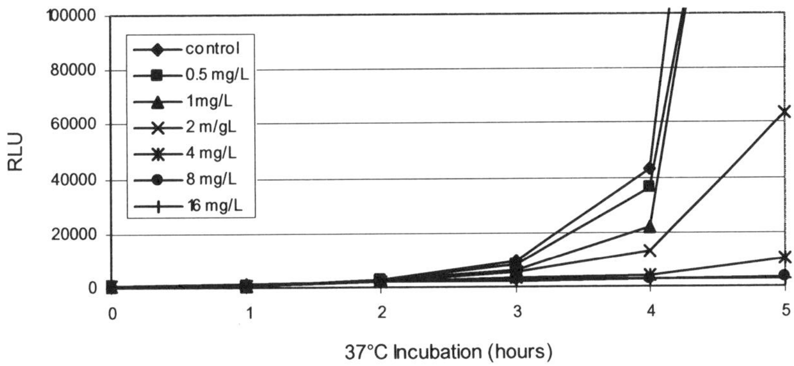

Fig. 2. Growth curves for two Gram negative bacteria showing effects of antibiotics A clear pattern emerged after 3-5 hours growth. *E. coli* proved to be sensitive to ceftazidime even at the lowest concentration tested. *P. aeruginosa* was resistant to

gentamicin at the lower concentrations but at = 2 mg/L the growth rate was affected, thereby giving a rapid estimate of the MIC (minimum inhibitory concentration).

MRSA screening. The BacLite *Rapid* MRSA system takes advantage of the "growth assay" approach to provide a 5-hour screening test to detect MRSA (methicillin resistant *Staphylococcus aureus*) from clinical swabs. The assay gains its specificity from three sources:[2]

1. Antibiotics that MRSA is resistant to, but inhibit the growth of other organisms;
2. Immunocapture on magnetic microparticles coated with antibodies to *S. aureus*;
3. AK release with lysostaphin, which preferentially lyses *S. aureus*.

The system comprises a Sample Processor to collect and wash the magnetic microparticles and a Reader (luminometer) to perform the AK assay. After an initial 2 h incubation and the immunomagnetic separation step, two sub-samples are tested, 2 h apart. If significant growth has taken place in the second incubation phase then the sample is judged to contain MRSA. The assay gave a diagnostic sensitivity of 93.4% and a specificity of 95.7% for 1377 nasal screening swabs tested by several different operators.

CONCLUSIONS

Measuring ATP derived from adenylate kinase provides a highly sensitive, rapid assay for the detection of micro-organisms, with potential applications in the direct testing of clinical specimens or surfaces in a hospital environment. The fact that AK correlates closely with cell numbers allows the technique to be used in a range of further applications where the need is to enumerate organisms rapidly or to measure the rate of microbial growth. The latter approach allows the rapid determination of antibiotic resistance, one particular application of which is the detection of MRSA directly from a screening swab. A similar approach can be applied to other important antibiotic resistant target organisms such as VRE (vancomycin resistant *Enterococcus*) or ESBL (extended-spectrum ß-lactamase) producers.

REFERENCES

1. Squirrell DJ, Murphy MJ, Leslie RL, Green JCD. A comparison of ATP and adenylate kinase as bacterial cell markers: correlation with agar plate counts. In: Stanley PE, Kricka LJ. eds. Bioluminescence and Chemiluminescence: Progress and Current applications. Singapore: World Scientific, 2002:487-90.
2. Leslie RL, Murphy MJ, Squirrell DJ, Cotterill SL, Matthews SCW, Skyrme M. Hospital testing of a rapid bioluminescent assay for MRSA. In: Tsuji A, Matsumoto M, Maeda M, Kricka LJ, Stanley PE. eds. Bioluminescence and Chemiluminescence: Progress and Perspectives. Singapore: World Scientific, 2004:417-20.

STANDARDIZATION OF BIOLUMINESCENT RAW MILK QUALITY ASSAY USING LUMTEK TEST-SYSTEMS

VG FRUNDZHYAN,[1] NN UGAROVA,[2] AA MAKARUSHIN,[3] GB GAVRILOV[3]
[1] Lumtek LLC, Vorobievy Gory, Moscow 119992, Russia; [2] Chemistry Faculty, Lomonossov Moscow State University, Moscow 119992, Russia, Email unn@enz.chem.msu.ru; [3] Yaroslavl State Laboratory for Raw Milk and Dairy Research, Moscovsky prospect, Yaroslavl 150030, Russia

INTRODUCTION

Bacterial cell count (BCC) and somatic cell count (SCC) are significant characteristics of raw milk. BCC and SCC indicate the hygiene of dairy farming, influence purchasing price value, determine raw milk processing into foodstuffs or further processing into casein or lactose. For rapid BCC and SCC count the Bioluminescent Assay (BA) is used often. We developed the Lumtek bioluminescent test-systems for BCC (detection limit – $5x10^4$ CFU/mL, duration of assay - 20-25 min per sample) and SCC (detection limit 10^4 cell/mL, duration of assay is 2-3 min per sample) raw milk assays. Lumtek test-systems comprise: 1) portable highly sensitive PMT-luminometer, 2) ready-to-use kits, long shelf life reagents and disposables, 3) assay protocol. Lumtek test-systems proposed were examined in an independent research centre (Yaroslavl State Laboratory for Raw Milk and Dairy Research). BCC Lumtek test-system was compared with Standard Plate Count (SPC) and Fluoro-Opto-Electronic Count (FOEC). SCC Lumtek test-system was compared with Direct Somatic Cell Count (DSCC), FOEC and Viscosimetric Assay (VA). Statistical analysis of the data obtained showed a good correlation between Lumtek test-systems and standard assays.

MATERIALS AND METHODS

Instrumentation. Luminometer LUM-1 (Lumtek LLC, Russia), fluoro-opto-electronic counter IBCm supplied both with Bactocount and Somacount modules (Bentley Instruments, USA), viscosimeter VNZh1,41 (Laverna-Lab, Russia) were used. Filtravettes™ (FV™) – luminometer polystyrene micro-cuvettes (h 13 mm, ∅ 10 mm, bottom made of bacterial membrane filter with pore size 0.45 μm), were from New Horizons Diagnostic Corp., USA.

Methods. BCC in raw milk (CFU/mL) was determined using BA (BCC Lumtek test-system), SPC [1] and FOEC. [2] SCC in raw milk (cell/mL) was determined using BA (SCC Lumtek test-system), FOEC,[2] DSCC,[3] and VA.[4]

Statistical analysis. Assessment of accuracy of BCC and SCC assay in milk was performed in compliance with established procedures.[5] The following statistical parameters were calculated: r (repeatability), sd_r (standard deviation of repeatability), sd_{xy} (residual standard deviation), R^2 (determination factor).

Protocol of BCC assay using Lumtek test-system. 1) ATP-reagent Lumtek and ATP-standard are reconstituted with 2 mL of Ultra-pure water and 1 mL of BCRA, respectively. FV^{TM} is placed into measuring cell of LUM-1 and 0.02 mL of ATP-standard solution, 0.5 mL of Ultra-pure water and 0.5 mL of ATP-reagent are pipetted into FV^{TM}. The content of FV^{TM} is stirred rapidly and bioluminescent signal (I_{st1}) is recorded. 2) Raw milk assayed (1 mL) is pipette to the vial with Milk pretreatment reagent followed by incubation in water bath (10 min, 45 °C). 3) FV^{TM} is placed on the Absorbent pad placed to the Rack and 0.2 mL of milk incubated is pipetted into the FV^{TM}. Milk is forced through FV^{TM} using Syringe supplied with rubber flange tip into Absorbent pad. Washing solution (2 drops) are added to the FV^{TM} and forced through FV^{TM} to the Absorbent pad. 4) FV^{TM} is placed into the measuring cell of LUM-1 and 0.02 mL of BCRA is pipetted to the FV^{TM}. 1 min later 0.5 mL of Ultra-pure water and 0.5 mL of ATP-reagent are pipetted into the same FV^{TM}. The content of FV^{TM} is stirred rapidly and bioluminescent signal (I_{milk1}) is recorded. 5) Bacterial ATP concentration (ATP_{bact}) in the milk is calculated by the equation (1) and BCC in milk is established using Table 2.

$$[\mathbf{ATP_{bact}, mol/mL}] = \mathbf{0.1 \times 10^{-11} \times (I_{milk1})' / (I_{st1})'} \qquad \mathbf{(1)}$$

Table 1. BCC/SCC raw milk assay kit composition
(Each kit designed for 10 samples assay in duplicate)

BCC-assay	SCC-assay
1. ATP-reagent (ATP detection limit 10^{-14} mol/cuvette), lyophilized	
2. ATP-control, lyophilized (ATP concentration in solution is 10^{-11} mol/mL)	
3. Ultra-pure water, sterile	
4. Milk pretreatment reagent, lyophilized	4. Somatic cell releasing agent (SCRA)
5. Washing solution, sterile	5. Plastic vials (2 mL), sterile
6. Bacteria cell releasing agent (BCRA)	6. Microcuvettes (0.35 mL), sterile
7. Absorbent pads, sterile	
8. Filtravettes^{TM}, sterile	
9. Rack for Filtravette^{TM}	
10. Syringe with rubber flange tip	

Table 2. BCC and ATP_{bact} concentration in raw milk*

ATP_{bact} in milk, mol/mL	BCC in milk, CFU/mL
less than $0.60 \cdot 10^{-12}$	less than $50 \cdot 10^3$
$(0.60\text{-}1.20) \cdot 10^{-12}$	$(50\text{-}100) \cdot 10^3$
$(1.21\text{-}2.70) \cdot 10^{-12}$	$(100\text{-}200) \cdot 10^3$
$(2.71\text{-}4.20) \cdot 10^{-12}$	$(200\text{-}300) \cdot 10^3$
$(4.21\text{-}7.30) \cdot 10^{-12}$	$(300\text{-}500) \cdot 10^3$
$(7.31\text{-}74.00) \cdot 10^{-12}$	$(500\text{-}4000) \cdot 10^3$
$(74.10\text{-}442.00) \cdot 10^{-12}$	$(4000\text{-}20000) \cdot 10^3$

*120 milk samples were assayed both by SPC and Lumtek test-system ($R^2 = 0.92$).

Table 3. SCC and ATP_Σ concentration in raw milk*

ATP_Σ in milk, mol/mL	SCC in milk, cell/mL	Cow's health status
less than $1.9 \cdot 10^{-10}$	less than $100 \cdot 10^3$	Healthy cow
$(1.9\text{-}4.4) \cdot 10^{-10}$	$(100\text{-}200) \cdot 10^3$	Mastitis suspected
$(4.4\text{-}7.4) \cdot 10^{-10}$	$(200\text{-}300) \cdot 10^3$	
$(7.4\text{-}11.0) \cdot 10^{-10}$	$(300\text{-}400) \cdot 10^3$	
$(1.1\text{-}1.6) \cdot 10^{-9}$	$(400\text{-}500) \cdot 10^3$	
$(1.6\text{-}3.3) \cdot 10^{-9}$	$(500\text{-}1000) \cdot 10^3$	Mastitis
more than $3.3 \cdot 10^{-9}$	more than $1000 \cdot 10^3$	

*80 milk samples were assayed both by DMCC and Lumtek test-system ($R^2 = 0.94$).

Protocol of SCC assay using Lumtek test-system

1) ATP-reagent Lumtek and ATP-standard are reconstituted with 2 mL of Ultra-pure water and 1 mL of SCRA, respectively. Microcuvette is placed into measuring cell of LUM-1 and 0.5 mL of ATP-standard solution and 0.5 mL of ATP-reagent are pipetted into Microcuvette. The content of Microcuvette is stirred rapidly and bioluminescent signal (I_{st2}) is recorded. 2) 0.05 mL of raw milk assayed and 0.45 mL of SCRA are pipetted to the Plastic vial and sample obtained is stored for 1 min before assay. 3) Microcuvette is placed into measuring cell of LUM-1. 0.5 mL of milk diluted and 0.5 mL of ATP-reagent are pipetted into Microcuvette. The content of Microcuvette is stirred rapidly and bioluminescent signal (I_{milk2}) is recorded. Non-bacterial ATP concentration (ATP_Σ) (sum of free ATP and somatic cell ATP) is calculated by the equation (2) and SCC in milk is established using Table 3.

$$[\mathbf{ATP\Phi_\Sigma, mol/mL}] = \mathbf{10 \times 10^{-11} \times (I_{milk2})' / (I_{st2})'} \quad (2)$$

BCC and SCC in milk samples were assayed by the methods foregoing and the data obtained was analyzed statistically (see Table 4).

Table 4. Precision of BCC and CSS raw milk assay

Assay	r, log	sd_r, %	sd_{xy}	R^2
BCC				
BA (BCC Lumtek test-system), n=28	0.30	2.0	0.92	0.24
SPC, n=31	0.18	1.2	-	-
FOEC, n=31	0.13	0.4	0.86	0.27
SCC				
BA (SCC Lumtek test-system), n=47	23.80	2.80	0.72	101.50
DSCC, n=47	22.40	2.90	-	-
FOEC, n=40	20.40	2.10	0.94	99.00
VA, n=30	24.00	3.00	0.61	111.00

RESULTS

Lumtek test-systems for BCC and SCC assay of raw milk are shown in the Fig. 1 and Table 1.

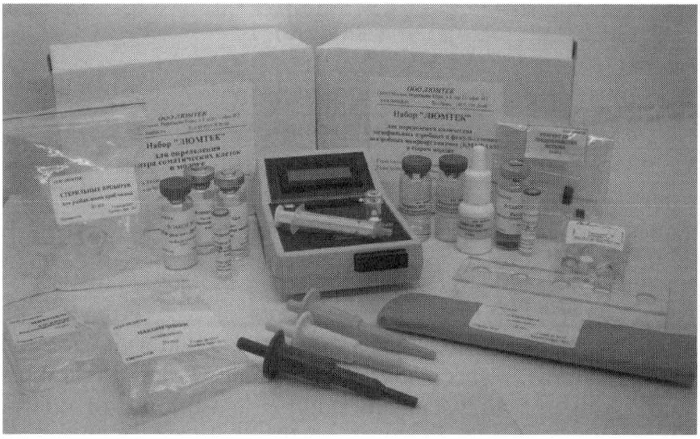

Fig. 1. Lumtek test-systems for raw milk assay of BCC and SCC

The most precise results, close to the results obtained by SPC or DSCC, were obtained by the FOEC, the least precise were obtained by VA. Lumtek test-systems showed good precision, especially in the case of BCC. In comparison with FOEC, Lumtek test-systems provide a less expensive and simpler assay. BCC Lumtek test-system was adopted by the Russia State Standard Committee (GOST 52415-2005. Fresh cow's milk – raw material. Quantitative determination of the total amount of mesophilic aerobic and elective anaerobic microorganisms. Chemiluminescent method.).

REFERENCES

1. State Standard 9225-84 (Russia). Milk and milk products. Methods of microbiological analysis.
2. Manual for bacteria and somatic cell count using Bactocount IBCm and Somacount IBCm modules (Bently Instruments).
3. ISO 13366-1:1997.
4. State Standard 23453-90 (Russia). Milk. Methods for determination of somatic cells quantity.
5. State Standard R ISO 5725 (Russia). Precision of methods and results of measurement. P.1-6.

BART – BIOLUMINESCENT ALTERNATIVE TO REAL-TIME PCR

OA GANDELMAN,[1,2] VL CHURCH,[1,2] CA MOORE,[1,2] C CARNE,[3]
H JALAL,[4] JAH MURRAY,[1,2] LC TISI[1]

[1] Lumora Ltd., c/o Institute of Biotechnology, Tennis Court Road, Cambridge, CB2 1QT, UK; [2] Institute of Biotechnology, Cambridge University, Tennis Court Road, Cambridge CB2 1QT, UK; [3] Dept of Genitourinary Medicine, Clinic 1A, Box 38, Addenbrooke's Hospital, Hills Road, Cambridge CB2 2QW, UK; [4] Clinical Microbiology & Public Health Laboratory, Box 236, Addenbrooke's Hospital, Hills Road, Cambridge CB2 2QW, UK E-mail: o.gandelman@lumora.co.uk

INTRODUCTION

Sensitive quantification of nucleic acids (NA) by real-time PCR (qPCR) relies on thermocycling samples in the presence of fluorophores that directly detect the amplicon produced. In practice, this requires a complicated and costly combination of a thermo-cycler and a fluorimeter. We have previously demonstrated that a viable alternative to qPCR is a combination of one of a number of isothermal NA amplification technologies with a bioluminescent reporter system called BART (Bioluminescent Assay for Real Time).[1] BART uses highly thermostable mutants of firefly luciferase (FFL) to detect inorganic pyrophosphate (PP_i) released as a by-product of specific NA target amplification, during the course of the amplification reaction itself. PP_i is used to generate ATP through a coupled reaction catalysed by ATP-sulphurylase (Fig. 1). Here we show that PP_i accumulates concomitantly with DNA synthesis, and that its detection in, real time, produces quantitative, highly sensitive and specific results that directly reflect the amount of DNA synthesized.

$$(DNA)_n + dNTP \xrightarrow{\text{polymerase}} (DNA)_{n+1} + PP_i$$

$$PP_i \xrightarrow{\text{ATP-sulphurylase}} ATP$$

$$h\upsilon \xleftarrow[\text{luciferase}]{} ATP$$

Fig. 1. Biochemistry of BART

Simple hardware requirements in combination with a closed-tube no-addition format make BART an ideal technology for *in vitro* diagnostics. We report here a pilot clinical evaluation of BART using isothermal Loop-Mediated Amplification (LAMP) for the detection of *Chlamydia trachomatis* (*CT*) in human urine samples. Our results demonstrated high specificity and sensitivity in assays completed within 1 h. This has significant implications for the development of non-invasive, clinic-based devices for the rapid detection of this pathogen as well as many others.

METHODS

BART reagent was formulated as described in the original patent with slight modifications.[2] In-house artificial construct of *CT* DNA sequence fragment (ChAT) was used as a template. Reactions were run at 50-65°C on Lumora's in-house hardware. PP_i was assayed using ELIDA[3] on a Luminoscan Ascent luminometer (Labsystems), DNA was assayed using picogreen (Molecular Probes) fluorescence on Cary Eclipse fluorimeter (Varian). DNA from urine samples was prepared using ChargeSwitch gDNA Mini Bacteria kit (Invitrogen) following a simplified protocol.

RESULTS AND DISCUSSION

We have demonstrated BART in combination with six different isothermal amplification technologies (data not shown). The light outcome in all cases followed a similar characteristic time-dependence which included:

- A background signal originating from the presence of dATP in the full reaction mixture (known to be a substrate for FFL) that gradually decayed;
- A bright flash followed by a rapid switch off of the signal to almost zero level, observed when the target NA was present;
- A very slow decay of the starting background signal observed throughout the reaction if there was no target NA (Fig. 2).

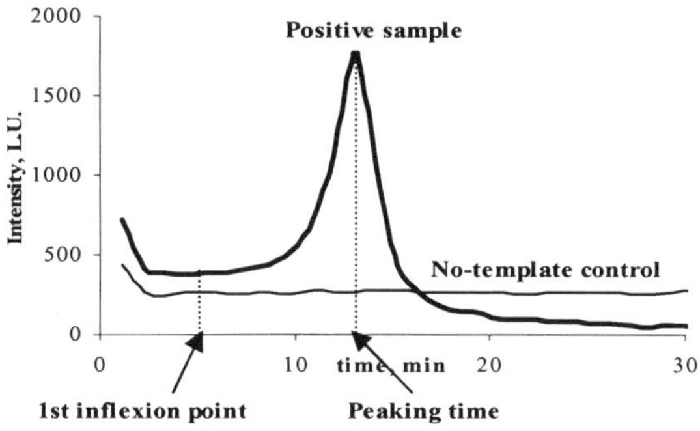

Fig. 2. Schematic representation of bioluminescent output in BART

Using LAMP as an example of isothermal amplification, we monitored in parallel the accumulation of PP_i and DNA for three ChAT concentrations: 10^{-12}, 10^{-11} and 10^{-10} g/µL (Fig. 3). A predicted synchronised increase in PP_i and DNA concentrations was confirmed by good linear correlation between the experimentally measured and calculated concentrations of PP_i. Calculations were based on independent fluorescent measurements of DNA. PP_i concentrations of 44±19 µM corresponded to the first inflexion point on the bioluminescent output, while 521±79 µM PP_i referred to the bioluminescent flash itself. The switch off of the signal was due to inhibition of FFL by PP_i accumulation during the amplification process.

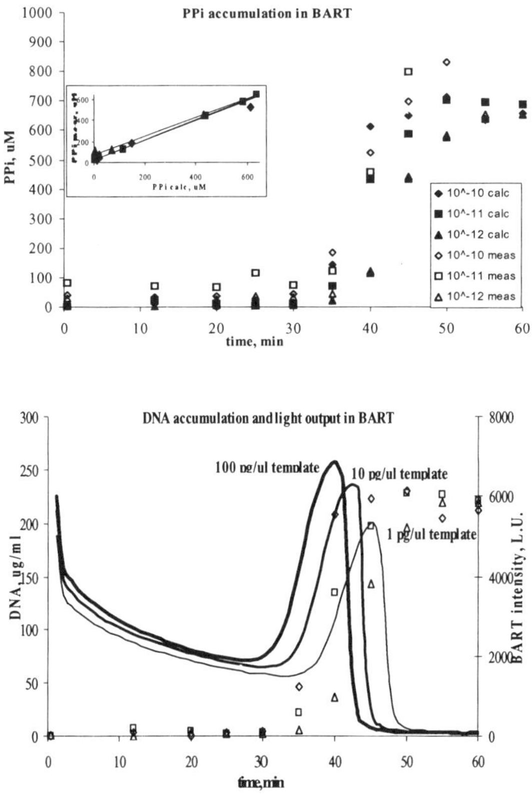

Fig. 3. Kinetics of PP_i and DNA accumulation and bioluminescent output in BART

Peaking time in BART generally depended on the reaction conditions: temperature and composition of the reagent. Under exactly the same conditions peaking time was a function of only the amount of target DNA. Amplification of three different templates using LAMP was monitored by BART in a wide concentration range (0.5-5x10^8 copies/10 μL reaction) and at different temperatures (Fig. 4). Linear correlation between the peaking time and the concentration of DNA target was observed across 6 orders of magnitude. Experimental errors and deviation from the linear fit became significant at DNA concentration below 100 copies/10 μL reaction. This could be explained by the lack of mixing in reaction causing non-homogenous spatial distribution of PP_i. The possibility to assay ≥100 copies/10 μL reaction of target DNA with high accuracy and speed (peaking time was less than 35 min) makes BART a perfect system for the point-of-care (POC) diagnostics.

We carried out a clinical evaluation of BART, using LAMP, for the detection of *CT* in human urine samples. Analytic specificity and microbial range of BART was assayed using DNA from 14 different strains of *CT*. All 14 were reliably detected by BART (data not shown). The specificity of the assay was further investigated using DNA from pathogenic bacteria and commensals of the oropharynx and genital tract, as well as high concentrations of salmon sperm DNA. Neither amplification nor bioluminescent flashing was observed in any of the samples containing DNA from pathogens (32 in total) or salmon sperm, while any samples spiked with ChAT came positive. BART displayed 100 % analytical specificity and selectivity towards *CT* DNA.

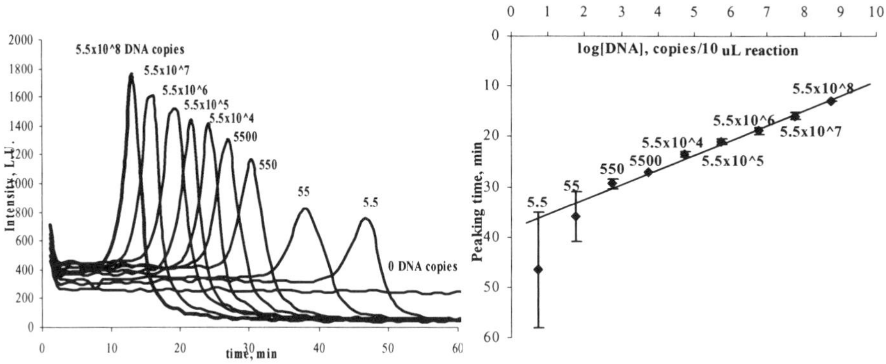

Fig. 4. Quantitative BART response to a wide range of target DNA concentration

DNA from a panel of 105 urine specimens presented to the GU clinic was prepared and assayed for the presence of *CT* DNA by home-made qPCR system and by BART in parallel. This resulted in finding 50 *CT*-positive and 55 *CT*-negative samples by qPCR and 44 *CT*-positive and 55 *CT*-negative samples by BART showing 100% clinical specificity of BART. Clinical sensitivity of BART, at the time of this trial was 88% (assuming that qPCR did not produce any false positives). However, more recent optimisation of BART and innovations in isothermal amplification methods are poised to significantly improve the speed and sensitivity. The great advantage of BART-testing for POC is that the hardware requirements are extremely simple and that results are available in less than 1 h, i.e. whilst the patient waits.

REFERENCES

1. Gandelman O, Church V, Murray J, Tisi L. BART-NAAT – a novel bioluminescent assay for real-time nucleic acid amplification. In: Tsuji A, Maeda M, Kricka L, Stanley P. eds. Bioluminescence & Chemiluminescence: Progress and Perspectives. Singapore: World Scientific, 2004:523-6.
2. Murray J, Tisi L. Method for determining the amount of template nucleic acid present in a sample. PCT/GB2004/000127.
3. Nyren P, Lundin A. Enzymatic method for continuous monitoring of inorganic pyrophosphate synthesis. Anal Biochem 1985;151:504-9.

ADVANCING THE DEVELOPMENT OF DUAL-LUCIFERASE ASSAYS

EM HAWKINS, D GARVIN, A PAGUIO, P STECHA,
B SWANSON, F FAN, KV WOOD
Promega Corporation, 2800 Woods Hollow Rd, Madison, WI 53711, USA

INTRODUCTION

Luciferases have long been used as genetic reporters because of their high sensitivity, wide dynamic range and resistance to interference by biologically active compounds. Repeated cell transfection is inconvenient and day-to-day changes in transfection efficiency can generate erratic results. This is a severe limitation in automated screening. Many researchers are creating and repeatedly using cell lines that stably express their genes of interest. Creating stable cell lines can be particularly challenging in receptor-mediated research since the receptor of interest and the responsive reporter gene must both be transfected into the cells. We have developed a method for constructing stable dual luciferase cell lines using two plasmids. The first plasmid expresses firefly luciferase genes under the control of a response element and a hygromycin selectable marker. The second plasmid expresses a target receptor and a *Renilla* luciferase-neomycin selectable marker fusion. Faster signal responses and shorter assay times were achieved by using destabilized luciferases, and the combined effect of these enhancements improved the data quality of this dual-luciferase assay.

MODEL FOR ANALYSIS OF G-PROTEIN COUPLED RECEPTORS

Much is known about the signal transduction mechanism of G-coupled protein receptors (GPCR's). In this paper, the human muscarinic 3 receptor (M3R) is used to demonstrate the analysis of reporter function using genetic reporters in stably transfected cell lines. The signaling pathway for M3R is normally coupled through the Gαs subunit with intracellular cAMP. We incorporated the gene encoding this receptor into a plasmid also containing a *Renilla* luciferase-neomycin protein fusion.

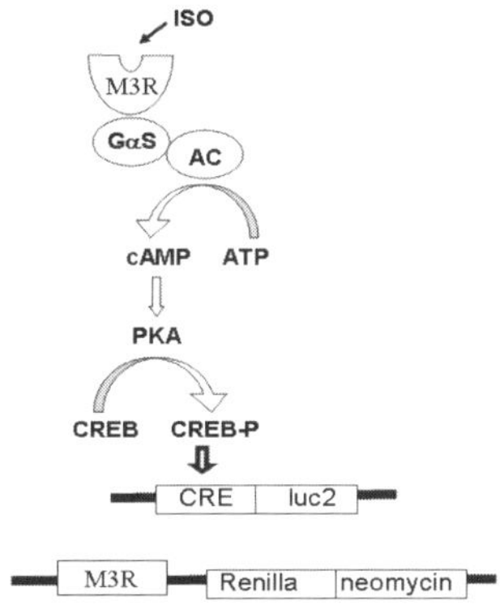

Fig.1. Diagram of CRE-induction pathway and transfected plasmid DNA

Activation of this receptor causes up regulation of the firefly luciferase gene controlled by the CRE-responsive element in the other plasmid (Fig. 1). All experiments used HEK 293 cells.

DUAL PLASMID STRATEGY FOR STABLE CELL LINES

In typical dual-reporter experiments, one reporter is used to track experimental changes and the other reporter is used as a control. The easiest way to express the two proteins in one cell is to place both genes on the same plasmid but control their expression with different promoters. When this type of plasmid was transfected into cells, however, expression levels of the two proteins were not independent (Fig. 2A). When the inducible gene was up regulated, expression of both proteins increased. This occurred even when the two genes were placed in the same or opposite orientations on the plasmid. However, placing the reporters on different plasmids separated the two responses so that each could reflect their independent regulation (Fig. 2B).

GENE CONSTRUCTS

The reporter genes and the selectable markers used to create stable cell lines were deliberately manipulated to be useful and perform consistently. Although the DNA sequence for the each luciferase was extensively modified, the protein sequence of firefly luciferase was completely unchanged and the protein sequence of *Renilla* luciferase was changed by only one amino acid. The DNA sequences were first codon optimized for expression in mammalian cells and then modified to remove as many putative consensus regulatory sequences as possible. As seen in Fig. 3, this process increased the expression of the new firefly luciferase (*luc2*) 4- to 12-fold over the parent gene (*luc+*), depending on the cell line. The firefly luciferase gene was also modified to contain a protein degradation sequence. The shorter protein half-life increased the rate and extent of change in protein expression after induction (Fig. 3). The *luc2P* gene was used to create our cell line. In the receptor-containing plasmid, the *Renilla* luciferase gene was fused to the neomycin gene. This protein fusion enabled antibiotic resistance for clone selection through the neomycin expression. It also simplified tracking cell health and baseline protein expression through *Renilla* luciferase expression.

SYSTEM TESTING

Creating stably transfected cell lines by simultaneously incorporating two plasmids can be difficult. It requires testing of many clones (often hundreds) to ensure that the experimental and control reporters and the membrane receptor all perform with the desired response characteristics. We have simplified the process by taking a step-wise approach. In our model, the CRE-controlled firefly luciferase was on the same plasmid as a hygromycin gene so that selection pressure could be used (Fig. 1). Once clones stably-expressing firefly luciferase were isolated, they were tested for the extent of CRE-regulation. Candidate clones were identified by stimulating either endogenously- or transiently-expressed GPCRs and looking for high induction and luciferase expression. The best-performing clones were designated as "master" cell lines, then modified to produce other transiently- or stably-expressed proteins.

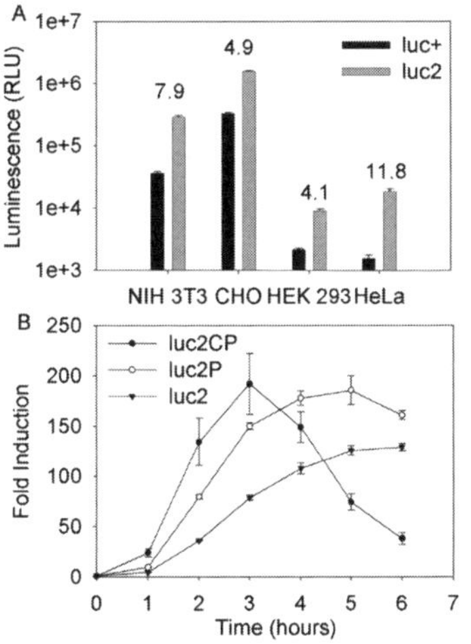

Fig. 2. Genetic modifications improved expression and response dynamics of firefly luciferase. Panel A. The luc+ and luc2 genes were expressed in the same backbone, under control of the CMV promoter in four different cell lines. Panel B. The luc2 gene encoding luciferase appended with hPEST or hCL1/hPEST sequences were stably transfected into HEK293 cells and induced with 1 μM isoproterenol/100 μM RO-20-1724.

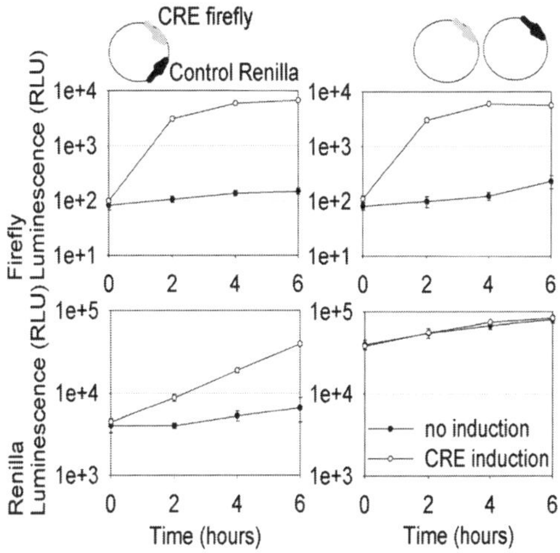

Fig. 3. Expression of control reporter genes on the same plasmid as inducible genes caused the expression of the control protein to change with induction. CRE-inducible firefly luciferase and *Renilla* luciferase under control of the TK promoter were transfected into HEK293 cells and luminescence was measured after induction (Fig. 2B). In the left 2 panels, the firefly and *Renilla* luciferases were expressed from the same plasmid, oriented in opposite directions. Results were similar when luciferases were oriented in the same direction. In the right panels luciferases were expressed from different plasmids.

Transfection with the second plasmid containing the M3R and the *Renilla* luciferase-neomycin fusion protein provided not only the receptor to be tested, but also a selection marker fused to an easily measured indicator of baseline transcription and cell viability. Cells stably-transfected with the second plasmid were selected for sensitivity to regulation and independently monitored for the retention of the experimental firefly luciferase-expressing plasmid and the control *Renilla* luciferase-expressing plasmid. Double-stable clones were tested and ranked based on a variety of characteristics: induction level, EC_{50} and IC_{50} values for known agonists and antagonists, rank potency of agonists and antagonists, etc. Our

HEK 293 cells stably transfected with CRE-controlled firefly luciferase and the muscarinic 3 receptor exhibited over 100-fold induction and generated an EC_{50} value of 2.6 x 10^{-6} M for muscarinic chloride, a known agonist, and an IC_{50} value of 6.3 x 10^{-9} M for atropine, a known antagonist, consistent with literature reports. In addition, titration of pilocarpine induced firefly luciferase expression by only a small factor, especially compared to other agonists, suggesting that both full and partial agonists were detectable with this cell line (Fig. 4A). Inclusion of the *Renilla* luciferase control is particularly valuable when measuring antagonists. Sanguinarine chloride downregulated firefly luciferase during a screen of the Sigma LOPAC1280™ library. Concurrent decrease of the *Renilla* luciferase signal during the screen flagged the compound for further investigation, and experiments showed that the compound was cytotoxic and not a specific M3R antagonist (Fig 4B).

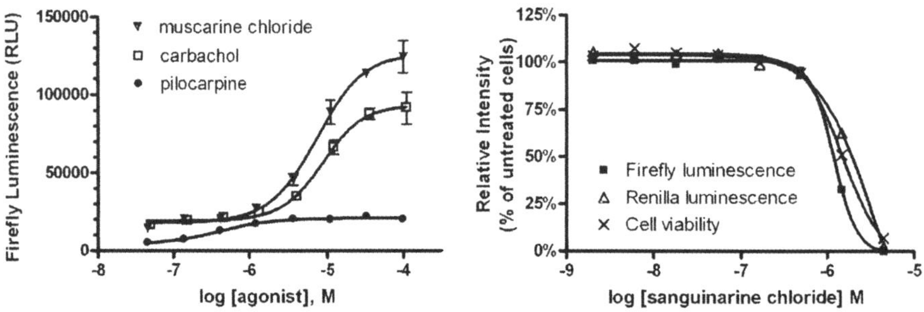

Fig. 4. Panel A. Known agonists of M3R stimulate luminescence when titrated onto the master cell line in 384-well plates. Pilocarpine, a known partial agonist, exhibited weak stimulation. Panel B. Sanguinarine chloride causes decreases in both firefly and *Renilla* luminescence. This parallels the measured effect on cellular viability (CellTiter-Glo™ Reagent, Promega).

CONCLUSION

We have successfully constructed a stable dual luciferase cell line using a simple stepwise approach, and the same method could be easily used to generate new lines. Other receptors that signal through the CRE pathway can be quickly evaluated by transfecting them into the CRE master cell line. Other response elements may also be transfected to create alternate master cell lines. The current stably-transfected master cell line exhibits characteristics that are desirable for daughter lines. Using the new generation of firefly luciferase gene (*luc2*) and its rapid response version (*luc2P*) provided brighter luminescence and improved responsiveness. Assays for the CRE pathway induced through the muscarinic 3 receptor showed good dynamic range and were useful in determining both agonists and antagonists. Finally, multiplexing using both firefly and *Renilla* luciferases improved data quality when screening compound libraries or performing other experiments by simultaneously monitoring cell viability using an internal control reporter.

SIMULTANEOUS BIOLUMINESCENT IMMUNOASSAY USING AEQUORIN LABELED FAB FRAGMENT AND BIOTINYLATED FIREFLY LUCIFERASE

KATSUTOSHI ITO,[1] WAKA NISHIMURA,[1] MASAKO MAEDA,[1] KEIKO GOMI,[2] SATOSHI INOUYE,[3] HIDETOSHI ARAKAWA[1]

[1] School of Pharmaceutical Sciences, Showa University, 1-5-8 Hatanodai, Shinagawa, Tokyo 142-8555, Japan; [2] Kikkoman Co., 399 Noda, Noda-shi, Chiba, 278-0037, Japan; [3] Chisso Co., Yokohama Research Center, 5-1 Okawa, Kanazawa-ku, Yokohama, 236-8605, Japan

INTRODUCTION

We established a simultaneous bioluminescent assay[1] using aequorin (Aq) and biotinylated luciferase (b-*Luc*).[2] These photo-proteins have totally different mechanisms of bioluminescence. Therefore, we proposed that Aq and b-*Luc* could be rapidly and continuously measured in a same assay medium. We applied the method to rapid and sensitive simultaneous bioluminescent immunoassay (BLIA) using Aq labeled Fab fragment and b-*Luc*/streptavidin complex.

MATERIALS AND METHODS

Reagents. Recombinant aequorin (Aq) which had a free cys residue was from Cisso Corporation (Yokohama, Japan). Thermostable biotinylated firefly luciferase (b-Luc) was obtained from Kikkoman Corporation (Chiba, Japan). N-Succinimidyl-6-maleimidohexanoate (EMCS) was purchased from Dojindo Laboratories (Kumamoto, Japan). Prostate specific antigen (PSA)-antichymotrypsin complex were purchased from Cortex Biochem, Inc. (San Leandro, CA, USA). Prostatic acid phosphatase (PAP) was gift from ASKA Pharmaceutical Co., Ltd. (Kawasaki, Japan). Mouse monoclonal anti-PSA IgG (clone no. 4D10 and 56) were purchased from Japan Clinical Laboratories (Kyoto, Japan). Mouse monoclonal anti-PAP IgG (clone no. 94-10045 and 94-10046) were purchased from OEM Concept Inc. (Toms River, NJ, USA). Sheep anti-digoxigenin (Dig) Fab fragment were purchased from Roche Diagnostics (Germany).

Preparation of aequorin labeled sheep anti-digoxigenin Fab fragment. We prepared Aq labeled sheep anti-Dig Fab fragment. Briefly, to introduce the maleimide group into anti-Dig Fab fragment, 1 mg of anti-Dig Fab fragment in 0.01 mol/L PBS (pH 7.0) was reacted with 10 μL of an EMCS (62 μg) solution in dimethyl sulfoxide. After incubation for overnight at 4°C, the excess EMCS derivatives were removed by gel filtration. The maleimide- anti-Dig Fab fragment was mixed with 0.6 mg of Aq which has a free sulfhydryl group in 0.05 mol/L PBS (pH6.0) containing 5 mmol/L EDTA. The mixture was purified by HPLC gel chromatography using a Superdex 200 10/300 GL column (10 mm i.d. x 300 mm). The fractions of Aq labeled sheep anti-Dig Fab fragment which have high immuno- and bioluminescent-reactivity were added to a water-soluble gelatin solution as a stabilizer, and then stored at -80°C prior to use. The Aq labeled sheep anti-Dig Fab fragment was stable for 1 week at 4°C.

Simultaneous sandwich BLIA for PAP and PSA. The principle of the proposed simultaneous BLIA is shown Fig. 1. The wells of the microtiter plate were coated with mouse monoclonal anti-PAP IgG (clone no. 94-10046) and mouse monoclonal anti-PSA IgG (clone no. 4D10). Each IgG were 5 μg/mL in 50 mmol/L sodium bicarbonate buffer (pH 9.5). The solution was added 100 μL to each well. The wells were then post-coated by adding 100 μL of a 1% water-soluble gelatin solution containing 0.05% NaN_3. The plates were stored at 4°C prior to use. After washing the plate, 20 μL of a standard or sample solution, 100 μL of Dig labeled anti-PAP IgG and biotinylated anti-PSA IgG mixed solution were added to each well. The Dig labeled anti-PAP IgG and biotinylated anti-PSA IgG were used 125 and 235 ng/mL, respectively. The plate were incubated for 2 h at 37 °C. After washing, 100 μL of b-*Luc*/streptavidin complex ($1x10^{-9}$ mol/L) and Aq labeled anti-Dig Fab fragment (495 ng/mL) mixed solution was added and allowed to stand for 1 h at room temperature. The microtiter plate was re-washed and Aq and b-*Luc* actibities were assayed by the simultaneous bioluminescent method.

The simultaneous bioluminescent method for Aq and b-Luc as follow. Fifty microliter of a 50 mmol/L $CaCl_2$ solution in 0.05 mol/L HEPES buffer (pH 7.0) was added. The bioluminescent intensity of Aq was measured for 1 s by a Centro LB960 luminescent reader (Berthold Technologies, Germany), immediately. Then, 50 μL of a bioluminescent reagent for b-*Luc* (containing ATP, D-luciferin and $MgSO_4$ in 50 mmol/L HEPES-KOH buffer, pH 7.0) were added to the same wells of the microtiter plate. The BL intensity of b-Luc was measured for 1 s after delay of 2 s.

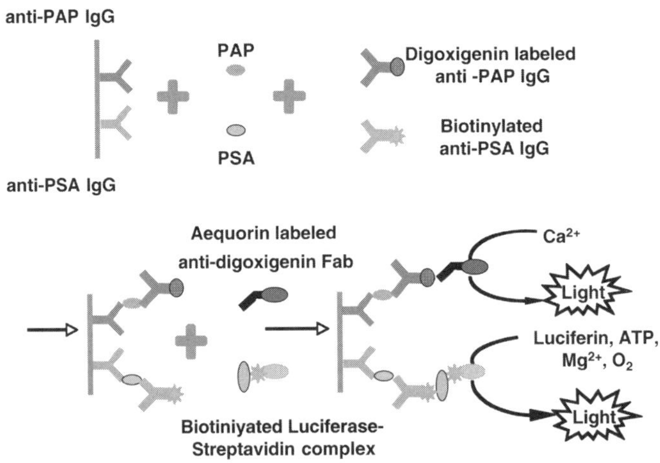

Fig. 1. Schematic illustration of the proposed simultaneous bioluminescent immunoassay

RESULTS AND DISSCUSSION

In the simultaneous BL assay of Aq and b-Luc, the sensitivity of Aq was 9.4 $\times 10^{-21}$ mol/assay (blank + 3SD), which is similar to the result of the individual assay for Aq. The intra-assay coefficients of variation (CV) for 8 replicates with each standard point of Aq were from 2.3 to 4.4%. The detection limit of b-Luc (at blank+3SD) was 3.6 x 10^{-19} mol/assay. The intra-assay CVs for 8 replicates with each standard points of b-*Luc* were from 3.6 to 5.8%.

The proposed simultaneous bioluminescent measurement for the Aq and b-Luc activities was applied to BLIA. We selected PAP and PSA, which are markers of prostate cancer as representative analytes. In this BLIA, the detection for PAP and PSA used Aq and b-Luc as labeled photo-protein, respectively. The measurable ranges of PAP and PSA were 0.04 - 100 and 0.2 - 200 ng/ml (Fig. 2), and the detection limits (at blank+3SD) of PAP and PSA were 1.9 x 10^{-18} and 4.6 x 10^{-18} mol/assay. The intra-assay coefficients of variation of PAP and PSA with each standard point were 2.9 - 4.8% (n=8) and 2.9 - 6.3% (n=8), respectively. The recovery of PAP and PSA from human serum were 75.6±9.0% and 85.6±12.0% (mean±SD, n=15), respectively.

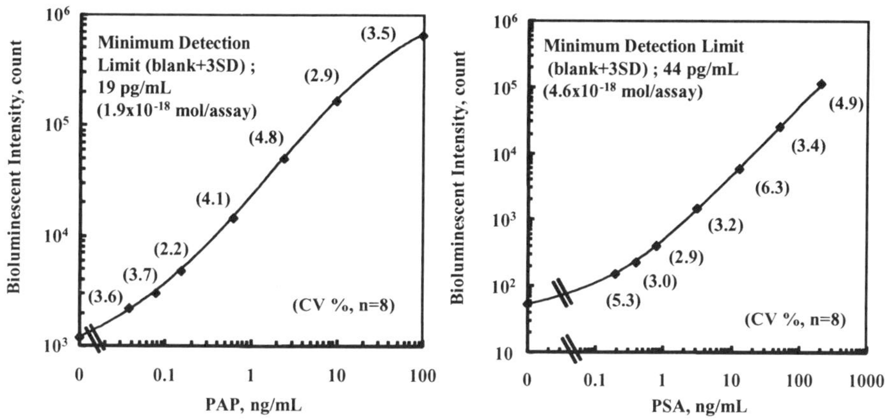

Fig. 2. Standard surves for prostatic acid phosphatase (PAP) and prostate specific antigen (PSA)

The concentrations of PAP and PSA in serum samples could be accurately measured by the BLIA. The values of PSA by the proposed BLIA were compared with an enzyme-linked immunosorbent assay (ELISA, Markit M-PA, Dainippon Sumitomo Pharma Co., Ltd., Osaka). The ELISA was carried out in the Clinical Laboratory, Showa University Hospital. The PAP levels were measured by Microwell ELISA PAP kit (Diagnostic Automation Inc., Calabasas, CA, USA). The regression equations determined for the results were y (BLIA) = 0.298 x (ELISA) -

0.087 (n = 34, r^2 = 0.988) for PAP and y (BLIA) = 2.043 x (ELISA) + 0.053 (n = 54, r^2 = 0.958) for PSA, respectively. The slopes of regression equations were differed with the conventional methods. But the measurement of PSA levels by Markit M-PA kit was already reported lower than other commercial kits.[3] Concentration of PAP standard of the ELISA kit was showed about 3-fold higher than our PAP standard. A high degree of correlation was observed between the values obtained using the BLIA and those by the conventional methods.

ACKNOWLEDGEMENTS
The authors would like to thank the Clinical Laboratory, Showa University Hospital, for kindly providing and analyzing the human plasma samples. We also thank Dr. Sasamoto, ASKA Pharmceutical Co., Ltd., for the gift of purified PAP standard. This research was partially supported by the Ministry of Education, Science, Sports and Culture, Grant-in-Aid for Scientific Research (C) No.17590039.

REFFERENCES

1. Tatsumi H, Fukuda S, Kikuchi M, Koyama Y. Construction of biotinylated firefly luciferases using biotin acceptor peptides. Anal Biochem 1996;243: 176-80.
2. Nishimura W, Ito K, Arakawa H, Maeda M, Inouye S, Tatsumi H. Rapid and simultaneous bioluminescent assay of aequorin and firefly luciferase. Luminescence 2004;19:167.
3. Kano S, Ishibashi M, Itoh Y. The present status of standardization in serum total PSA measurement--a result of "Survey 2000" conducted by PSA ad hoc Committee of Japanese Urological Association. Rinsho Byori 2001;49:967-73.

DESIGNING ENHANCED THERMOSTABLE LUCIFERASE FOR PROTEOLYTIC ASSAY

AP JATHOUL,[1] OA GANDELMAN,[2] GHE LAW,[1] JAH MURRAY,[1,2] LC TISI[2]
[1]*Institute of Biotechnology, Tennis Court Road, University of Cambridge, CB2, IQT, UK;* [2]*Lumora Ltd., Tennis Court Road, Cambridge, CB2 IQT, UK*

INTRODUCTION

In bioluminescent protease assays, proteolysis is detected via the bioluminescence of the firefly (FF) luciferin (LH_2) analogue 6'-amino-*D*-luciferin (ALH_2) with FF Luciferase (*Luc*) after hydrolysis from a peptide-conjugated form ('proluciferin').[1] The amount of light emitted in such assays depends not only on the activity of a specific protease, but also on the activity and substrate range of *Luc* with ALH_2. Yet, little is known about various aspects of ALH_2 activity with wild-type (WT) or thermostable *Lucs*.[2] Here, we report kinetic parameters for WT *Luc* and a series of existing thermostable mutants, defining Km values for LH_2 and ALH_2, but also ATP in the presence of either LH_2 or ALH_2. This study clarifies unpublished aspects of ALH_2 substrate range and activity, and may help shed light on aspects of *Luc* reactivity. These results form a benchmark for further comparison with other *Luc* mutants, which may be selected for enhanced activity with ALH_2.

MATERIALS AND METHODS

D-LH_2 potassium salt was purchased from Europa Bioproducts Ltd. (Cambridge, UK) and *D*-ALH_2 from Marker Gene Technologies Inc. (Oregon, USA). x2*Luc* was kindly donated by Dr. Peter White (Dstl, UK), x4*Luc* by Dr. Laurence Tisi (Lumora Ltd, UK), and, x5*Luc* and x12*Luc* by Dr. Erica Law (University of Cambridge, UK; Lumora Ltd., Cambridge, UK). UG was purchased from Promega Corporation (Madison, WI, USA). Kinetic measurements were carried out in a Labsystems Ascent Luminoskan Luminometer. Individual experimental conditions are indicated in the figure legends.

RESULTS AND DISCUSSION

Substrate range and activity with thermostable *Lucs*. First, activities of two of the most thermostable *Lucs* available, Lumorase™ (x12*Luc* – mutations: F14R, L35Q, A105V, V182K, T214C, I232K, D234G, F295L, D357Y, E354R, S420T and F465R) and Ultra-Glo™ (UG) were tested with ALH_2, and compared to WT. LH_2 Michealis-Menton constant and Vmax were determined at concentrations between 0.2mM-500mM (in 0.1M Tris-acetate at pH 7.8, 10mM $MgSO_4$ and 2mM EDTA) and for ALH_2 0.1mM-600mM, at saturating ATP (1mM-5mM). For ATP, a 10mM-1mM or 4mM concentration range was used, at saturating LH_2 (*ca.*150mM) or ALH2 (ca.90mM). All enzyme concentrations were 67 nM. Flash heights measured in triplicate at (or corrected to) 550v PMT.

Table 1. Km and Vm Values for WT, x12*Luc* and UG with LH_2 ALH_2, ATP in the presence of LH_2 and ATP in the presence of ALH_2

Mutant	Residues	Km/Vm LH_2 (μM)	Km/ Vm ATP (LH_2) (μM)	Km/ Vm ALH_2 (μM)	Km/ Vm ATP (ALH_2) (μM)
WT	-	14.8±0.2 / 5000±0	66.0±6.0 / 5000 ±0	1.6±0.2 / 588±15	83.0±20.0 / 667±84
x12	F14R, L35Q, A105V, V182K, T214C, I232K, D234G, F295L, D357Y, E354R, S420T, F465R	7.0±2.0 / 314±14	55.0±6.0 / 301±14	6±0.2 / 357±0	297.5±46.8 / 345±0
x4	T214A, I232A, F295L, E354K	15.8±0.3 / 1250±0	56.0±3.0 / 1429±60	11.0±1.2 / 388±28	260.2+8.1 / 400 ±0
x5	F14R, L35Q, V182K, I232K, F465R	12.0±0.0 / 5000±0	87.0±2.0 / 5000±0	2.3±0.0 / 556 ±0	193±16 / 769 ±0
RY	E354R, D357Y	5.8±0.5 / 2500±0	52.0±4.0 / 2500±0	1.4±0.8 / 518±12	210.0±24.0 / 526±23
UG	20+ mutations of *P. pennsylvanica*	3.8±0.1 / 107±4	4.8±0.7 / 111±11	6.8±0.2 / 24±1	19.7±2.9 / 21±1

For WT and x12*Luc*, the Km for ALH_2 (14.8+0.2 μM and 7.0+2.0 μM, respectively) and was found to be lower than that of their respective LH_2 Kms, resulting in higher Kcat/ Km values with ALH_2 than LH_2. x12*Luc* was found to have a markedly higher Kcat (proportional to Vm - not shown) with ALH_2 than UG, displaying flash kinetics more akin to WT than UG (data not shown). The Kcat of x12*Luc* with ALH_2 was found to be approximately 14 times higher than that of UG suggesting the latter is catalytically compromised compared to x12*Luc*. The thermostable *Lucs* tested had higher ATP Kms in the presence of ALH_2 than in the presence of LH_2 (termed the 'ATP-effect' - discussed below). WT *Luc* had no or little increase of $ATP_{(ALH2)}$ Km relative to its $ATP_{(LH2)}$ Km (ca.1.2-fold), while UG had a 4-fold increase in $ATP_{(ALH2)}$ Km above its $ATP_{(LH2)}$ Km. However, x12*Luc* had the highest such increase of 5.4-fold. This may indicate mutations are present in x12*Luc* that affect the coordination of ATP specifically in the presence of ALH_2 during the reaction.

To dissect which thermo-stabilising mutations in x12*Luc* affected ALH_2 Vm, and an increased Km for ATP in the presence of ALH_2 (see Fig. 1), kinetic parameters were tested for three further in-house thermostable *Lucs* containing subsets of the mutations in x12*Luc*: x2*Luc* (E354R and D357Y) (3), x4*Luc* (T214A, I232A, F295L and E354K) (4) and x5*Luc* (F14R, L35Q, V182K, I232K and F465R)

(5). All *Lucs* tested except UG were originally engineered from *Photinus pyralis* (*Ppy*). Their kinetic data are also displayed in table 1. x2*Luc* and x5*Luc* had similar Kms for ALH_2 to WT (ca.1-2 μM), whereas for x12*Luc* it was 6 μM, while x4*Luc* had the highest Km for ALH_2 at 11μM. WT and x5*Luc* showed the largest decreases in ALH_2 Km compared to their LH_2 Kms, whereas x4*Luc* and x12*Luc* have similar Kms for both substrates. As x12*Luc* is comprised of x4*Luc* and x5*Luc* mutations (among others), it is feasible that a particular mutation in x4*Luc* is responsible for the increased Km for ALH_2 of these two *Lucs*. Also, while the Vm was halved for x4*Luc* with ALH_2 relative to LH_2, the x12*Luc* Vm (and therefore Kcat) remained the same for both. Both x2*Luc* and x5*Luc* were seen to have Vm values closer to WT with ALH_2, while x4*Luc* x12*Luc* were lower (x4*Luc* the lowest). This again suggests a common mutation in them that reduces their Vm with ALH_2.

ATP effect. $ATP_{(ALH2)}$ Kms were elevated for all further thermostable *Luc* mutants tested, with x5*Luc* being elevated ca.2.2-fold relative to its $ATP_{(LH2)}$ Km, and x2*Luc* showing a 4-fold increase. However, the ATP Km of x4*Luc* increased 4.6-fold (second highest after x12*Luc*). This pattern of increase was seen to match the degree of thermal inactivation of the Ppy-based *Lucs* (Fig. 1), suggesting thermo-stabilising mutations may either reduce ATP coordination in the presence of ALH_2, or reduce the ability to undergo necessary conformational changes for reaction with ALH_2. Mutations in x12*Luc* and x4*Luc* appear to exacerbate this effect. As mutations at positions 214 and 295 are only present in x4*Luc* and x12*Luc* and not in any other *Ppy*- based *Luc* tested, they are prime candidates for conferring the increased Kms for ALH_2 and the $ATP_{(ALH2)}$ (testable by back mutation).

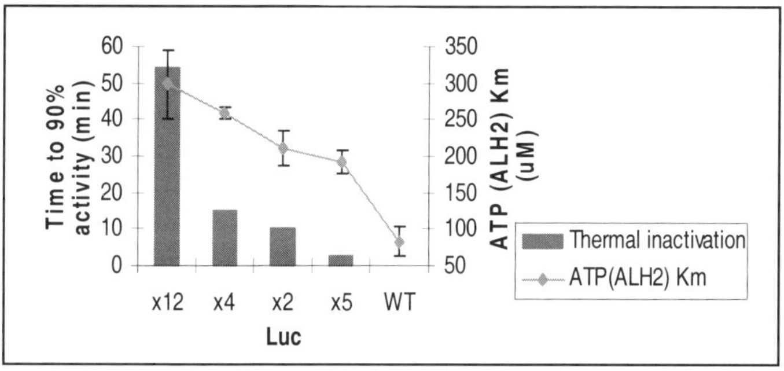

Fig. 1. Time to 90% thermal inactivation of *Luc* at 40°C. Each enzyme (67 nM) incubated at 40°C, activity measured at intervals up to 80 min in Luminometer by injection of 10x Km of LH2 and ATP and measurement of flash heights in triplicate

It was seen that, by mixing both LH_2 and ALH_2 together in varying percentages at concentrations 10x their respective Kms and using them to measure ATP Kms, the dynamic range of ATP could be increased to intermediate levels between the Kms of $ATP_{(LH2)}$ (low Km) and $ATP_{(ALH2)}$ (high Km) for x2*Luc*, x5*Luc* (data not shown) and x12*Luc* (Fig. 2).

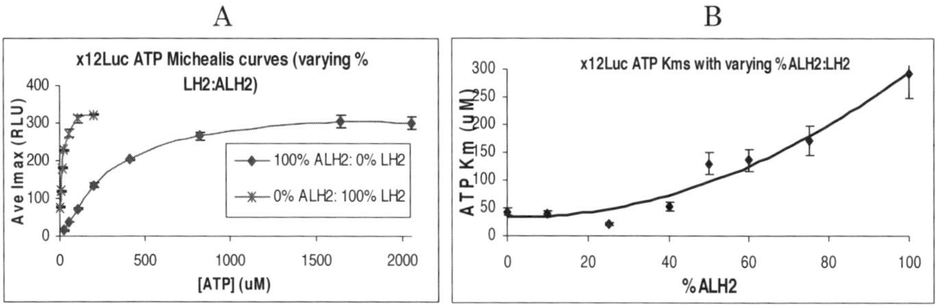

Fig. 2. Effect of varying ALH_2:LH_2 mixture ratio when determining ATP Km A. Michealis-Menton curves B. Km relationship to ratios. ATP Km measurements carried out as described in table 1, using differing ratios of ALH_2:LH_2 mixtures

CONCLUSIONS

It was found that the Km for ALH_2 is lower than that of LH_2 for *Ppy* WT *Luc* and *Ppy*-based thermostable *Luc*, but not a *Photuris pennsylvanica*-based *Luc* (UG). x12*Luc* has the highest ALH_2 activity concurrent with highest thermostability of all the *Luc* mutants tested. Another effect observed was the increase of $ATP_{(ALH2)}$ Km for all *Luc* mutants. This will allow ATP assays with variable substrate range with just one type of *Luc*. This work may also impact on the planning of bioluminescent protease assays. The Km of ALH_2 may be relevant to the optimal usage of proluciferin, and to the determination of levels of proteolytic activities required in assays for detection effective by *Luc*. Higher Vm for ALH_2 of thermostable *Lucs* may allow faster assays. Work is underway to further enhance x12*Luc* ALH_2 activity by back mutation of residues common to x4*Luc*.

REFERENCES

1. Monsees T, Miska W, Geiger R. Synthesis and characterisation of a bioluminogenic substrate for alpha-chymotrypsin. Anal Biochem 1994;221: 329-34.
2. White EG, Worther H, Seliger HH, McElroy WD. Amino analogs of firefly luciferin and their biological activity thereof. JACS 1966;88:2015-8.
3. White PJ, Squirrell DJ, Arnaud P, Lowe CR, Murray JAH. Improved thermostability of the North American firefly luciferase: saturation mutagenesis at position 354. Biochem J 1996;319:343-50.
4. Tisi LC, White PJ, Squirrell MJ, Murphy MJ, Lowe CR, Murray JAH. Development of a thermostable firefly luciferase. Anal Chim Acta 2002; 457:115-23.
5. Law GHE, Gandelman OA, Tisi LC, Lowe CR, Murray JAH. Mutagenesis of solvent-exposed residues in *Photinus pyralis* luciferase improves thermostability and pH-tolerance. Biochem J 2006;397:305-12.

BIOSENSORS FOR THE NON-INVASIVE EVALUATION OF BACTERIAL QUORUM SENSING IN GI DISORDERS

ANJALI KUMARI,[1] PATRIZIA PASINI,[1]
SAPNA K. DEO,[1] DEBORAH FLOMENHOFT,[2]
HAROHALLI SHASHIDHAR,[2] SYLVIA DAUNERT[1]
[1]Department of Chemistry; [2]Department of Pediatrics,
University of Kentucky, Lexington, KY 40506-005, USA

INTRODUCTION

Bacteria are thought to be involved in many clinical disorders ranging from infections to chronic inflammation. There are numerous reports in both animal models and humans where bacteria play a significant role in gastrointestinal (GI) diseases, such as inflammatory bowel disease (IBD).[1] Inflammatory bowel disease, which includes Crohn's disease (CD) and ulcerative colitis, is a chronic and relapsing inflammatory condition of the GI tract. The detailed mechanisms and causes of IBD are not fully understood. However, there is evidence to suggest that environmental, genetic, immunologic and microbial factors predispose patients to, as well as inducing IBD. To support the involvement of bacteria in IBD the identification of serologic markers against microbial antigens, and the successful use of antibiotic therapy in patients with Crohn's disease, as well as the observation that colitis cannot be induced in animals without bacterial flora.[2]

A key component of bacterial behavior and colonization is communication between bacteria, which is based on a phenomenon known as quorum sensing (QS). Quorum sensing enables bacteria to communicate with each other and control the expression of certain specialized genes by producing and responding to extracellular signal molecules in proportion to cell density.[3] Bacteria have elaborate chemical signaling systems that enable them to communicate within and between species.[4] Quorum sensing has been demonstrated to be involved in the pathogenesis of several diseases. For example, *N*-acyl homoserine lactones (AHLs), the signal molecules employed by most Gram-negative bacteria, have been detected in sputum samples of cystic fibrosis patients.[5] They have also been shown to be responsible for biofilm formation, both, in the lungs and on inanimate surfaces.[6] Interestingly, AHL molecules have been proven to modulate animal and human immune system.[7]

All these observations have led us to hypothesize that quorum sensing signal molecules (QSMs) may be employed as biomarkers to measure intestinal inflammatory activity. To that end, we have developed genetically engineered whole-cell sensing systems for the detection of QSM levels in body fluids such as saliva and stool. We envision that these biosensors could provide a useful tool for the non-invasive monitoring of inflammatory activity in IBD.

EXPERIMENTAL

Two whole-cell sensing systems were employed for detection of *N*-acyl homoserine lactones. The whole-cell sensors were based on *Escherichia coli* JM109 bacterial cells harboring a plasmid containing a bioluminescent reporter gene under control of an AHL-inducible promoter. Specifically, the plasmids pSB406 and pSB1075 were constructed by taking advantage of the LasI/LasR and RhlI/RhlR quorum sensing systems of the bacterium *Pseudomonas aeruginosa*.[8] LasR and RhlR are regulatory proteins that recognize and bind AHLs, then activate gene transcription, while LasI and RhlI are synthases of the signal molecules. Plasmid pSB406 contains the *rhlR* gene and the promoter of the *rhlI* gene (P_{rhlI}) fused with the reporter *luxCDABE* gene cassette. Similarly, plasmid pSB1075 contains the *lasR* gene and the promoter of the *lasI* gene (P_{lasI}) fused with the reporter *luxCDABE* gene cassette, which is responsible for production of the bacterial bioluminescent protein luciferase and the enzymes that form the substrate for luciferase. Upon binding AHLs, RhlR and LasR proteins activate expression of the reporter gene, which is under transcriptional control of the promoters P_{rhlI} and P_{lasI}, respectively. Consequently, a bioluminescent signal is produced whose intensity is directly proportional to the amount of signal molecule present in the cell. The AHL assay was performed as previously described.[9] Briefly, bacteria containing plasmid pSB406 or pSB1075 were grown in Luria Bertani medium containing ampicillin, at 37°C, with constant shaking at 250 rpm, until the optical density at 600 nm reached a 0.5 value. Subsequently, either solutions of standard AHLs at different concentrations or biological samples were added in triplicate to cell suspensions in culture tubes, and incubated for 3 h at 37°C with constant shaking at 250 rpm. Aliquots of cell suspensions were then transferred to a 96-well microtiter plate, and the bioluminescence signals were measured using a standard microplate luminometer. Light intensity was expressed in relative light units (RLU).

RESULTS AND DISCUSSION

The initial step in the development of our method was the characterization of the living whole-cell sensing systems, which was performed by incubating the bacterial cells with different commercially available standard AHL compounds. Indeed, individual bacterial species produce AHLs that differ in the length (4-18 carbon atoms) and substitution in C-3 of the molecule acyl side chain. An increase in bioluminescence intensity was observed with increasing concentrations of the AHLs tested. The response of each sensing system varied depending on the length of the acyl chain of the AHL analyzed. The sensing system harboring plasmid pSB406 showed a signal response to short acyl chain AHLs that was higher than that to long acyl chain ones. For instance, *N*-butyryl-DL-homoserine lactone (C4-HSL) and *N*-hexanoyl-DL-homoserine lactone (C6-HSL) were more effective than *N*-octanoyl-DL-homoserine lactone (C8-HSL), *N*-dodecanoyl-DL-homoserine lactone (C12-HSL) and *N*-tetradecanoyl-DL-homoserine lactone (C14-HSL) in inducing the sensor response (Fig. 1). On the contrary, the sensing system harboring plasmid pSB1075 was more responsive towards long acyl chain AHLs than short acyl chain ones (data not shown). The ability of these sensing systems to respond to different AHLs allow us to detect the whole range of AHLs produced by bacteria present in nature. The detection limits achieved with both whole-cell sensing systems were 1×10^{-9} M, which is sufficiently low to detect AHLs produced by bacteria. This method of detection proved to be precise and reproducible, with coefficients of variation less than 7%.

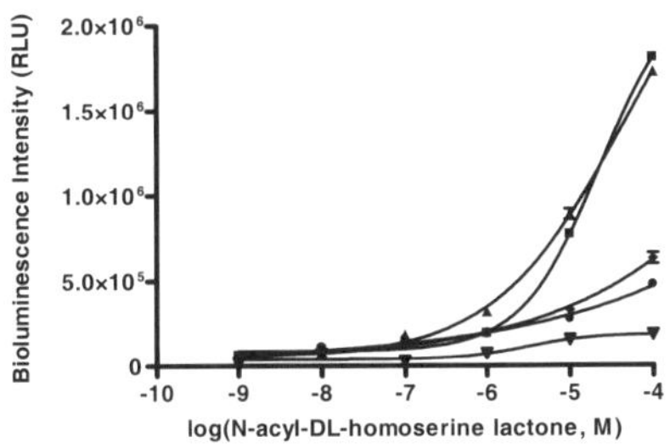

Fig. 1. Calibration curves for the biosensing system employing plasmid pSB406 obtained with C6-HSL (▲), C4-HSL (■), C8-HSL (●), C12-HSL (♦), and C14-HSL (▼). The signals have been corrected with respect to the blank. Data shown are the average ± one sd (n=3).

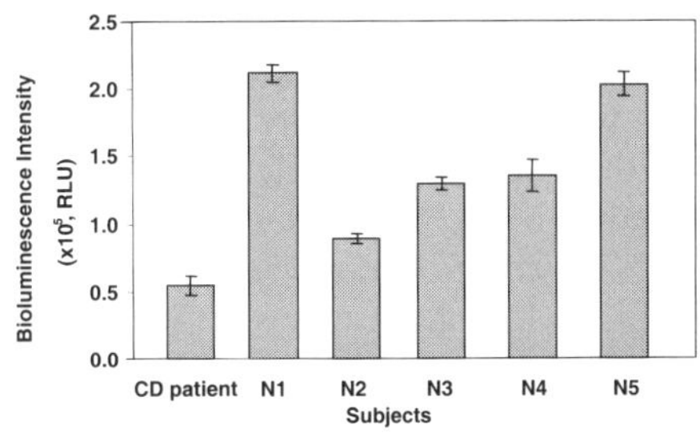

Fig. 2. Detection of AHLs in saliva samples from healthy volunteers and a Crohn's patient using the sensing system harboring plasmid pSB406. The signals have been corrected with respect to the blank. Data shown is the average ± one sd (n=3).

In order to validate our method for detection of AHLs in saliva and stool, we evaluated the effect of the biological sample matrix on the whole-cell sensing system response. To that end, we generated dose-response curves in pooled saliva and stool samples, respectively. Calibration curves were also obtained in the absence of the biological matrix as a reference within the same analytical run. The slopes of the dose-response curves generated either in saliva or stool were similar to those of the dose-response curves generated in the absence of matrix, thus demonstrating that the biological sample matrix components did not affect the sensor response, in the experimental conditions chosen. This was achieved by simple sample preparation, i.e., centrifugation of saliva samples to remove the viscous components, and suspension and dilution of stool samples in distilled deionized water. The whole-cell sensing systems were then employed to detect the presence of AHLs in human saliva and stool. Saliva samples were collected from healthy volunteers and a Crohn's patient. Stool samples were obtained from infants in the Newborn Intensive Care Unit of the University of Kentucky Children's Hospital. The results achieved in real samples showed that AHLs could be detected reproducibly in saliva and stool from both healthy and diseased individuals (Fig. 2).

The levels of these molecules varied among different subjects, while no significant variations were observed for a single normal subject over a period of time.

In conclusion, we have developed a biosensor that can be employed as an effective simple tool for detection of quorum sensing molecules in biological matrixes such as saliva and stool of healthy individuals and Crohn's patients. This biosensing system should be ideal for non-invasive monitoring of disease activity, and diagnosis and management of bacteria-related disorders of the GI tract and other apparatuses, thus supporting the potential use of QSMs as biomarkers of disease.

ACKNOWLEDGMENTS

This work is supported by the National Science Foundation (Grant CHE-0416553) and the Children's Miracle Network. We thank the Office of the Vice President of Research of the University of Kentucky for a University Research Professorship to SD. AK and SD acknowledge support from a Gill Fellowship and a Gill Eminent Professorship, respectively. We also thank Dr. Paul Williams (University of Nottingham, Nottingham, UK) for providing the plasmids pSB406 and pSB1075.

REFERENCES

1. Sartor RB. Microbial influences in inflammatory bowel diseases: role in pathogenesis and clinical implications. In: Sartor RB, Sandborn WJ. eds. Kirsner's Inflammatory Bowel Diseases. Elsevier, 2004:138-62.
2. Chandran P, Satthaporn S, Robins A, Eremin O. Inflammatory bowel disease: dysfunction of GALT and gut bacterial flora (II). Surgeon 2003;1:125-36.
3. Miller MB, Bassler BL. Quorum sensing in bacteria. Annu Rev Microbiol 2001;55:165-99.
4. Federle MJ, Bassler BL. Interspecies communication in bacteria. J Clin Invest 2003;112:1291-9.
5. Middleton B, Rodgers HC, Camara M, Knox AJ, Williams P, Hardman A. Direct detection of N-acylhomoserine lactones in cystic fibrosis sputum. FEMS Microbiol Lett 2002;207:1-7.
6. Kjelleberg S, Molin S. Is there a role for quorum sensing signals in bacterial biofilms? Curr Opin Microbiol 2002;5:254-8.
7. Rumbaugh KP. Convergence of hormones and autoinducers at the host/pathogen interface. Anal Bioanal Chem 2006; DOI 10.1007/s00216-006-0694-9.
8. Winson MK, Swift S, Fish L, Throup JP, Jorgensen F, Chhabra SR, et al. Construction and analysis of luxCDABE-based plasmid sensors for investigating N-acyl homoserine lactone-mediated quorum sensing. FEMS Microbiol Lett 1998;163:185-92.
9. Kumari A, Pasini P, Deo SK, Flomenhoft D, Shashidhar H, Daunert S. Biosensing systems for the detection of bacterial quorum signaling molecules. Anal Chem 2006; in press.

IMPROVED CA2^{+}-ACTIVATED PHOTOPROTEINS FOR HTS APPLICATIONS

N MASTROIANNI, T FLAK, S CORAZZA
Axxam srl, San Raffaele Biomedical Science Park, Via Olgettina 58, 20132 Milan, Italy. E-mail: nadia.mastroianni.nm@axxam.com

INTRODUCTION

The Ca^{2+} bivalent ion is a critical second messenger involved in many physiological and signal transduction processes within the cell. Transient changes in intracellular Ca^{2+} concentrations are mainly triggered either by ligand-mediated activation of GPCRs (G-protein coupled receptors) or by opening of Ca^{2+} permeable ion channels. Both are transmembrane proteins and belong to the most preferred target classes within the pharmaceutical industry. The development of precise Ca^{2+} mobilization assays for HTS (High Throughput Screening) applications is challenged by the nature of the Ca^{2+} signal which is rapid, short, transient and sometimes weak.[1] The use of a Ca^{2+} activated photoprotein with its inherent property of flash-type luminescence represents a valid tool for analyzing all aspects of Ca^{2+} mediated signal transduction processes in mammalian cells. The advantages of using photoproteins as Ca^{2+} indicators include: low background levels resulting in a good signal-to-noise ratio; monitoring of reaction kinetics, and possibility to measure Ca^{2+} concentrations at specific cellular sites. While the miniaturization of the HTS process for testing chemical compound libraries has allowed the scaling up of the number of wells/plate from 96 to 384 and up to 1536 and a saving on the volumes of compounds used, the lower number of cells/well means that the reporter system must necessarily generate a strong and reproducible signal that can be accurately detected. To face these challenges, we are engaged in finding and developing new photoproteins with novel or improved properties, such as Ca^{2+} affinity, higher light release, slower kinetics.

METHODS

Generation of a randomly mutagenized library and screening. The GeneMorph II Random Mutagenesis kit (Stratagene) was used following supplier's instructions in order to achieve a high mutation rate. Directional cloning of the PCR products was obtained using the pET-30 Ek/LIC vector, engineered to express the target protein immediately downstream of an enterokinase cleavage site. A library of approximately 84,000 colonies expressing the randomly mutated photoprotein was generated. Functional screening was performed after IPTG induction and charging with 10 μM coelenterazine. Bioluminescence was assayed by detection of light emission upon Ca^{2+} stimulation. The best colonies were picked, grown in LB liquid medium and re-tested using the same experimental conditions described before.

RESULTS AND DISCUSSION

In order to select a pool of very sensitive photoproteins, several approaches, including construction of chimeric molecules, site-directed and random mutagenesis were evaluated. Considering both intensity of light release and sensitivity to low Ca^{2+} stimulation as selecting criteria, a high throughput functional screening for new photoproteins was developed, using a CCD camera-based plate reader for the detection of flash luminescence.

Functional screening of randomly mutagenized libraries. The mutation-based approach produced a huge number of putative new photoproteins that needed to be analyzed for their functionality. Random mutagenesis was performed on the cDNA sequences of different natural photoproteins and the obtained pool of molecules was cloned in bacterial expression vectors. Typically, after transformation in an appropriate bacterial strain, the library was plated, induced for protein expression, charged with coelenterazine and then tested for light emission upon Ca^{2+} stimulation. Functional screening was performed using a robotic system based on the CCD camera-based plate reader in combination with a dispensing unit. The selection of the positive clones was performed by marking on the luminescent images the positive luminescent colonies and then comparing these files with the corresponding visible-light images.

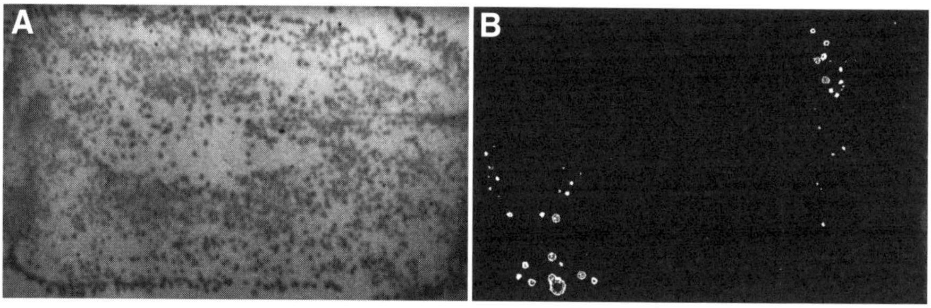

Fig. 1. High throughput functional screening with a CCD camera
A: Visible image of bacterial colonies
B: Light emission upon Ca^{2+} stimulation of the same filter

Out of 84,000 colonies expressing the randomly mutated photoprotein, 1089 were positive and therefore had bioluminescent characteristics. The best colonies were picked, for a total of 289 colonies, grown in LB liquid medium and re-tested using the same experimental conditions. Finally 8 clones were chosen, tested for light emission upon stimulation with low Ca^{2+} concentrations (as shown in Fig. 2) and further analyzed after small scale protein purification.

Cell-based functional assays. The expression and functionality of a few mutants was further evaluated in mammalian cell lines frequently used in HTS, such as CHO and HEK 293. The gene encoding for these photo-mutants was targeted to various cell compartments in order to generate cell lines allowing a refined detection of Ca^{2+}-dependent signaling pathways in response to activation of different targets.[2] All the transfected cells displayed stable expression of the photoprotein over time, without affecting neither cell growth nor viability.

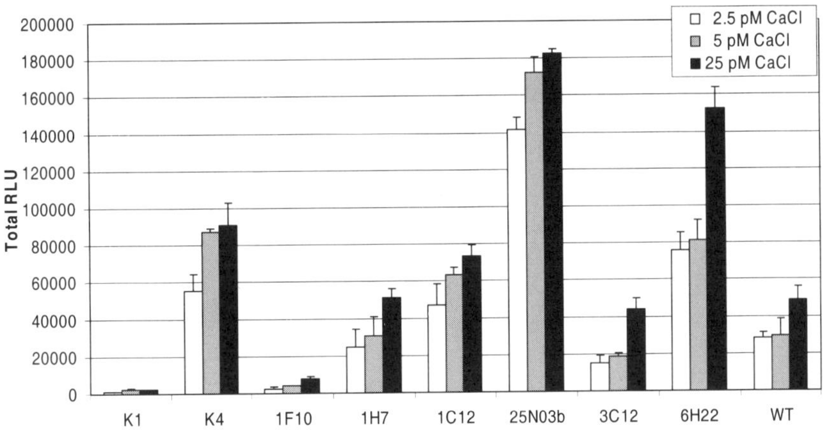

Fig. 2. Re-test of mutant clones; light release from bacterial cultures

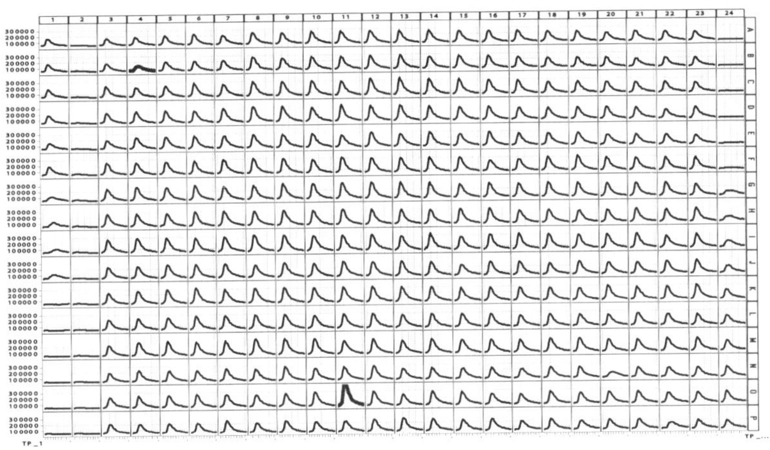

Fig. 3. Plate layout of primary HTS in 384 MTP - kinetics curves

One specific mutant, namely 25N03b, showed a higher light release compared to the others and to the cell line transfected with the wild type photoprotein, thus reconfirming the indications obtained in other expression systems. Sustained, reproducible signals could be detected using a range of plate readers (among them FLIPR^{TETRA} and CyBi®Lumax) and with different MTP formats (96, 384 and 1536). Moreover a reliable light signal could be obtained with 100 cells/well, a considerably low cell density. This feature made 25N03b particularly attractive as reporter system for HTS where miniaturization and reproducibility are essential requirements. As a matter of fact HTS campaigns involve testing of large compound collections (reaching 1,200,000 chemical entities) and the use of a recombinant cell line expressing a brighter photoprotein could reduce significantly the total amount of cellular material required. Thanks to a strong signal-to-noise ratio, the kinetics of Ca^{2+} movement upon stimulation with compounds can be easily followed in cell-based functional assays for HTS. An example of a 384-plate layout from primary screening is given in Fig. 3:

- Column 1 and 24: ANTAGONIST dose response (col 24 in reverse orientation)
- Column 2: ANTAGONIST EC_{80} (MIN response)
- Column 23: TYRODE (MAX response for hit reference)
- Column 3 to 22: COMPOUNDS.

CONCLUSIONS

In conclusion, the selected mutated photoproteins exhibit a robust flash luminescence signal in different expression systems. The most interesting mutants show enhanced bioluminescence, high affinity for Ca^{2+} and long-lasting light emission. They are conveniently used in cell-based assays to determine variations of intracellular calcium concentration, particularly in assays for the screening of molecules with high and ultra-high-throughput techniques. The strong light signal obtained with 25N03b mutant makes it quite attractive for exploring new applications for this mutated photoprotein like *in vivo* cell imaging and biosensor arrays for toxicity.

REFERENCES

1. Mattheakis L, Ohler LD. Seeing the light: Calcium imaging in cells for drug discovery. Drug Disc Today, HTS suppl, 2002;1:15-9.
2. Brini M, Pinton P, Pozzan T, Rizzuto R. Targeted recombinant aequorins: tools for monitoring Ca^{2+} in the various compartments of a living cell. Microsc Res Tech 1999;46:380-9.

DEVELOPMENT OF A MULTIPLEXED BIOLUMINESCENT CELL-BASED ASSAY WITH THE *LUC* GENE FROM *LUCIOLA ITALICA* FOR HIGH THROUGHPUT SCREENING OF CHOLESTEROL-LOWERING DRUGS

E MICHELINI,[1] TL SOUTHWORTH,[2] D ABLAMSKI,[2]
BR BRANCHINI,[2] A RODA[1]
[1] *Department of Pharmaceutical Sciences, University of Bologna, Bologna, Italy;*
[2] *Department of Chemistry, Connecticut College, New London, Connecticut, USA*
Email: elisa.michelini8@unibo.it

INTRODUCTION

The cDNA encoding the luciferase from the Italian firefly *Luciola italica* has been recently cloned and expressed in *E. coli*.[1] The bioluminescence emission of the Italian firefly luciferase has a maximum at 566 nm at pH 7.8 and shows extended light emission decay kinetics. This new luciferase has been cloned in suitable vectors and expressed in *E. coli* and mammalian cells (HepG2) to explore its applicability as a reporter protein in whole-cell biosensors, non-invasive *in vivo* imaging and multiplexed assays. Diverse mutants have been also obtained by site-directed or random mutagenesis.[2]

In the present study, a new cell-based assay with two luciferases emitting at different wavelengths has been developed to *in vivo* monitor the two pathways of bile acid biosynthesis. Bile acid biosynthesis is a key step of intracellular cholesterol homeostasis and, in turn, affects the rate of cholesterol synthesis in hepatocytes. Bile acids are synthesized via the classic pathway initiated by cholesterol 7 alpha-hydroxylase (CYP7A1) and via alternate pathways, one of which is initiated by sterol 27-hydroxylase (CYP27A1). Both the two pathways seem to be regulated via Farnesoid X receptor (FXR) activation.

We report here the development of a recombinant HepG2 cell-based luciferase assay that enabled us to evaluate the ability of natural and synthetic bile acids and other compounds to activate or inhibit the two bile acid synthesis pathways.

MATERIALS AND METHODS

***Luciola italica* luciferase mutagenesis, cloning and expression.** The QuikChange Multi Site-Directed Mutagenesis kit (Stratagene, Heidelberg, Germany) was used to perform site-directed mutagenesis using the *L. italica* luciferase wild-type DNA sequence in the pGEX-6P-2 vector as a template. Green- and red-light emitting thermostable (TS) *L. italica* mutants were made and the peroxisome targeting sequences were removed. The green mutant (Lit GR-TS) contains the mutations V243I, G248A and F250S; whereas the red enzyme (Lit RE-TS) has the S286T change. Both luciferases contain the TS mutations G216A, T217L, S234A, and E356K. The mutants were subsequently cloned downstream of the CYP7A1 and CYP27A1 promoters in the pGL3 vector.

First, the expression of the red and green *L. italica* thermostable mutants were evaluated in *E. coli* after 2 h IPTG induction at 37°C. Fig. 1 shows the normalized emission spectra obtained from JM109 *E. coli* cells transformed with the plasmids pQE30 expressing Lit WT, Lit RE-TS and Lit GR-TS.

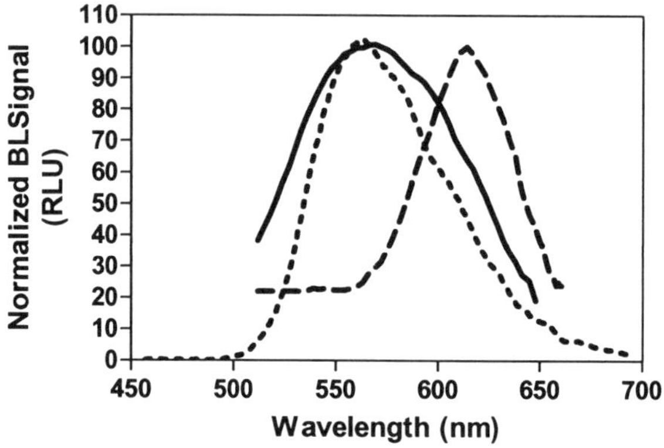

Fig. 1. Normalized bioluminescence emission spectra obtained after addition of 100 µL of 1 mM D-luciferin in 0.1 M Na-citrate-buffer at pH 5 to JM109 *E. coli* cells expressing Lit WT (dotted line), Lit RE-TS (dashed line) and Lit GR-TS (straight line) in pQE30 vector. Luminescence measurements were performed using a Cary Eclipse fluorescence spectrophotometer (Varian, Palo Alto, CA, USA) set on the chemi/bioluminescence mode

Mammalian expression and development of bioassay for cholesterol-lowering drugs. The human hepatoblastoma cell line, HepG2, was stably transfected with pcDNA3.1-FXR expression vector containing the cDNA encoding human Farnesoid X receptor in order to obtain clones that stably expressed the receptor. Clones stably expressing FXR were transiently transfected with reporter plasmids containing CYP7A1 (-1887/+24) and CYP27A1 (-531/+1) promoters driving the expression of Lit GR-TS and Lit RE-TS, respectively. One day before transfection, approximately 1.5 x 10^5 HepG2 cells were seeded per well in 24-well microtiter plates. Cells were washed with PBS, then transiently transfected using Exgen500 (MBI Fermentas, Vilnius, Lituania) according to the manufacturer's instructions. All transfections were performed in triplicate. Twenty-four h after transfection, cells in individual wells were incubated with different concentrations of bile acids or other compounds for 20 h, then washed in PBS and lysed with 1% Triton® X-100. Then, after centrifugation, 70 µL of surnatant were transferred to a black microtiter 96-well microplate. After addition of 100 µL luciferase assay system (Promega, Madison, WI, USA) the signal was acquired for 30 s with a Luminoskan luminometer (Labsystem). Several parameters were optimized in order to increase the analytical performance of the bioassay, including cell number, analyte incubation time and transfection conditions.

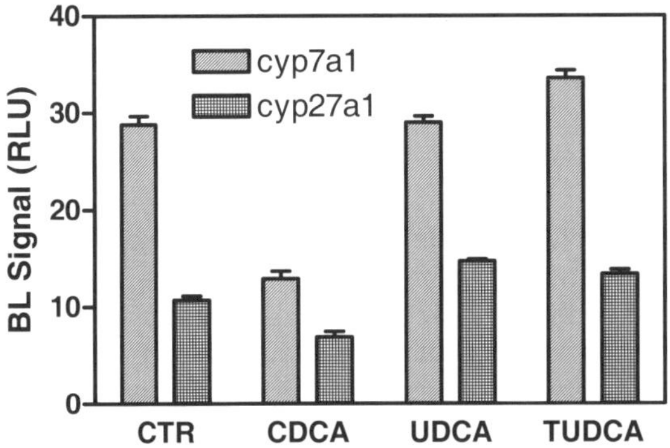

Fig. 2. Validation of the system. Effect of known bile acids at a final concentration of 50 μmol/L (CDCA, UDCA and TUDCA) on CYP7A1 and CYP27A1 transcriptions. HepG2-FXR were transfected with pGL3-CYP7A1-Lit GR-TS or pGL3-CYP27-Lit RE-TS vectors and treated with bile acids for 20 h. Each value represents the mean ± SD of three determinations

Fig. 2 shows the typical data obtained with the bioassay, in which measurements of the two bioluminescent emissions were used to monitor and compare the transcriptional regulation of bile acid biosynthesis. In particular, repression of CYP7A1 and, in a minor proportion, of CYP27A1 are mediated by chenodeoxycholate (CDCA), which is the main actor of negative feedback regulation of bile acid synthesis. Other bile acids, known for their hepatoprotective effect, (e.g. ursodeoxycholate, UDCA, and its taurine-derivative tauroursodeoxycholate, TUDCA) were not able to repress CYP7A1 nor CYP71.

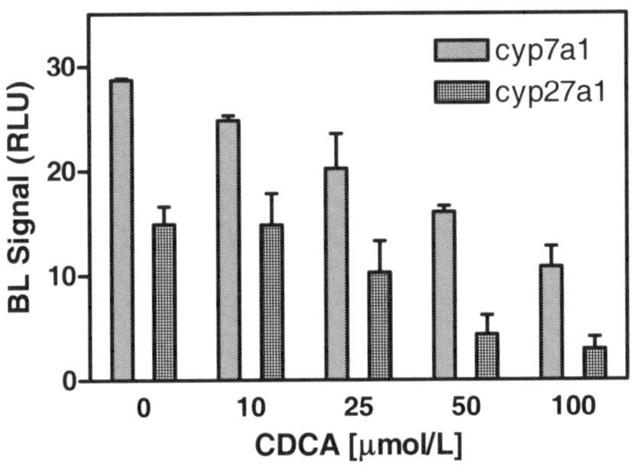

Fig. 3. Dose-dependent effect of CDCA on CYP7A1 and CYP27A1 transcriptions. HepG2-FXR were transfected with pGL3-CYP7A1-Lit GR-TS or pGL3-CYP27-Lit RE-TS reporter vectors, treated with increasing concentrations of CDCA for 20 h

To investigate whether the CDCA-induced inhibition was dose-dependent, dose-response curves were obtained for CDCA concentrations ranging from 10 to 100 μmol/L (Fig.3). Treatment of the transfected cells with CDCA caused a dose-dependent decrease of luciferase activity.

CONCLUSIONS

We developed a stable HepG2 cell line overexpressing human FXR for studying the role of FXR in the control of gene expression by monitoring the transcriptional state of the two main enzymes involved in bile acid biosynthesis. We demonstrated the FXR agonist activity of some natural bile acids such as chenodeoxycholate, ursodeoxycholate (UDCA) and its taurine-derivative tauroursodeoxycholic acid (TUDCA). The newly developed cell-based assay allows the monitoring of the two distinct biosynthetic pathways in a high throughput format (96-well microtiter plate) and is highly suitable for the screening of new cholesterol-lowering drugs. In addition, the two *L. italica* luciferase mutants obtained by site directed mutagenesis of *L. italica* luciferase both showed good thermostability and good expression in mammalian cells, and are therefore suitable for the development of novel multiplex bioassays by using optical filters for spectral resolution.[3,4]

REFERENCES

1. Branchini BR, Southworth TL, De Angelis JP, Roda A, Michelini E. Luciferase from the Italian firefly *Luciola italica*: Molecular cloning and expression. Comp Biochem Physiol B Biochem Mol Biol 2006; 145:159-67.
2. Branchini BR, Southworth TL, Khattak NF, Michelini E, Roda A. Red- and green-emitting firefly luciferase mutants for bioluminescent reporter applications. Anal Biochem 2005; 345:140-8.
3. Gammon ST, Leevy WM, Gross S, Gokel GW, Piwnica-Worms D. Spectral unmixing of multicolored bioluminescence emitted from heterogeneous biological sources. Anal Chem 2006;1:1520-7.
4. Nakajima Y, Kimura T, Sugata K, Enomoto T, Asakawa A, Kubota H, et al. Multicolor luciferase assay system: one-step monitoring of multiple gene expressions with a single substrate. Biotechniques 2005;38:891-4.

NOVEL BRET-BASED BIOSENSORS WITH A NEW BIOLUMINESCENT DONOR, *GAUSSIA* LUCIFERASE, FOR ESTROGEN RECEPTOR LIGANDS

E MICHELINI,[1] M MAGLIULO,[1] M BARALDINI,[2] A RODA[1]
[1] *Department of Pharmaceutical Sciences, Via Belmeloro 6, University of Bologna, Bologna, Italy;* [2] *Department of Sciences of Metals, Electrochemistry and Chemical Techniques, Via San Donato 15, Bologna, Italy, Email: aldo.roda@unibo.it*

INTRODUCTION

Protein-protein interactions play a fundamental role in a the majority of biochemical systems. The most commonly used method to identify protein-protein interactions relies on the conventional two-hybrid system. In this method, the interaction must occur in the yeast nucleus, this means interactions that strictly depend upon cell-type specific processing and compartmentalization will not be detected. Among alternative methods, such as fluorescence resonance energy transfer (FRET) and protein mass spectrometry, Bioluminescence Resonance Energy Transfer (BRET) is emerging as a promising tool to monitor protein-protein interactions in living cells and in real time.

BRET is a natural non-radiative energy transfer phenomenon occurring in marine organisms between a bioluminescent donor protein and a fluorescent acceptor protein. In order to study an interaction between two target proteins, labelled proteins are generated by genetic fusion with either a bioluminescent donor or fluorescent acceptor. Recently, several *in vivo* and *in vitro* applications of BRET assays have been reported, ranging from membrane-bound and cytosolic receptors dimerization studies to nuclear factors interactions monitoring.[1] The majority of BRET assays use *Renilla* luciferase (Rluc) as energy donor; in the presence of coelenterazine, Rluc instead of emitting blue light, emits blue light which is able to excite a variant of the green fluorescent protein (enhanced yellow fluorescent protein, EYFP) acting as fluorescent acceptor.

The recent cloning of several reporter proteins with different spectral properties has enlarged the number of possible bioluminescent donor proteins to be used in BRET assays.[2] We investigated the suitability of a new bioluminescent donor, Gaussia luciferase (Gluc) from *Gaussia princeps*, a copepod living in the deep that is able to produce a brilliant blue bioluminescent display when disturbed.

MATERIALS AND METHODS

The plasmid pECFP-19-EYFP (provided by Dr. Marko Virta, Turku, Finland) was used as starting vector to construct the fusion proteins Gluc-EYFP, hERa-EYFP and hERß-Gluc. All materials for cell culture and Platinum Pfx DNA Polymerase were supplied by Gibco-BRL Life Technologies (Gaithersburg, MD, USA).

Coelenterazine, pGLuc vector and pCDNA3.hGLuc were kindly provided by Bruce Bryan (Prolume, Pinetop, AZ, USA). Restriction endonucleases, T4 DNA ligase, and other modifying enzymes were from MBI Fermentas (Vilnius,

Lithuania). All other chemicals were of analytical grade and were used as received. Luminescence measurements were performed using a Cary Eclipse fluorescence spectrophotometer (Varian, Palo Alto, CA, USA) equipped with a microtiter plate reader. The instrument was set on fluorescence mode for performing fluorescence excitation or emission scans and fluorescence measurements at fixed wavelengths. For bioluminescence or BRET measurements, the instrument was set on chemi/bioluminescence mode.

One day before transfection, approximately 1.5 x 10^5 HepG2 cells were seeded per well in 24-well microtiter plates. Cells were washed with PBS, then transiently transfected using Exgen500 (MBI Fermentas, Vilnius, Lithuania) according to the manufacturer's instructions. All transfections were performed in triplicate.

RESULTS AND DISCUSSION

Gluc expression in mammalian cells. At first the expression of Gaussia luciferase was determined in human hepatoblastoma cell line HepG2 by transient transfection of the plasmid pcDNA3.hGluc containing the humanized form of Gaussia luciferase with secretion signal. Forty-eight hours after transfection, cells were detached and an aliquot of intact viable cells, the conditioned cell-free medium, or both was assayed for Gluc activity using a luminometer by introducing coelenterazine to a final concentration of $1x10^{-5}$ mol/L and integrating the signal over the first 10 s.

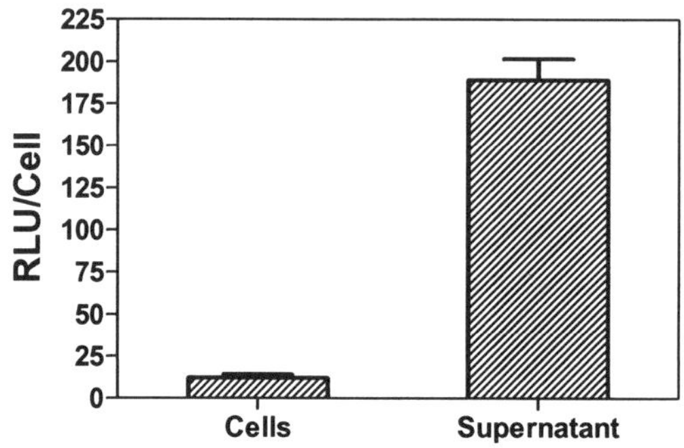

Fig. 1. Signals obtained from viable cells normalized to cell number. Viable cells versus conditioned medium. RLU, relative light units

Forty-eight hours after transfection, we observed over 10 times higher bioluminescence in the cell-free, conditioned medium compared to live cells themselves or lysates prepared from them, confirming secretion of active hGLuc into the medium (Fig. 1).

Intramolecular BRET between Gluc and EYFP. To explore the applicability of Gluc and EYFP as BRET partners, a fusion construct directly linking Gluc to EYFP with a linker of 19 aminoacids was developed. HepG2 cells were transiently transfected with an expression plasmid carrying the fusion protein (pCDNA.3hGluc-EYFP) and luminescence spectra recorded as shown in Fig. 2.

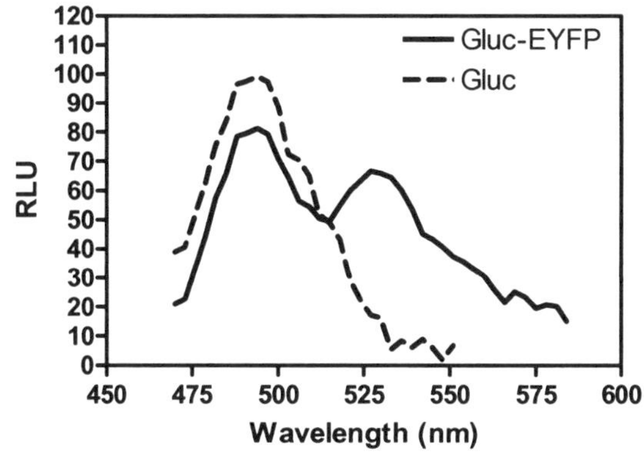

Fig. 2. HepG2 cells transfected with the plasmid pCDNA3Gluc-EYFP. Bioluminescence emission spectra were recorded in the range 450-600 nm, upon addition of coelenterazine at a final concentration of $1x10^{-5}$ mol/L (solid line). The dotted line represents the bioluminescent emission spectrum recorded for cells transfected with pCDNA3.hGLuc

BRET assay for human Estrogen Receptor α-β hetero-dimerization (hERα-hERβ). By fusing both hERα and hERβ to either Gluc or EYFP it was possible to study hetero-dimerization and compare it with homo-dimerization α–α determined with the previously developed *in vivo* BRET assay.[1] Two plasmids were developed for the co-expression of the two fusion proteins ERβ-Gluc and ERα- EYFP in HepG2 cells.

The BRET signal was calculated as follows using emission intensities recorded when the two fusion proteins are expressed together for the Gluc construct expressed alone in the same experiment:

$$\frac{(emission_{530nm} - emission_{485nm})C_f}{emission_{485nm}} \quad \text{where } C_f \text{ is: } \frac{\text{emission}_{530\text{nm}}}{\text{emission}_{485\text{nm}}}$$

HepG2 cells co-transfected with the plasmids encoding for the fusion proteins ERα-EYFP and ERβ-Gluc were incubated for 30 min with different concentrations of 17-βestradiol (range 10^{-10}-10^{-6} mol/L). Bioluminescence emission spectra were recorded upon addition of coelenterazine at a final concentration of $1x10^{-5}$ mol/L. The obtained dose response curve was compared to the curve obtained with the in vivo BRET assay for hERα homo-dimerization previously developed (Fig. 3).

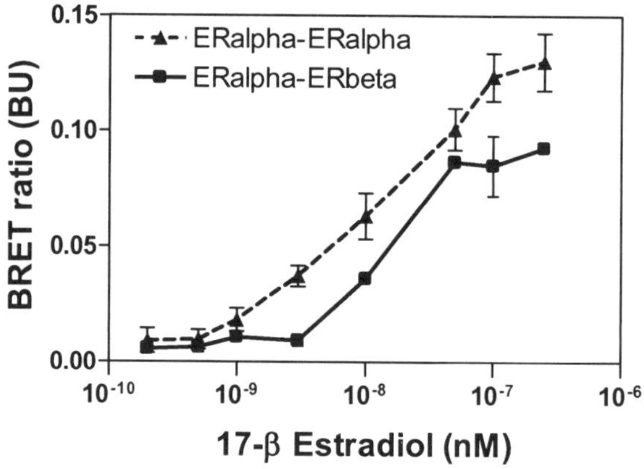

Fig. 3. Dose response curves for 17-βestradiol obtained with two different BRET based assays. HepG2 cells transfected with cDNA coding for ERa-EYFP and ERa-Rluc were incubated for 30 min with increasing concentrations of E_2. Then coelenterazine was added and light emission acquisition performed (solid line). The dotted line represents the same assay performed by co-expressing ERa-EYFP and ERβ-Gluc. Data are means ± SD of at least three independent experiments each performed in sextuplicate

CONCLUSIONS

The use of a new bioluminescent donor, Gaussia luciferase, has been investigated as an alternative to Rluc. This protein has the advantages of smaller size, higher bioluminescent signal emission when it is codon humanized and flash kinetics (λ_{max} 485 nm), thus making it suitable for an efficient energy transfer to EYFP. A novel bioluminescence resonance energy transfer (BRET) homogeneous assay for estrogen compounds was developed and optimized based on human estrogen receptor homo- and hetero-dimerization.

The assay fulfils all the standard requirements of accuracy and intra- assay and inter-assay precision with a limit of detection comparable to those of whole-cell biosensors or related bioassays and the specific advantages of homogeneous assays.

REFERENCES

1. Michelini E, Mirasoli M, Karp M, Virta M, Roda A. Development of a bioluminescence resonance energy-transfer assay for estrogen-like compound in vivo monitoring. Anal Chem 2004;76:7069-76.
2. Bakhos A, Tannous BA, Kim DE, Fernandez JL, Weissleder R, Breakefield XO. Codon-optimized Gaussia luciferase cDNA for mammalian gene expression in culture and *in vivo*. Mol Ther 2005;11:435-43.

THE USE OF *LUX* GENES FOR MONITORING UPTAKE OF *PSEUDOMONAS AERUGINOSA* BY RESPIRATORY EPITHELIAL CELLS

GM ROBINSON, SK JACKSON, VC SALISBURY, SM NELSON
University of the West of England, Coldharbour Lane, Frenchay, Bristol, BS16 1QY, UK

INTRODUCTION

Pseudomonas aeruginosa is found ubiquitously in the environment and is an important cause of respiratory infections in individuals with impaired immunity. Increasing antibiotic resistance in *P. aeruginosa* coupled with the danger of nosocomial transmission to susceptible individuals illustrates the importance of fully understanding the means by which the organism persists within the human lung. Numerous bacterial virulence determinants have been implicated in respiratory adhesion and a number of host cell receptors have been identified as potential targets. It has been suggested that a defective host-cell receptor mechanism leads to uptake of bacteria without clearance from the lung and that such an intracellular reservoir might contribute to the persistence of *P. aeruginosa* lung-infections.

We have used bioluminescent *P. aeruginosa* to observe bacterial uptake by respiratory epithelial cells and showed that light from intracellular bacteria remained clearly distinguishable 24 h post-inoculation. Internalisation was mitigated by temperatures of 10°C indicating that the process of bacterial uptake was mediated by the human cell-line rather than by the bacteria, and we have begun receptor blocking experiments in order to identify key host-cell receptors.

Bioluminescent *P. aeruginosa* has the potential to resolve key difficulties in the understanding of how *P. aeruginosa* persists in susceptible individuals.

METHOD

Preperation of inoculum. *Pseudomonas aeruginosa* PAO1-lite transformed with the *Photorhabdus luminescens lux* operon (plasmid borne)[1] was grown overnight in nutrient broth containing 10 mg/L gentamicin (for plasmid maintenance) at 37°C. The bacteria were washed twice in pre-warmed phosphate-buffered saline (PBS) and resuspended in pre-warmed RPMI-1640 (5% foetal bovine serum, 2 mM glutamine). Optical density was adjusted to OD_{590} 1.0 in RPMI-1640.

Maintenance and propagation of cell lines. Adherent human lung carcinoma cell line A549 was cultured in RPMI-1640 (5% foetal bovine serum, 2 mM glutamine) at 37°C in an atmosphere with 5% CO_2. A549 cells were trypsinised on a weekly basis and subcultured 1:10 into fresh RPMI-1640. Twenty-four h prior to use, a flask of confluent A549 cells was trypsinised and the cells resuspended in fresh RPMI-1640, and 200 μL added to a 96-well black-walled, clear-bottom microplate.

Experimental setup. Once A549 cells in microplates were confirmed confluent, they were washed twice in 200 μL pre-warmed PBS and 200 mL fresh RPMI-1640 was added to each well. 10 μL of inoculum (prepared as described above) was added to each well as appropriate and the plate was transferred to a Tecan GENios Pro automated luminometer where bioluminescence was monitored at 37°C over time. After 2 h incubation, the plate was withdrawn from the luminometer and colistin (50 mg/L) added to appropriate wells to eliminate extracellular bacteria. Following another 2 h of monitoring, the plate was withdrawn and saponin (0.1%) was added to appropriate wells to lyse A549 cells and the plate returned to the luminometer. When antibody blocking was used, the cells were preincubated with anti-TLR4 antibody for 45 min prior to the addition of the inoculum. When low temperatures were used, the cells and bacteria were cooled to 10°C prior to adding the inoculum. Following inoculation, the microplate was incubated at 10°C for 2 h, then colistin (50 mg/L) was added to appropriate wells and the microplate transferred to the luminometer and incubated at 37°C.

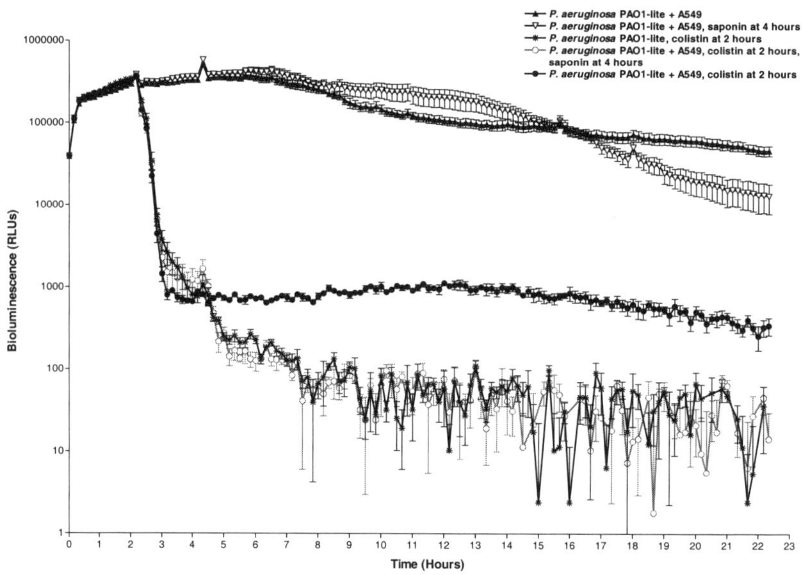

Fig. 1. Internalisation of *P. aeruginosa* by cell line A549. Light output measured in relative light units (RLUs)

RESULTS

Adding colistin to *P. aeruginosa* PAO1-lite resulted in a drop of bioluminescence down to background levels (Fig. 1). In the presence of A549 cells, a level of protection from the antimicrobial agent was apparent, with light levels dropping to around 1,000 RLUs. This protection was mitigated by the addition of saponin which lysed the A549 cells and exposed the bacteria to the colistin, followed by a drop in bioluminescence down to background levels.

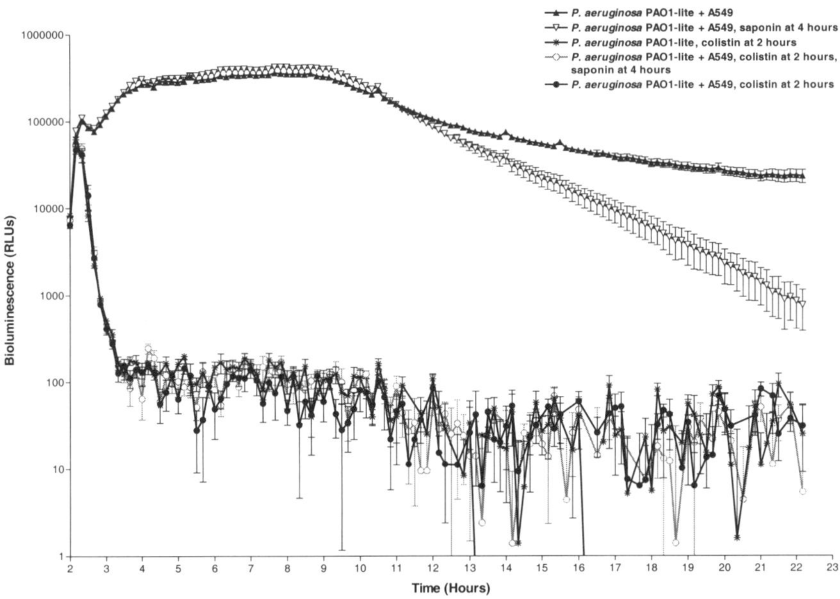

Fig. 2. As Fig. 1, with 2 hour pre-incubation at 10°C

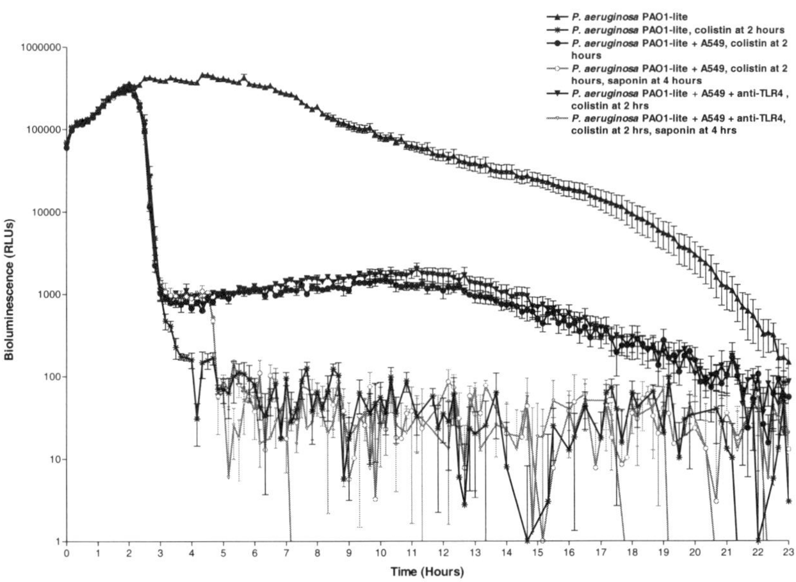

Fig. 3. Internalisation of *P. aeruginosa* by cell line A549 following incubation with anti-TLR4. Light output measured in relative light units (RLUs)

The protective effect was also mitigated by pre-incubation at 10°C (Fig. 2), suggesting that no invasion/internalisation took place. The result of the 10°C pre-incubation temperature suggests the internalisation of the A549 cells by *P. aeruginosa* is more of a cell-mediated than bacteria-mediated process, as a temperature of 10°C would be more likely to be inhibitory to the A549 cells than the bacteria. To investigate whether Toll-like receptor 4 acts as a potential binding site on A549 cells for *P. aeruginosa*, the A549 cells were pre-incubated with anti-TLR4 antibody prior to the addition of the bacteria. As shown in Fig. 3, the pre-incubation with anti-TLR4 did not inhibit bacterial internalisation, indicating that TLR4 is not involved in *P. aeruginosa* adhesion to A549 cells.

DISCUSSION

P. aeruginosa expressing the *P. luminescens lux* genes has illustrated that bioluminescence is a powerful reporter in terms of confirming the intracellular location of bacteria. Bacterial uptake was clearly observable by monitoring light output following the elimination of extracellular bacteria. Intracellular light output was stable, consistent and repeatable and was detectable in excess of 24 h post-inoculation. The process of bacterial uptake by respiratory epithelial cells is most likely a receptor mediated process. Using blocking antibodies such as anti-TLR4, cellular receptors can be explored or eliminated as targets for bacterial binding and the use of bioluminescent reporter strains can simplify and expedite the process.

REFERENCES

1. Nelson SM, Marques CNH, Greenman J, Lewis RJ, Bowker KE, Salisbury VC. Real-time monitoring of metabolic activity in biofilms. In: McBain A, Allison D, Brading M, Rickard A, Verran J, Walker J. eds. Biofilm Communities: Order from Chaos? Cardiff: Bioline, 2003:257-68.

BIOLUMINESCENCE CHARACTERISTICS OF AN OBELIN MUTANT IN VARYING SOLVENT CONDITIONS

LAURA ROWE, ANNA ROTHERT, SAPNA K. DEO,
MARK ENSOR, SYLVIA DAUNERT
University of Kentucky, Department of Chemistry, Lexington, KY 40509, USA

INTRODUCTION

Obelin is a globular, 22 kDa bioluminescent photoprotein composed of four EF-hand domains, and a hydrophobic core cavity in which coelenterazine is located. Upon binding calcium, obelin undergoes a conformational change which leads to the oxidation of coelenterazine to an excited coelenteramide. This excited coelenteramide subsequently relaxes, and releases the characteristic 485 nm bioluminescent light of native obelin.[1] Obelin was isolated from the bioluminescent hydroid *Obelia longissima*, and is structurally analogous to aequorin, but offers certain advantages over aequorin in terms of calcium sensing within a cell. These advantages include the fact that obelin's bioluminescence is less sensitive to competing intracellular divalent cations, such as Mg^{2+}, and that the calcium induced bioluminescence kinetics are faster in obelin.[1] This allows for less cross-reactivity and an ability to monitor ultra-fast calcium fluctuations *in vivo*.

Native obelin contains five cysteine residues, at positions 51, 67, 75, 151, and 158. Obelin has a high ratio of cysteine residues, and these residues are thought to be important for bioluminescent activity.[1] However, recent X-ray crystal structures have been resolved for both apoobelin and obelin, and no cysteine residues are found to be close enough to the active site of the photoprotein to be directly involved in the bioluminescence reaction, nor are any of the five cysteines involved in intramolecular disulfide bonds.[1] All cysteine residues are, however, located in or very near the a-helices of the EF-hand domains.[2] To the best of our knowledge no mutational studies have been undertaken concerning the cysteine residues in obelin, with the exception of mutations of C158 to serine and alanine.[2] Therefore, in order to more fully understand the role of cysteine in the bioluminescence of obelin we undertook a mutational study on all five cysteine residues in obelin. The results of this study are currently still under analysis, however, we selected one mutant, in which the cysteine residues at positions 67 and 75 were both mutated to serines, and analyzed the change in bioluminescence properties when the photoprotein was exposed to a range of conditions that may be encountered in a cellular environment. Different areas and organelles within cells exhibit specific microenvironments whose physical properties may differ than the cell in general. Such pH and solvent changes may affect the bioluminescent properties of obelin in general, and our C65-76 mutant specifically. Therefore, herein we describe the analysis of the relative activity, decay half-life time, and emission maximum of obelin C67-75 in buffers varying in pH and buffer composition.

MATERIALS AND METHODS

Expression and Purification of Obelin C67-75. The cysteine residues at positions 67 and 75 were mutated to serine using a Qwikchange site directed mutagenesis kit. This obelin was then ligated into a pin4 plasmid containing an OmpA signaling peptide, and this plasmid was transformed into *E. coli*. The mutant was grown overnight in LB Broth to an OD 650 nm of 0.6, and 1 mM IPTG (Isopropyl ß-D-1-thiogalactopyranoside) was added to induce expression. Obelin C67-75 was expressed into the media, and purified by successively passing through a DEAE and butyl Sepharose column. Purity was verified using SDS-PAGE and Western blotting.

Bioluminescence Activity, Decay Half Life, Emission Spectrum Determination. Bioluminescence activity was determined using an Optocomp 1 test tube luminometer, by injecting 100 μL of calcium containing buffer into 10 μL of sample and integrating the signal over 6 s. Decay half life times were determined using a micro titer plate Polarstar Fluorometer by injecting 100 μL of calcium containing buffer into 50 μL of sample and integrating for 10 s at 0.1 s intervals. The decay kinetics and half lives were determined using GraphPad Prism. Emission spectrum were determined using a custom built luminometer that scans the emission wavelengths from 400-700 nm simultaneously for 5.0 s following the injection of a calcium containing buffer. All samples were measured in triplicate. Obelin was diluted in a Tris buffer containing 30 mM Tris, 2 mM EDTA, 1 mg/mL BSA, 2 mM DTT, pH:7.0 buffer. The pH of the buffer was altered using HCl or NaOH in order to produce buffer pH levels of 4.5, 5, 6, 7, 8, 9, and 9.5. The buffer composition was altered by excluding DTT or BSA in the Tris buffer, and by using deionized water as a solvent that contained only DTT.

Table 1. Summary of activity, emission maxima, and half-lives of obelin C67-75 when in different solvents

	Activity (%)	Emission Maxima (nm)	Half-life (s)
pH: 4.5	26	493	1.25
pH: 5	35	493	1.13
pH: 6	57	491	1.06
pH: 7	100	491	1.12
pH: 8	75	491	1.34
pH: 9	67	491	1.27
pH: 9.5	35	493	1.23
Tris buffer no BSA	45	493	1.11
Water with DTT	30	493	0.95
Tris buffer no DTT	7	491	1.25

RESULTS AND DISCUSSION

The obelin mutant C67-75 exhibited high activity and good stability in a standard Tris buffer. In order to study the effects that changes in pH and solvent composition have on this mutant obelin was charged overnight at 4°C with coelenterazine while in ten different solvents. Seven solvents were Tris buffer with pH values of 4.5,5,6,7,8,9, and 9.5, one solvent was deionized water with 2 mM DTT, one solvent was Tris buffer without BSA (pH:7), and the final solvent was a Tris buffer (pH:7) without DTT. We found that the bioluminescent intensity of obelin C67-75 progressively decreased as the pH deviated farther and farther from neutral (Fig. 1).

Additionally, obelin C67-75 dissolved in the Tris solvent containing both BSA and DTT yielded the most activity, with the activity of the obelin in the Tris buffer not containing BSA, the deionized water containing DTT, and the Tris buffer not containing DTT decreased successively (Table 1).

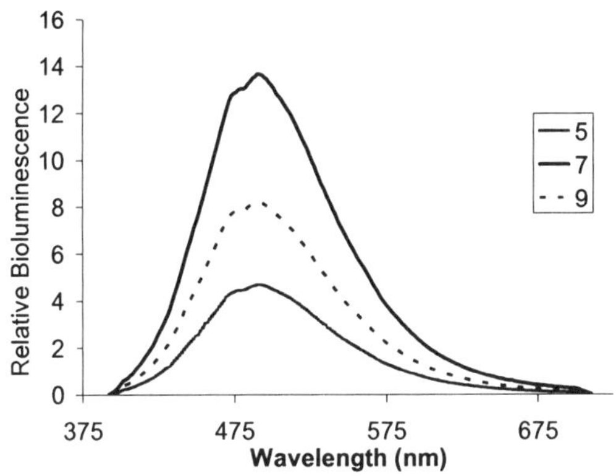

Fig. 1. Emission spectra of Obelin C67-75 when at different pH's

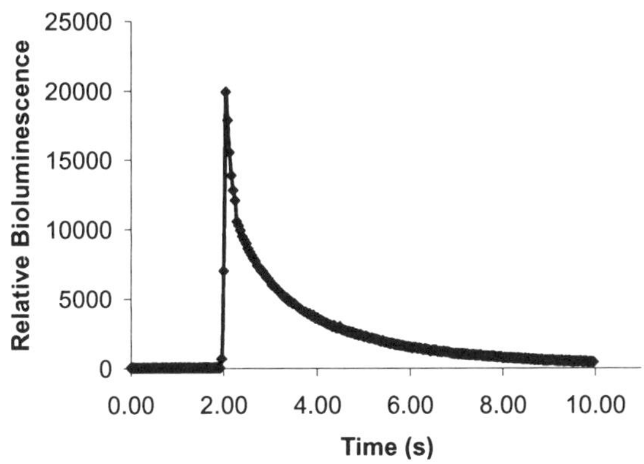

Fig. 2. Decay half-life of obelin 67-75 at pH of 7.0

Although activity was lost with extremes in pH and varying solvents, this obelin still exhibited high bioluminescence under all the tested conditions, with the exception of the DTT free solution. This indicates that obelin C67-75 should be viable for use in a variety of cellular microenvironments. Next, we determined if emission maxima or decay half-lives were altered under these varying solvent conditions. The shape and peak of the emission spectra from the obelin in all solvents were very similar, with emission maxima ranging from 491-493 nm (Fig. 1).

The spectra varied only in their relative intensity, as was indicated in the bioluminescence activity study. Decay half-lives were also very similar in all solvents, with half-lives ranging from 0.95 -1.34 s (Table 1). The data from this study suggests that the bioluminescent properties such as emission maxima and half-life do not vary significantly, regardless of the environment that the obelin is exposed to. The high activity and stable bioluminescent properties exhibited by obelin C67-75 over a wide range of solvent conditions suggests that this obelin mutant may be a practical tool for *in vivo* cellular analysis in the future.

ACKNOWLEDGEMENTS
This work was supported by the National Institutes of Health. Laura Rowe would like to thank NSF-IGERT for a pre-doctoral fellowship and the NIH for a pre-doctoral support grant.

REFERENCES

1. Deng L, Markova SV, Vysotski ES, Liu ZJ, Lee J, Rose J, et al. Crystal structure of a Ca2+-discharged photoprotein: implications for mechanisms of the calcium trigger and bioluminescence. J Biol Chem 2004;279:33647-52.
2. Deng L, Vysotski ES, Markova SV, Liu ZJ, Lee L, Rose J, et al. All three Ca+-binding loops of photoproteins bind calcium ions: the crystal structures of calcium-loaded apo-aequoriun and apo-obelin. Prot Sci 2005;14:663-75.
3. Bondar VS, Frank LA, Malikova NP, Inzhevatkin EV, Illarionova VA, Vysotski ES. Bioluminescent activity of the recombinant obelin after chemical modification of histidine and cysteine residues. In: Roda A, Pazzagli M, Kricka LJ, Stanley PE. eds. Bioluminescence and Chemiluminescence: Perspectives for the 21st Century. Chichester: Wiley, 1999:400-403.
4. Liu Z, Stepanyuk GA, Vysotski ES, Lee J, Markova SV, Malikova NP, et al. Crystal structure of obelin after Ca2+-triggered bioluminescence suggests neutral coelenteramide as the primary excited state. Proc Natl Acad Sci USA 2006;103:2570-5.
5. Vysotski ES, Lee J. Ca2+-regulated photoproteins: structural insight into the bioluminescence mechanism. Acc Chem Res 2004;37:405-15.
6. Bondar VS, Purtov KV, Malikova NP, Frank LA, Illarionov BA. Role of conservative residue Cys158 in the formation of an active photoprotein complex of obelin. Biochemistry (Moscow) 2001;66:1014-8.

A SIMPLE CHEMILUMINESCENCE FLOW INJECTION ANALYSIS FOR THE DETERMINATION OF SULPHIDE IN AQUEOUS SAMPLES

SAKCHAI SATIENPERAKUL,[1] SAISUNEE LIAWRUANGRATH,[2] TERENCE J. CARDWELL[3]

[1] Department of Chemistry, Faculty of Science, Maejo University 50290 Thailand; [2] Department of Chemistry, Faculty of Science, Chiang Mai University 50200 Thailand; [3] Department of Chemistry, La Trobe University, Bundoora, Victoria 3086, Australia, E-mail address: sakchais@mju.ac.th

INTRODUCTION

Hydrogen sulphide is widely distributed among a variety of man-made and natural settings where sulphur-containning matter may decompose anaerobically. Sulphide salts may also be added to industrial wastewater for precipitation and control the level of some toxic metals. Levels of hydrogen sulphide in the workplace environment are one of the important factors for work safety, especially in mines or underground tunnels where acute intoxication is problematic.[1] Thus, the determination of sulphide is usually included to comply with the standards set by many environmental agencies.

The standard accurate spectrophotometric procedure that is suitable for quantitative determination of low concentration of hydrogen sulphide in wastewater has been recommended by the American Public Health Association.[2] This technique requires fairly expensive instrumentation, skill in operation and is time consuming, which makes their routine implementation problematic, although this method gives a reasonably high detection limit, 2 mg/L.

Few studies have utilised chemiluminescence flow injection (CL-FI) as an analytical procedure for sulphide ion.[3-5] According to the recent review by Hindson and Barnett,[6] work on the application of chemiluminescence for the analysis of sulfide in aqueous solution with permanganate has also not yet been reported. This led us to investigate a new flow based method combined with sensitive chemiluminescence detection for examining sulphide in our environment with respect to reliability, robustness and sensitivity. We now report the first use of this common reagent for the sensitive determination of sulphide and its application for environmental samples.

MATERIALS AND METHODS

Reagents, Solutions and Sample. All reagents were of analytical reagent (AR) grade and were used as received. NANOpure deionised water (18.0 MΩ, Millipore, USA) was used for all solution preparations.

Sulphide stock solution (1000 mg/L) was prepared by dissolving 0.750 g of $Na_2S.9H_2O$ (Sigma, USA) in 100 mL of boiled, cooled deionised water. Intermediate standard solutions of sulphide (10.0 mg/L) were made as needed by making appropriate dilutions of the stock sulphide solution (1000 mg/L) with the deionised water. Working solutions were prepared by appropriate dilution of this intermediate solution with the carrier solution and analysed immediately due to the instability of very dilute sulphide in acid media. Potassium permanganate stock solution (1.0 x 10-2 M) was prepared by dissolving 0.1587 g of $KMnO_4$ (BDH, UK) in 100 mL deionised water. The carrier stream solution containing 1% (m/v) of sodium hexametaphosphate was prepared daily by dissolving the appropriate amount of sodium hexametaphosphate (Sigma-Aldrich, USA) in sulphuric acid solution (pH 2.0). The reagent stream solution, acidic potassium permanganate (5 x 10-5 M), was prepared fresh daily by making appropriate dilution of the $KmnO_4$ stock solution in carrier solution.

Apparatus. The flow injection manifold used in all experiments is illustrated in Fig. 1. The experimental setup consisted of a two-channel peristaltic pump with rate selector (Minipuls 3, Gilson, France) equipped with silicone rubber tubing, a sample injection valve (Type 50, Rheodyne Inc., CA, USA) and PTFE connection tubing (0.5 mm i.d.). The chemiluminescence signal was monitored in a custom built flow-through luminometer, which consisted of a flat spiral glass flow cell (glass tubing i.d. 1.5 mm, spiral coil diameter 25 mm) mounted flush against a red sensitive photomultiplier tube (PMT) (Thorn-EMI 9878SB, Electron Tubes Ltd, UK). The operational potential for the PMT was provided by a stable power supply (Thorn-EMI model PM28SB, Electron Tubes Ltd, UK). The flow cell, photomultiplier tube and voltage divider were encased in light tight housing. The detector output was recorded on a strip chart recorder (Kipp & Zonen, Netherlands).

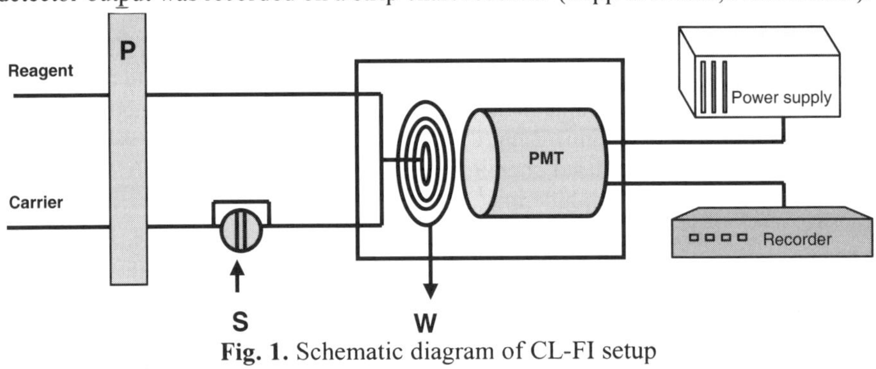

Fig. 1. Schematic diagram of CL-FI setup

Procedure. The peristaltic pump was employed to propel the reagent and carrier streams at equal flow rates with a total flow rate of 5.0 mL/min. Analyte or standard solutions were injected manually into a 1% (m/v) sodium hexametaphosphate in 0.02 M H_2SO_4 carrier stream (pH 2.0), which then merged at a T-piece with the reagent stream of potassium permanganate (5 x 10^{-5} M) made up in the acidic sodium hexametaphosphate carrier solution. The combined reaction mixture was passed through the flat spiral coil flow cell, where the chemiluminescence intensity was detected at a photomultiplier tube operated at a voltage of 1.00 kV. The output of the PMT which was proportional to the CL intensity was monitored continuously. The analytical signal of the CL-FI system was the maximal output potential corresponding to peak maximum.

RESULTS AND DISCUSSION

Optimisation of chemical and physical variables. Optimisation of the CL-FI system was conducted. The univariate optimisation results are essential in assessing the influence of individual system parameters on its performance. The reliability of the results obtained by this approach could be highly dependent on their degree of interrelation. The system's parameters optimised in this study were the: (i) PMT voltage; (ii) flow rates of the carrier and reagent streams; (iii) sample volume; (iv) concentration of $KMnO_4$ in the reagent stream; (v) concentration of H_2SO_4 in both the carrier and reagent streams (pH); and (vi) concentration of sodium hexametaphosphate in both the carrier and reagent streams. The working ranges within which these parameters were optimised and the corresponding optimal values are presented in Table 1.

Analytical figures of merit. A series of standard solutions (from 2.0 – 250 μg/L) were used to study the linear range of the calibration graph under the optimised conditions. The calibration line was linear in the concentration range 5.0–50.0 μg/L. In this dynamic range, the linear regression equation is $I = 0.092\ C + 0.256$ ($R^2 = 0.992$) where I is the chemiluminescence signal (mV) and C is the concentration expressed in μg/L. The limit of detection, calculated at 2σ of the noise was 2.0 μg/L. The sample throughput was estimated to be > 150/h.

Table 1. Flow system parameters optimised in this study

Parameters	Range studied	Optimal value
$KMnO_4$ concentration (M)	1.25×10^{-5} - 5×10^{-3}	5.0×10^{-5}
H_2SO_4 concentration (M)	0.005 - 0.1	0.02
Sodium hexametaphosphate concentration (% m/v)	0.25 - 3.0	1.0
Sample volume (μL)	50 - 500	200
Flow rate (mL/min)	3.0 - 6.0	5.0
PMT voltage (kV)	0.7 - 1.15	1.00

Analysis of real samples by CL-FI. The applicability of this CL-FI method to the analysis of 'total sulphide' in natural water samples and wastewater was evaluated by external calibration and standard addition methods. The three water samples were analysed after appropriate dilution and *t*-tests indicated that they were no significant differences at 95% confidence level, compared with those obtain from methylene blue spectrometric method.[2]

CONCLUSIONS

A new, simple, sensitive and inexpensive CL-FI method has been developed for the determination of sulphide in aqueous samples based on the oxidation of the analyte by acidic potassium permanganate in the presence of sodium hexametaphosphate. Samples could be analysed at a rate of *ca.* 150 sample per hour with a detection limit of 2.0 μg/L.

REFERENCES

1. Committee on Medical and Biologic Effect of Environmental Pollutants. Hydrogen Sulfide. Baltimore: University Park, 1979.
2. American Public Health Association. Standard Methods for the Examination of Water and Wastewater, 20th ed. APHA AWWA WEF: Washington DC, 1998.
3. Klockow D, Teckentrup J. Investigation and analytical application of the sulphide–hypobromite chemiluminescence reaction. Talanta 1981;28:653-63.
4. Townshen A, Burguera JL. Determination of ng/mg levels of sulphide by a chemiluminescence reaction. Talanta 1980;27:309-14.
5. Safavi A, Karimi MA. Flow injection chemiluminescence determination of sulfide by oxidation with chlorinated isocyanurates. Anal Lett 2002; 35: 2023-37.
6. Hindson BJ, Barnett NW. Analytical application of acidic potassium permanganate as a chemiluminescence reagent. Review. Anal Chim Acta 2001;445:1-19.

USING RECOMBINANT ANTIGENS FOR THE DETECTION OF ANTIBODIES TO *TRYPANOSOMAS CRUZI* ON A FULLY AUTOMATED CHEMILUMINESCENCE ANALYZER

D SHAH, C-D CHANG, K CHENG, L JIANG,
V SALBILLA, A HALLER, G SCHOCHETMAN
Abbott Laboratories, Abbott Park, IL 60064, USA, Email: Dinesh.shah@abbott.com

INTRODUCTION

Acridinium (Ac) derivatives have been utilized for highly sensitive immunoassays because of their stability and high chemiluminescence yield. Abbott PRISM® is a high throughput, fully automated serological analyzer to screen plasma or sera for Hepatitis B surface antigen, antibodies to Hepatitis B core, HCV, HIV, and HTLV.[1] The instrument uses acridinium labels to tag analytes captured on microparticles, buffer to wash nonspecific binding, H_2O_2 to trigger chemiluminescence, and a photon multiplier to collect photon counts. Data are processed by a computer to align with sample barcodes and to sort out the positive samples. The total assay time is under 1 hour and 160 samples/hour can be processed.

The Chagas' disease (American trypanosomiasis), caused by the protozoan *Typanosomas cruzi*, is endemic to most regions of the Latin Americas. Of the estimated 12-14 million infected people, approximately 25,000 die from this disease yearly typically due to cardiac rhythm disturbances or congestive heart failure.[2] Transfusion of blood from infected donors has become a major route for contracting the disease. Concerns on the safety of blood in the U.S. have been raised because over 12 million immigrants from the endemic countries and they represent a growing donor population but also a largely "silent" reservoir of *T. cruzi*.[3]

Laboratory diagnosis of Chagas' disease is complex, primarily because of the genetically diverse and polymorphous parasite. PCR is not always able to detect the specific DNA because of intermittent or low levels of parasites in the blood stream during the chronic stage. In contrast, serologic tests detecting antibodies to *T. cruzi* are well-suited for fast and inexpensive diagnosis of the disease. A *T. cruzi* lysate based Chagas assay on the PRISM® analyzer was reported previously.[4] However, due to many advantages with recombinant antigens (rAg) over crude lysate, such as quality control and reproducibility, we switched to the use of rAg of *T. cruzi* for the development of a Chagas assay.[5]

MATERIALS AND METHODS

Assay Reagents, Controls, and Samples. Solid-phase: A blend of several different rAg coated microparticles, each kind of microparticle was coated separately with a rAg of *T. cruzi.*

Acridinium Conjugate. Prepared from a Mab anti-human IgG conjugated with 10-(3-sulfopropyl)-N-tosyl-N-(2-carboxyethyl)-9-acridinium carboxamide via Nhydroxysuccinimide ester in a phosphate buffer. The conjugate was sized by HPLC to a pool of fractions with Ac/IgG molar ratios ranging from 3 to 8.

Specimen Diluent Buffer (SDB). SDB contained surfactants, blockers, and buffer salts to enhance specific binding and minimize nonspecific binding.

Controls. Negative Calibrator (NC) is re-calcified negative plasma. Chagas Positive Control (PC) was diluted from pooled plasma of Chagas' patients; each patient was confirmed serologically by at least 2 different tests.

Chagas Positive Specimens. 377 Chagas positive specimens were obtained from American Red Cross, Boston Biomedica, Inc. (BBI, West Bridgewater, MA, USA), BioClinical Partners (Franklin, MA, USA), Teragenix (Ft. Lauderdale, FL, USA) and Goldfinch (Iowa City, IO, USA). The human serum or plasma specimens were collected from donors covering many regions, including Central and Southern American countries, as well as United States. These positive specimens were confirmed by RIPA or 2 to 3 different immunotests.

Assay Format. Assay steps and reactions are illustrated in Fig. 1. In Step-1, the assay starts with 100 μL of sample that is incubated with 50 μL of SDB and 50 μL of *T. cruzi* rAg coated microparticles in the sample well of a PRISM reaction tray. After the first incubation, the tray is moved to the transfer station where the reaction mixture is flushed into the sandwich reaction well by transfer wash buffer and excess fluid is absorbed by a blotter underneath. In Step-2, 50 μL of the acridinium anti-human IgG conjugate is dispensed to the reaction well at the conjugate dispensing station. After the 2nd incubation, unbound conjugates are washed into the blotter. Fifty μL of an alkaline hydrogen peroxide solution is then injected at the trigger/read station to trigger chemiluminescence from acridinium labels captured in the reaction well. The chemiluminescence reaction mechanism is illustrated in Fig. 2. The intensity of chemiluminescent signal is proportional to the amount of anti-*T. cruzi* antibodies in the sample.

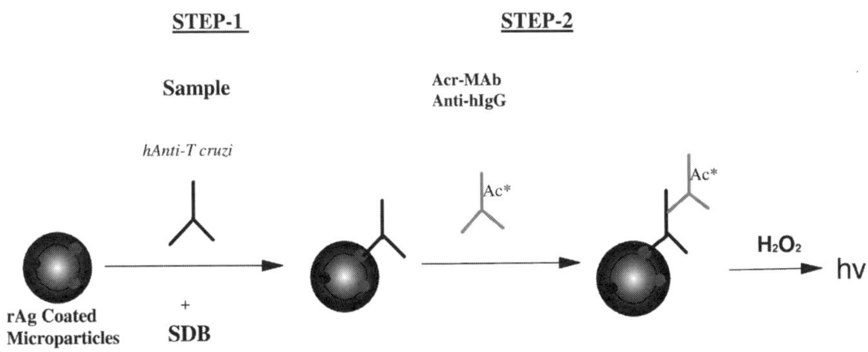

Fig. 1. ABBOTT PRISM® Chagas Assay Format

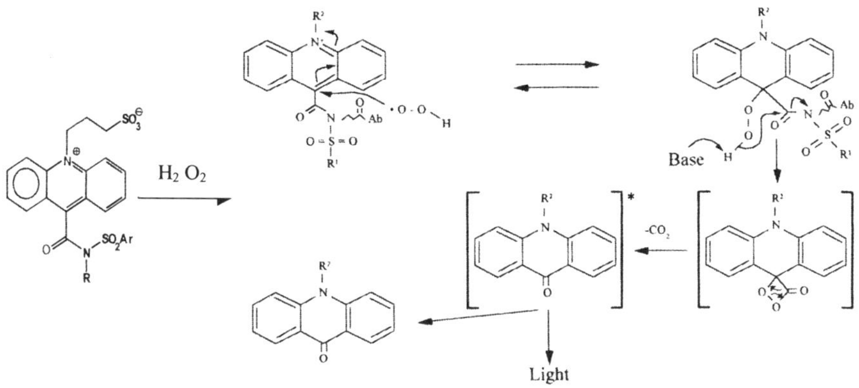

Fig. 2. Sulfopropyl acridinium and the chemiluminescent reaction mechanism

RESULTS AND DISCUSSION

The cut-off calculation is based on mean net NC counts plus 0.15 times the mean net PC counts. If a sample response with net counts is equal to or greater than the cut-off, i.e. S/CO value $\geq$ 1.0, it is reactive; if less than the cut-off, it is negative. However, if a sample S/CO value is between 0.9 and 1.0, it is considered "gray zone" and will be re-tested in duplicate.

The Chagas assay was tested with 7,911 serum and plasma specimens from random U.S. donors in Texas, Florida, and California through Gulf Coast Regional Blood Center (Houston, TX, USA) and ProMedDx (Norton, MA, USA). Six (0.08%) specimens were repeatedly reactive. Sample ID #5060 was also reactive in 2 reference ELISAs and further confirmed positive in RIPA. The remaining 5 were non-confirmable; hence, the prototype Chagas assay had a resolved specificity of 99.94% (7,905/7,911).

In further assessment on the specificity, several panels of human sera or plasma specimen (n = 115) with other diseases (e.g. Leishmaniasis, auto-immune, multi-myeloma, toxoplasmosis, and syphilis) or potentially interferencing substances (e.g. bilirubin and triglycerides) were tested and found non-reactive in this assay.

Due to the genetically diverse and polymorphous parasite, sensitivity of the prototype assay was assessed with various positive specimens (n = 377) from donors covering a wide region of Latin America, Mexico, and southern US. All 377 specimens were detected as reactive or 100% sensitivity. Optimization to further improve the assay performance is in progress. The Abbott PRISM® Chagas Assay should function well as a screening test to improve the safety of the blood supply by reducing the risk of *T. cruzi* transfusion.

ACKNOWLEDGEMENTS
We thank D. Leiby of American Red Cross and L. Kirchhoff of Goldfinch Inc for providing RIPA confirmed Chagas positive human specimens for our evaluation. Furthermore, we appreciate their help in using RIPA to confirm true positive samples with antibodies to T. *cruzi*. We also thank Drs. R. Badaro and E. Martins Netto of Federal University of Bahia, Brazil, for their suggestions in Chagas' disease and collaborative research.

REFERENCES

1. Kirchhoff LV. American trypanosomiasis (Chagas' disease). In: Guerrant RL, Walker DH, and Weller PF. eds. Tropical Infectious Diseases: Principles, Pathogens, and Practice. New York: Churchill Livingstone, 2006:1082-94.
2. Shah DO, Stewart J. Automated panel analyzers PRISM®. In: Wild D. ed. The Immunoassay Handbook - 2nd Edition. Landon: Nature Publishing, 2001: 297-303.
3. Leiby DA. Threats to blood safety posed by emerging protozoan pathogens. Vox Sang 2004;87(Suppl 2):S120-2.
4. Dubovoy N, Munson S, Shah DO. A feasibility of Chagas assay on a fully automated chemiluminescence based immunoassay analyzer. In: Roda A, Pazzagli M, Kricka LJ, Stanley PE. eds. Bioluminescence and Chemiluminescence: Perspectives for the 21st Century. Chichester: Wiley, 1998:95-8.
5. Chang CD, Cheng KY, Jiang L, Salbilla VA, Haller AS, Yem AW, et al. Evaluation of a prototype *T. cruzi* antibody assay using recombinant antigens on a fully automated chemiluminescence analyzer for blood donor screening. Transfusion 2006;46:1737-44.

QUANTITATIVE BIOLUMINESCENT RNA ASSAY

YE SUN, K. BRUCE JACOBSON, VAL GOLOVLEV
Sci-Tec, Inc.,10425 Cogdill Rd. Suite 300, Knoxville, Tennessee 37932, USA

INTRODUCTION

Gene expression is a measure of gene function and provides a better understanding of diseases. While microarrays have emerged as a dominant tool for high-density gene expression research, the alternative non-array gene expression technologies can be superior to microarrays in accuracy and sensitivity. The Real Time quantitative PCR (RT-qPCR, polymerase extension reaction) is often considered to be the method of choice for mRNA quantification.[1] Several factors contribute to the transformation of RT-qPCR technology into a mainstream research tool: (1) as a homogeneous assay, it does not need post-PCR processing; (2) a wide (>10^7-fold) dynamic range allows straightforward comparison between RNAs that differ widely in their abundance; and (3) the assay realizes the inherent quantitative potential of the PCR, as well as its qualitative uses. Yet, RT-qPCR is not pitfall-free.[1] The nature and extent of the unreliability of quantitative RT-PCR data is not widely appreciated or acknowledged. One example is the threshold cycle C_t, which records the cycle when sample fluorescence exceeds a chosen threshold of the background fluorescence. The C_t is used for quantifying target copy number. Yet C_t value is subjective and can be altered at will, making it difficult to compare results with different platforms or experimental conditions. Another issue is the reverse transcription step of RT-qPCR process. The step of converting RNA into a cDNA template is an important contributor to the variability often observed in RT-PCR. Most cDNA priming in RT-PCR assays use random primers or oligo-dT.[1] These create problems if the mRNA targets vary significantly in size, since a single mRNA specie can be represented by multiple transcripts, which are proportional to mRNA length. It has been demonstrated that random hexamers can overestimate mRNA copy numbers up to 19-fold compared with a sequence-specific primer.[2] An alternative approach uses oligo-dTs to bind to the poly(A) RNA tail, but this is not an effective choice for transcribing RNA that is likely to be fragmented by partial degradation.

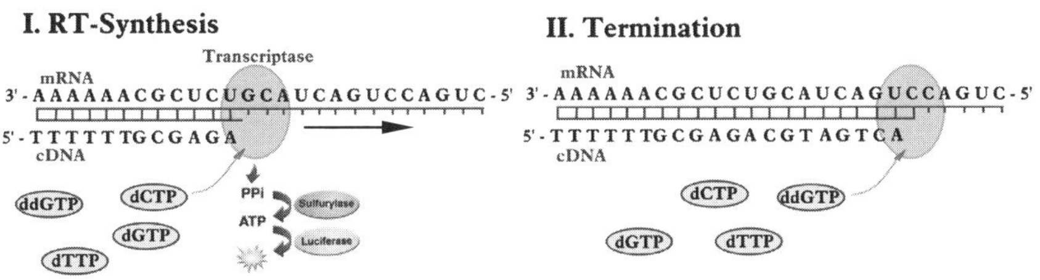

Fig. 1. Detection concept of the proposed reverse-transcription quantitative bioluminescence assay (RT-qBLA)

To address the drawbacks of the existing technologies, we developed RT-qBLA, a nucleic acid detection technique based on detection of released inorganic pyrophosphate (PPi) during oligonucleotide synthesis by PCR (Fig. 1). The target nucleic acid molecule can be RNA or DNA. In a series of three consecutive enzymatic reactions, the visible light generated is directly proportional to the number of incorporated nucleotides. During the nucleic acid polymerization reaction, the inorganic pyrophosphate PPi is released as the result of nucleotide incorporation by polymerase. When the released PPi is converted into ATP by ATP-sulfurylase, it provides the energy to luciferase to oxidize luciferin and generate light.

The PPi identification technique is extremely sensitive and can detect a single target DNA/RNA molecule. This is possible because the synthesis of a single large cDNA can require incorporation of thousands of deoxynucleotides, consequently producing thousands of ATP molecules and resulting in emission hundreds of photons per each cDNA copy synthesized. Yet, currently the PPi detection technique is used mostly for qualitative detection in applications such as pyrosequencing, i.e., for determination of DNA sequence and for SNP detection.[3,4] Indeed, bioluminescent signal is proportional to the number of incorporated nucleotides rather than to the number of target molecules. This makes it difficult to measure the number of DNA or RNA copies based on the intensity of bioluminescent signal alone. To make the bioluminescent signal proportional to the number of target copies and thus make it suitable for RNA quantification, we have developed an approach for programmable termination of reverse transcription (RT) reaction by using a mixture of dNTP and ddNTP. The ratio of ddNTP:dNTP defines the lengths of a cDNA transcript to be synthesized. The self-termination of reaction allows generation of the same amount of bioluminescence for all target RNAs, regardless of length. In combination with the recently reported improvements for reducing background and increasing the sensitivity of PPi detection, the RT-qBLA can be adopted for highly sensitive and accurate quantitative assay for RNA analysis. Another advantage of the RT-qBLA is that the assay can be performed either in 96 well microplates for multiple samples or in tube format for fewer samples. Because single tube and microplate luminometers are commonly available, they can easily be adapted for RT-qBLA assay with virtually no instrument setup cost.

RESULTS

DNA synthesis reaction by polymerase was performed using a single stranded M13mp18 phage DNA (Sigma, Cat. No. D-8410) and –40 forward primer (5’-gttttcccagtcaggacgttgta-3’). The reaction solution prepared by addition of 5.0 μL of 10x PCR buffer, 5 μL of 25 mM $MgCl_2$, 5 μL each of 2 mM dGTP, dCTP, dTTP, 5 μL of 2 mM dATPaS (Trilink Technologies, San Diego, CA, USA) 5 μL of ATP-sulfurylase (Sigma-Aldrich, St. Louis, MO, USA), 1.5 μL of 50 pmol/μL of primer, 1 μL of the M13 template, and 20 μL of ATP Detection Solution (FAAL, Sigma-Aldrich). To reduce the background, dATPaS was substituted for dATP.[5] A low-mass endogenous PPi from M13 DNA stock solution was removed by ultrafiltration on a YM100 Microcon column (Millipore). This 12 min clean-up procedure greatly reduces background caused by break-down products and low-mass contaminants.

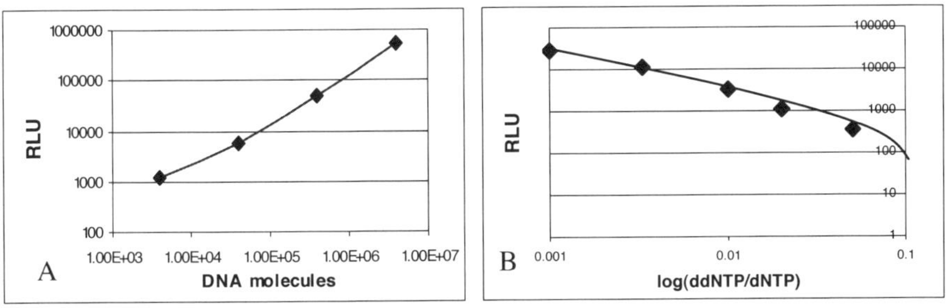

Fig. 2. (A) Bioluminescent signal vs. the amount of M13 DNA template. (B) PCR termination: intensity of bioluminescent signal vs. the ratio of ddNTP:dNTP in PCR solution

Before adding to the reaction solution, the M13 DNA was annealed at 95°C for 1 min and placed on ice for 3 min. The reaction solution with no Taq Polymerase added was placed in the luminometer to measure background. Next, 1 μL of 1.25 U/μL of Taq Polymerase was injected into the solution to record luminescence signal (see Fig. 2) for four different reaction solutions carrying different amount of M13 DNA template: $6.6\text{x}10^{-21}$ mole, $6.6\text{x}10^{-20}$ mole, $6.6\text{x}10^{-19}$ mole, and $6.6\text{x}10^{-18}$ mole (see Fig. 2A). The detection sensitivity is defined as three times the noise level of background luminescence, which is about 200 RLU/s in this experiment, thus corresponding to the detection of 180 DNAs (notice, the size of DNA replicate is likely exceeding 1,000 nt in this experiment). Fig. 2B shows decrease in bioluminescent signal due to addition of ddNTP into PCR solution. In this case, incorporation of ddNTP terminates synthesis, producing smaller sizes transcripts and reducing the amount of PPi released in solution. The RT-qBLA dynamic range was the ratio of the maximum luminescent signal to the minimum detectable signal. At about 6% of 3,200 RLU background signal, or 200 RLU/s in this experiment, the dynamic range of the assay performed on Sirius luminometer (Bethold Detection Systems, Pforzheim, Germany) is (7,996,800RLU/200RLU) = 39,984 or 4.6 $\log_{10}$.

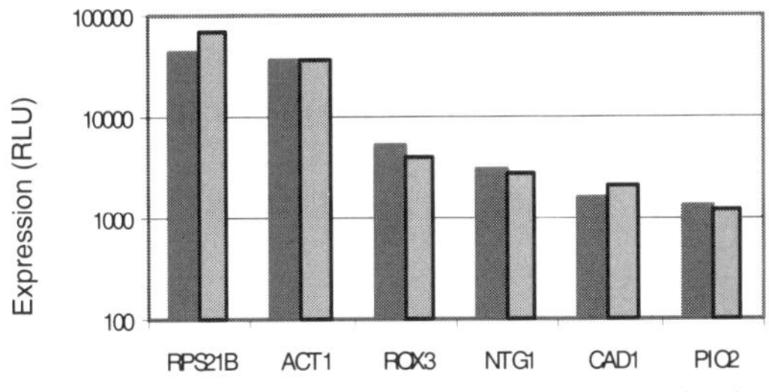

Fig. 3. RT-qBLA (left) and microarray detection (right) of selected genes in yeast. High (RPS21B, ACT1), Medium (ROX3, NTG1), Low (CAD1, PIO2)

Fig. 3 shows six selected genes in yeast detected by RT-qBLA (left) and measured by expression levels reported in a microarray experiment.[6] The expression level of ACT1 was used to normalize data across two platforms, although RT-qBLA used approximately three times less of total RNA than in most microarray experiments.

DISCUSSION

The RT-qBLA has higher sensitivity and dynamic range than procedures using microarrays, and therefore can be used for validation of microarray results from RT-qPCR. It allows RNA quantification in a broad dynamic range; can detect long (mRNA) and short (miRNA) targets; requires fewer reagents per reaction; is normally performed at a constant temperature and is amendable to various replication strategies including "constant temperature" PCR and rolling circle amplification; it can be implemented using significantly less costly equipment (<25%) than RT-qPCR systems. The RT-qBLA is expected to outperform RT-qPCR by: 1) better potential accuracy due to linear scale measurement by RT-qBLA vs. logarithmic scale measurement by RT-qPCR; 2) quicker assays since no temperature cycling is needed for RT-qBLA; and 3) less expensive reagents and equipment since only one transcriptase and one primer per reaction are needed for RT-qBLA vs. a typical RT-qPCR reaction which requires transcriptase, DNA Polymerase, RT primer and two PCR primers. The RT-qBLA also provides an open platform for various RNA and DNA replication strategies. It can be used in commercial assays for life science research, drug discovery and environmental analysis.

ACKNOWLEDGEMENTS

Supported by SBIR grant RGM080037 from NIH/NIGMS. This study used a Sirius single-tube luminometer (Berthold Detection Systems, Pforzheim, Germany).

REFERENCES

1. Bustin SA, Benes V, Nolan T, Pfaffl MW. Quantitative real-time RT-PCR - a perspective. J Mol Endocrinol 2005;34:597-601.
2. Bustin SA, Nolan T. Pitfalls of quantitative real-time reverse-transcription polymerase chain reaction. J Biomol Techniques 2004;15:155-66.
3. Ronaghi M, Uhlen M, Nyren P. A sequencing method based on real-time pyrophosphate. Science 1998;281:363, 365.
4. Hassibi A, Contag C, Vlad MO, Hafezi M, Lee TH, Davis RW, et al. Bioluminescence regenerative cycle (BRC) system: theoretical considerations for nucleic acid quantification assays. Biophys Chem 2005;116:175-85.
5. Zhou GH, Gotou M, Kajiyama T, Kambara, H. Multiplex SNP typing by bioluminometric assay coupled with terminator incorporation (BATI) Nucleic Acids Res 2005;33:e-13.
6. Ball CA, Jin H, Sherlock G, Weng S, Matese JC, Andrada R, et al. Saccharomyces Genome Database provides tools to survey gene expression and functional analysis data. Nucleic Acids Res 2001;29:80-1.

A NOVEL MULTI-CHANNEL CHEMILUMINESCENCE-IMMUNOASSAY TO DETECT HIV ANTIBODIES AS A MORE SPECIFIC SUPPLEMENTAL TEST

XUDONG SUN[1] AND XIAOLIN YANG[2]
[1]Beijing Yuande Bio-Medical Engineering, Beijing, 100176, P. R.China;
[2]People's Hospital of Peking University, Beijing, 100044, P.R. China,
Email: yangzhng@public3.bta.net.cn

INTRODUCTION

AIDS is a worldwide public health crisis, and HIV (Human Immunodeficiency Virus, or former HTLV-III) has been identified as its pathogen.[1, 2] Today, the only available way to prevent and control its spread is to find the HIV infection early and to practice careful personal protection. According to the guidelines of WHO (World Health Organization) and the authorities in most countries[3- 6] for diagnosis on a HIV infection, the specimens should be tested at first by more than two kinds of screening tests, such as EIA, RIA and LIA.[3,4,6] For confirmation of infection, the sera that are positive must be tested again by a more specific supplemental assay, such as WB (Western-Blot).[3-6] WB results can only be judged visually by an experienced technician. WB is a high-cost assay that is very difficult to implement in most kinds of medical service, especially in undeveloped areas and at the point of care (e.g., first aid). We have developed a new multi-channel chemiluminescence-immunoassay (MCCLIA) that is cheap and easy to use, in order to provide a more practicable confirmatory test for HIV infection.

MATERIALS AND METHODS

The MPC-1 luminometer and software were supplied by Beijing Yuande Bio-Medical Engineering Co. Ltd. (Beijing, China). The recombinant HIV antigens were supplied by National Center for AIDS/STD Control and Prevention of Chinese Center for Disease Control and Prevention (NCAIDS/STD, CHINA CDC). Microwells were purchased from NUNC (Denmark). The seroconversion panel for WB quality control was supplied by National Institute for the Control of Pharmaceutical and Biological Products (NICPBP) of State Food and Drug Administration (SFDA) of China. The enhanced chemiluminescence substrates and HRP labeled anti-human antibody were prepared as described previously.[7] The WB kit (HIV BLOT 2.2) to confirm the HIV infection was purchased from GeneLabs Diagnostics Pte Ltd. (Singapore). The normal sera were supplied by health volunteers. The sera that underwent the screen and supplemental test were supplied by the AIDS Confirmation Lab of the People's Liberation Army (PLA) (Beijing, China), AIDS Confirmation Lab of Shaanxi Province of China, AIDS Confirmation Lab of Sichuan Province of China respectively. All other reagents were commercial products at AR grade.

The procedure of MCCIA was as follows: 7 recombinant proteins gp41, gp120, p66, p31, p55 and p24 of HIV-1,[1-7] as well as gp36 of HIV-2[7,9] were separately coated on 7 microwells, while the bovine albumin and mouse anti-human antibody were also coated on another 2 microwells for negative and positive control respectively. All of their concentrations ranged between 0.1-2 μg/mL. To perform the test, 1:1000 diluted samples were separately added into 7 test and 2 control wells (100 μL per well). After 120 min incubation, all unbound materials were removed by a washing procedure. Then 100 μL of HRP labeled anti-human antibody was added, followed by 60 min incubation. Finally, the 100 μL enhanced chemiluminescence substrate was added after the second washing. The signal was detected in a luminometer. A S/Co (Sample/Cutoff) index higher than 1.0 was regarded as a positive response to corresponding HIV protein. The WB was performed as per the instruction manual. For WB or MCCIA, HIV infection was confirmed using the same WHO guidelines for the diagnosis of a HIV infection, which interpreted the results as positive, negative and indeterminate through the response pattern of each sample to the 7 proteins. All of the operations relevant to the sera were performed according to the guidelines of Bio-Safe for an HIV laboratory.[3,4,6]

RESULTS AND DISCUSSION

To acquire background response of normal human serum to the proteins, 500 normal sera were tested, and then each of the cutoff ratios to individual proteins against negative control was setup separately as shown in Table 1, based on a 99% confidence interval. MCCIA also interpreted all of the normal samples as negative using these cutoffs. To confirm sensitivity and specificity, the MCCIA was tested using the seroconversion panel for WB quality control, including 12 negative, 10 positive and 5 indeterminate calibrators. The MCCIA gave a correct response for all 27 samples (Table 2).

Table 1. The cutoff ratios of individual protein

Proteins	P24	P55	P31	P66	GP41	GP120	GP36
Cutoff	16	9	27.6	24	7.3	10.8	36

Table 2. MCCIA results for the seroconversion panel for WB quality control

Response	Seroconversion panel (27)		
	Negative (12)	Indeterminate (5)	Positive (10)
Negative	12	0	0
Indeterminate	0	5	0
Positive	0	0	10

MCCIA was compared with WB using 1070 specimens, which underwent WB as a supplemental test, obtained from 3 of authorized local AIDS Confirmation Labs of China. The results showed high concordance between WB and MCCIA (Table 3).

Table 3. The results of comparison experiment between MCCIA and WB

MCCIA Interpretation	WB Interpretation			Total
	Positive	Indeterminate	Negative	
Positive	594	8	0	612
Indeterminate	4	12	5	21
Negative	2	32	413	447
Total	600	52	418	1070

Using WB as a reference, the sensitivity and specificity of MCCIA were 99% and 98.8%, respectively. Furthermore, 3 of 51 WB indeterminate specimens were correctly identified as positive, because they were reconfirmed by a PCR assay during a follow-up procedure. These results may mean that MCCIA could provide higher sensitivity in detecting HIV in some kinds of samples. However, there was discordance in some samples, including 4 positive-indeterminate, 2 positive-negative, 8 indeterminate-positive, 12 of 32 indeterminate-negative, 5 negative-indeterminate.

The source of the differences between WB and MCCIA may relate to detectability or cross-reaction to individual proteins. Based on our data, MCCIA may play a very important role for quick confirmation of HIV infection at low cost.

ACKNOWLEDGEMENTS
Special thanks to NCAIDS/STD. CHINA CDC, AIDS Confirmation Lab of PLA of China, AIDS Confirmation Lab of Shaanxi Province of China, AIDS Confirmation Lab of Sichuan Province of China, as well as the volunteers.

REFERENCES

1. Schupbach J, Popovic M, Gilden RV, Gonds MA, Sarngadharan MG, Gallo RC. Serological analysis of subgroup of human T-lymphtropic retroviruses (HTLV-III) associated with AIDS. Science 1984;224:503-5.
2. Sarngadharan MG, Popvic M, Bruch L, Schupbach J, Gallo RC. Antibodies reactive with human T-lymphotropic retroviruses (HTLV-III) in the serum of patients with AIDS. Science 1984;224:506-8.
3. Tsang VCW, Hancock K, Wilson M, Palmer DF, Whaley S, Mcdougal JS, et al. Developmental procedure: enzyme-linked immunoelectro-transfer blot technique for HTLV-III/LAV antibodies. CDC March 1985; Atlanta.
4. CDC. Provisional public health service inter-agency recommendations for

screen donated blood and plasma for antibody to the virus causing acquired immune deficiency syndrome. United States morbidity weekly report 1985; 43:1-5.

5. WHO collaborating group on HIV-2. WHO Weekly Epidem Rec 1990; 10: 74-5.
6. CHINA CDC. National Guideline for Detection of HIV/AIDS. October 2004
7. Bottiger B, Karlsson A, Andreasson F. Envelop cross-reactivity between human immunodeficiency virus type 1 and 2 detected by different serological methods: correlation between cross-neutralization and reactivity against the main neutralizing site. J Virol 1990;64:3492-9.
8. Yang X. An improvement of enhanced chemiluminescence and its application to immunoassay. In: Roda A, Pazzagli M, Kricka LJ, Stanley PE. eds. Bioluminescence and Chemiluminescence: Perspectives for the 21st Century. Chichester: Wiley, 1998:138-41.
9. Clavel F. HIV-2, the West African AIDS virus. AIDS 1987;1:135-40.

TIME-RESOLVED MEASUREMENT OF FIREFLY BIOLUMINESCENCE USING PHOTOLYSIS OF CAGED-ATP

MASATOSHI TANAKA,[1] RYOMA IGARASHI,[1] NAOHISA WADA[2]

[1] *Department of Physics, Faculty of Engineering, Yokohama National University, 79-5 Tokiwadai, Hodogaya-ku, Yokohama 240-8501, Japan;*

[2] *Department of Life Sciences, Toyo University, 1-1-1 Izumino Itakura-machi, Gunma 374-0113, Japan, E-mail: mtanaka@ynu.ac.jp*

INTRODUCTION

Recently, the local structures of the luciferase complexed with a luciferyl adenylate intermediate analogue and that complexed with AMP plus oxyluciferin were determined by means of X-ray diffraction to clarify the mechanism of the luminescence color change.[1] However, questions concerning reaction dynamics of bioluminescence still remain; e.g., when does luciferase catalyze luciferin with ATP and Mg ion, and how long does luciferase take to exhaust the product? To answer these questions, time-resolved measurement overcoming diffusional delays is required. Until now, "caged" compounds, which change from photosensitive precursor molecules to effector ones by an intense pulse of radiation, were developed to study the mechanisms of such kinetic biological processes such as muscle contraction and ion transport ATPase.[2]

In this study, a new method using a caged compound to study reaction dynamics of bioluminescence has been developed. The solution involving caged-ATP, instead of ATP, and other ingredients necessary for the luminescence was irradiated by a laser pulse. ATP released from caged ATP triggered instantly the reaction. Temporal variation of the luminescence intensity was successfully observed in the time scale of ms. The time course of the luminescence reveals two component processes and they are identified from the obtained decay constants.

EXPERIMENTAL

Caged ATP (Dojindo Lab; $8x10^{-4}$M) was added to luciferin ($5x10^{-7}$M to $5x10^{-5}$M), luciferase (Sigma; $5x10^{-6}$M) and $MgSO_4$ ($5x10^{-3}$M) in HEPES buffer in the dark at room temperature. The experimental setup is shown in Fig. 1. The solution (1.5 mL) was put in a quartz cell (10x10x50 mm^3) and irradiated by a pulse of third harmonic wave (355 nm) of a Q-switched Nd:YAG. The pulse was purified in wavelength with a Pellin Broca prism and a dielectric mirror (pulse width, 7 ns, fluence 20 mJ/cm^2 per pulse). It is estimated from the fluence that every laser pulse transforms 4 % of caged ATP to ATP. This ATP triggered instantly the reaction, and generated luminescence that was passed through a band-pass filter (530-610 nm) and detected with a photomultiplier. Pulses were counted for 1 ms every 3 ms or for 8 ms every 10 ms. A mechanical shutter was closed for 2 ms after the laser shot so as to prevent output saturation due to the luminescence of luciferin itself which decays within about 1 μs. Temporal variation of the luminescence intensity was averaged for 30 laser shots.

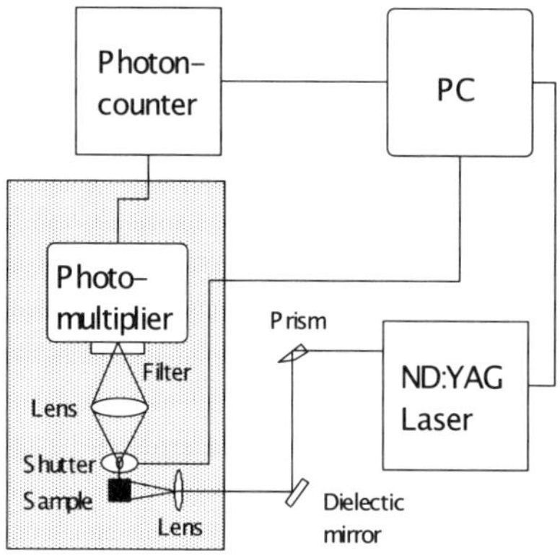

Fig.1. Schematic diagram of the experimental setup

RESULTS AND DISCUSSION

Temporal variations of the luminescence intensity at different pH after a laser shot with luciferin concentration of $5x10^{-6}$M are shown in Fig. 2. A prominent point at 2.5 ms and subsequent monotonous decay are clearly observed. This is the first report of time-resolved measurement of bioluminescence in the ms range as far as the authors know. The luminescence intensity at pH 7.8 is twice as large that at pH 7.6 or 8.0. However, the quantum efficiency of the luminescence should be estimated considering the decay constants described later. The inset of Fig. 2 magnifies the temporal variations at longer time periods.

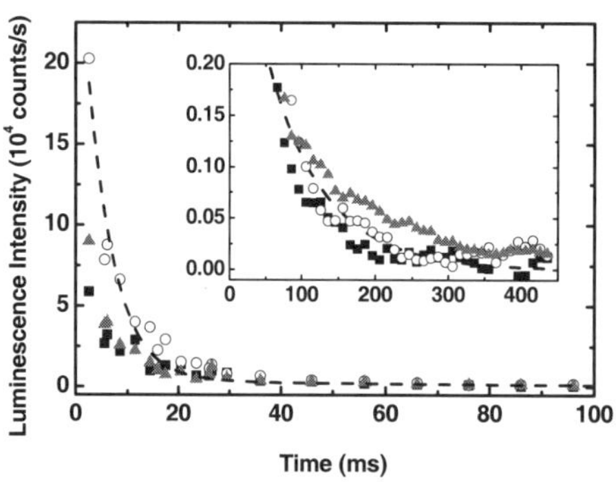

Fig. 2. Luminescence decay at pH 7.6, 7.8 and 8.0 (represented by squares, circles and triangles, respectively)

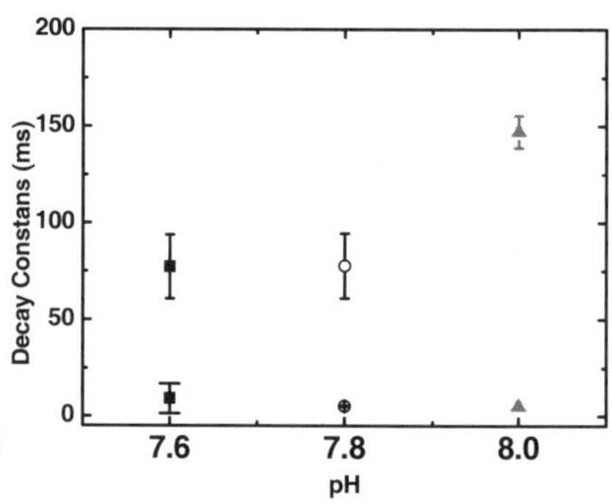

Fig. 3. Decay constants at various pH

Since fluctuation of background luminescence scatters the data in this range, each data point is obtained by smoothing five data points. The luminescence intensity at pH 8.0 decays slower than those at pH 7.6 and 7.8 at longer time range. Each time course thus consists of two decay constants with slow and fast phases, and is well reproduced by a sum of two exponential functions (dashed curve for the time course at pH 7.8). The decay constants for the two components are 5 ms and 75 ms (constants compared with those at different pH in Fig. 3). The decay constant of the first component depends weakly on pH, whereas that of the second component increases with pH, from 75 ms at pH 7.6 to 140 ms at pH 8.0. Thus obtained pH dependence of the decay constants qualitatively agrees with that of the rate constants of ATP formation from caged ATP.[3] If ATP is bound to luciferase > ms time scale, the luminescence intensity is proportional to the concentration of ATP, and the decay constant will agree with the rate constant of ATP formation. It is therefore natural to assign the second component to when ATP is generated from caged ATP binds the active site of luciferase and successively reacts to emit light. The first component which decays much faster than the second component can be ascribed to the ATP generation process from the caged ATP trapped before photolysis in the proximity of the active site of the luciferase. Although the nitrobenzyl moiety of the caged-ATP hinders binding at the active site, it may be possible to attract the caged ATP through a weak interaction such as Van der Waals. This interaction may weaken the bond strength between the nitrobenzyl moiety and ATP part of the caged-ATP to make the photochemical release of ATP faster than that of free caged ATP in the buffer.

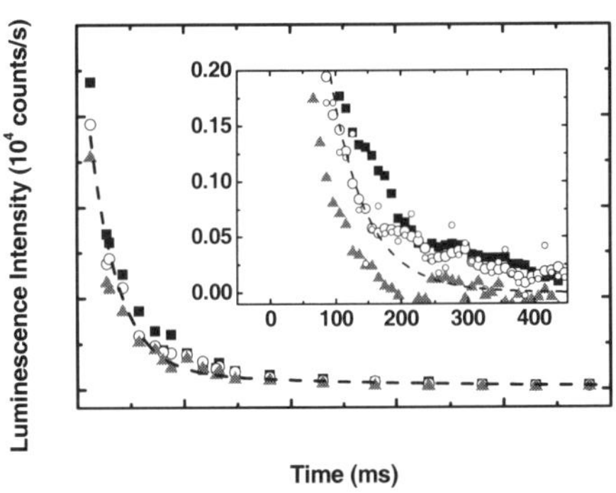

Fig. 4. Luminescence decay at various luciferin concentrations. Squares, circles and triangles stand for $5x10^{-7}$, $5x10^{-6}$ and $5x10^{-5}$ M

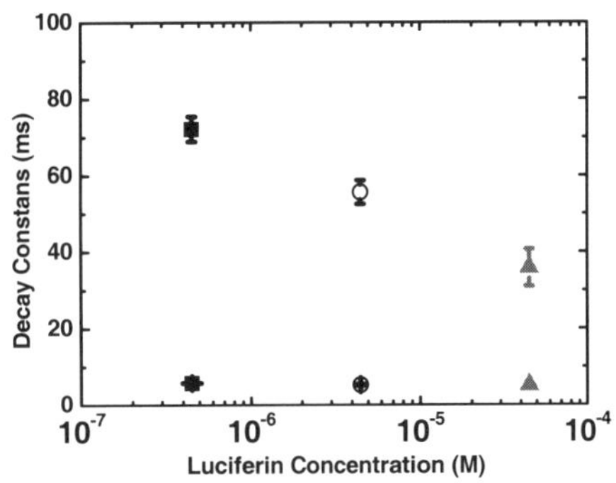

Fig. 5. Decay constants at different luciferin concentrations

Fig. 4 shows the temporal variations of luminescence intensity at several luciferin concentrations at pH 7.8. The concentration of $5x10^{-6}$M is equal to that of luciferase. Since luciferin molecules absorb incident laser light, spatially averaged light intensity at high luciferin concentration is lower than that at low concentration. Accordingly, the luminescence intensity is normalized to that in the case without absorption. The luminescence intensity does not depend on luciferin concentration. However, in the longer time range as shown in the inset, the luminescence intensity decays faster as the luciferin concentration increases. Dashed curves represent the best-fit double exponential functions at 5×10^{-6}M. The decay constants at different luciferin concentrations are shown in Fig. 5. The decay constant of the first component slightly depends on luciferin concentration, but that of the second component decreases with luciferin concentration. Probably, at low luciferin concentration, free ATP generated in the buffer takes more time to encounter the luciferase-luciferin complex than it does at higher concentration.

Time-resolved measurement reported in this paper reveals reaction dynamics of bioluminescence triggered by the photolysis of caged-ATP: ATP generated in the proximity around the active site of the luciferase yields light emission within several ms, whereas free ATP in the buffer yields emission several ten ms later with the intensity reflecting the ATP concentration.

ACKNOWLEDGEMENTS

This work was supported by a Grant-in-Aid from the Ministry of Education, Culture, Sports, Science and Technology of Japan (No.17656004). The authors are indebted to F. Arita and Prof. T. Sekiya for their contribution in the initial stage of this study, and Dr. S. Ohno and Prof. K. Shudo for their cooperation in the experiments.

REFERENCES

1. Nakatsu T, Ichiyama S, Hiratake J, Saldanha A, Kobashi N, Sakata K, et al. Structural basis for the spectral difference in luciferase bioluminescence. Nature 2006;440:372-6.
2. Wooton JF, Trentham DR. 'Caged' compounds to probe the dynamics of cellular processes: synthesis and properties of some novel photosensitive p-2-nitrobenzyl esters of nucleotides. In: Nielsen PE. ed. Photochemical probes in biochemistry. Dordrecht: Kluwer, 1989:277-96.
3. Walker JW, Reid GP, McCray JA, Trentham DR. Photolabile 1-(2-nitrophenyl)ethyl phosphate esters of adenine nucleotide analogues. Synthesis and mechanism of photolysis. J Am Chem Soc 1988;110:7170-7.

RED-CHEMILUMINESCENT PROBES FOR DETECTING SUPEROXIDE ANIONS

KATSUNORI TERANISHI

Faculty of Bioresources, Mie University, Tsu, Mie 514-8507, Japan

INTRODUCTION

Various imidazopyrazinone-type chemiluminescent probes have been developed, including 6-(4-methoxyphenyl)-2-methylimidazo[1,2-a]pyrazin-3(7H)-one (MCLA),[1,2] have been widely used in detecting superoxide anions in biological systems. The imidazopyrazinone compound presumably reacts with superoxide anions to form a singlet-excited amide to generate an amide compound and light. In the case of FCLA (the fluoresceinyl analog of MCLA),[2] the singlet-excitation energy that is generated by the reaction between the MCLA moiety and the superoxide anions transfers to the fluorescein moiety, which in turn emits green light. The luminescence intensity of a green-chemiluminescent probe (called Green Chemiluminescent CD), which has an MCLA and a fluorescein moiety covalently attached to a γ-cyclodextrin on its secondary and primary hydroxyl groups, respectively, is higher than that of FCLA. Although, FCLA and Green Chemiluminescent CD emit light with the longest wavelength under aqueous neutral conditions, luminescence at longer wavelengths is desirable. We present the synthesis and luminescence properties of novel red-luminescent probes (Fig. 1, 1-3) that consist of the MCLA and sulforhodamine 101 (Texas red) moieties.

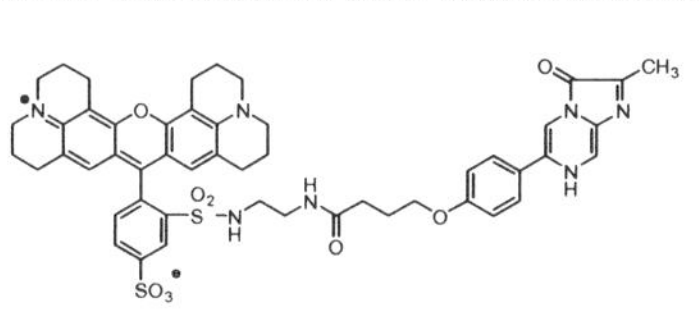

1

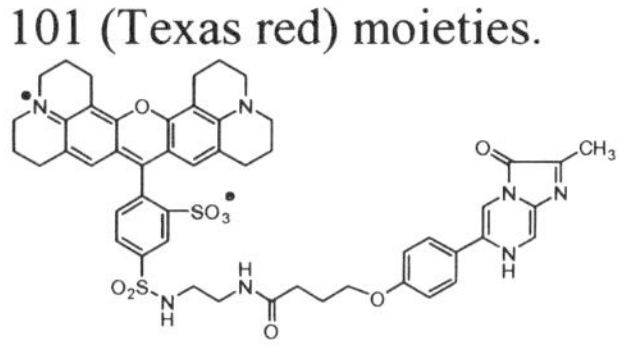

1

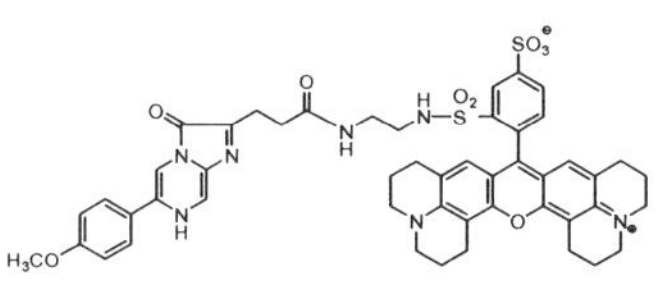

2

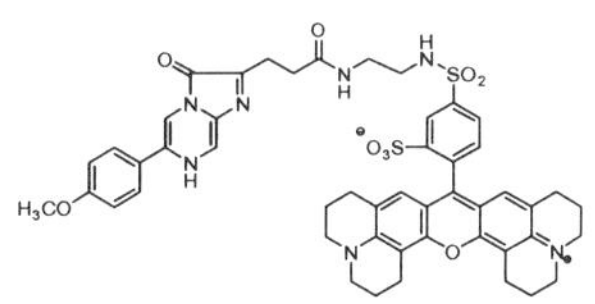

2

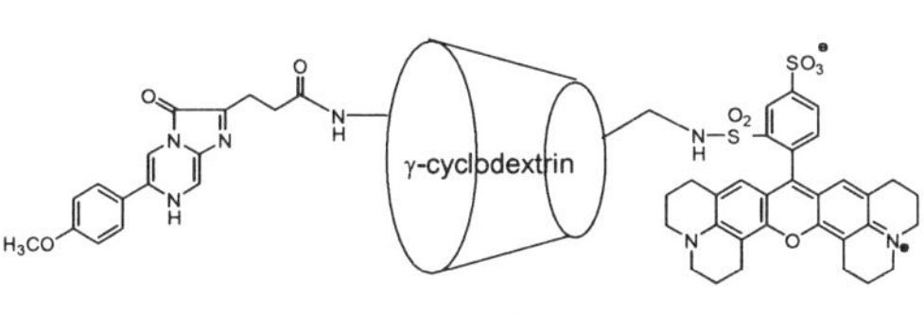

3

Fig. 1. Chemical structures of red-chemiluminescent probes 1-3

MATERIALS AND METHODS

General measurements of chemiluminescence intensities and spectra. To a mixture of 20 mmol/L Mops / 0.2 mol/L KCl (pH 7.2, 0.5 mL), 0.3 mmol/L hypoxanthine (0.5 mL), and probe at 20°C was added xanthine oxidase and then the reaction mixture was placed in the photometer. Chemiluminescence spectra were obtained immediately after the addition of xanthine oxidase.

RESULTS

Red-chemiluminescent probes 1 (a mixture of 1a and 1b) and 2 (a mixture of 2a and 2b) were successfully synthesized from the corresponding MCLA derivatives (6 and 7) by condensation with sulforhodamine 101 derivative 5 (a mixture of 5a and 5b, Scheme 1). The chemical structure of red-chemiluminescent probe 3, shows that the MCLA and sulforhodamine 101 are covalently attached on a γ-cyclodextrin through its secondary and primary hydroxyl faces, respectively. Previously, we showed that the chemiluminescence intensity of Green Chemiluminescent CD was greater than that of FCLA. Analogously, the chemiluminescence intensity of red-chemiluminescent probe 3 was expected to greater that those of probes 1 and 2. As shown in Scheme 2, red-chemiluminescent probe 3 was successfully synthesized, as a mixture of regioisomers.

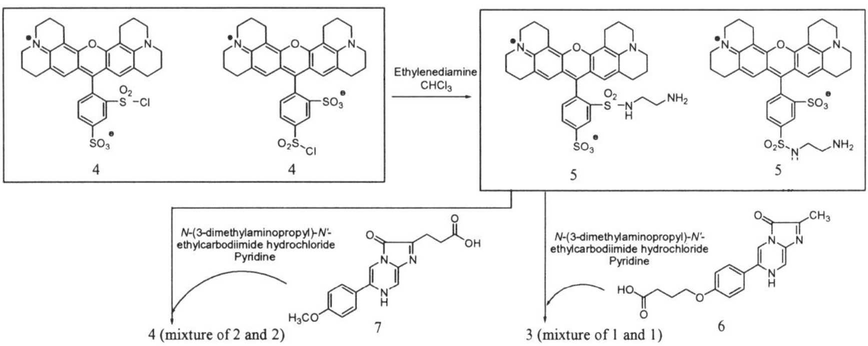

Scheme 1. Synthesis of red-chemiluminescent probes 1 and 2

Superoxide anion-induced chemiluminescences of probes 1-3, MCLA, and FCLA were investigated using the hypoxanthine-xanthine oxidase system as the source of superoxide anions. MCLA exhibited superoxide anion-induced chemiluminescence peaks at around 460 nm in the luminescence spectrum. In contrast, chemiluminescence spectra of probes 1-3 revealed a maximum at only 610 nm. The absence of chemiluminescence due to the MCLA moiety indicated that the superoxide anion-induced chemiluminescences of probes 1-3 are generated from the sulforhodamine 101 moiety. These results clearly indicate that the excitation-energy generated from the MCLA moiety is efficiently transferred to the sulforhodamine 101 moiety.

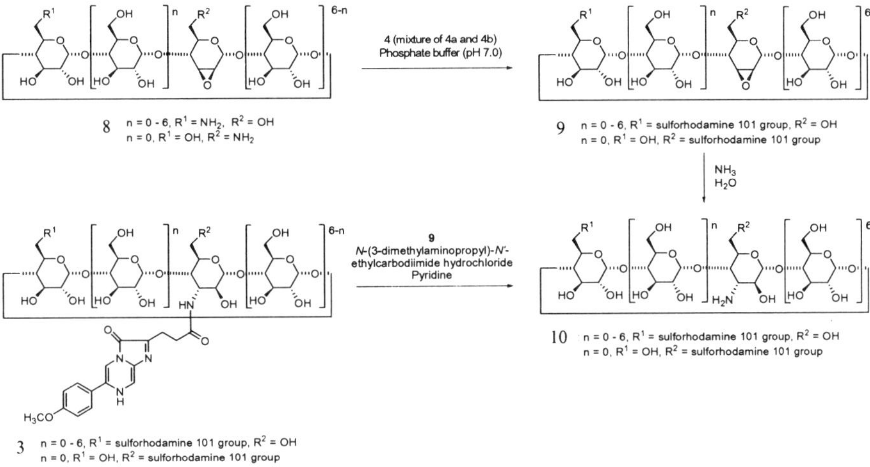

Scheme 2. Synthesis of red-chemiluminescent probe 3

Concentration effects of the probe on the intensity of the superoxide anion-induced chemiluminescence are shown in Fig. 2. For probes 1 and 2, at concentrations less than 0.5 μmol/L, the chemiluminescence intensities increased with higher probe concentrations; this behavior was similar to those of MCLA and FCLA. For probe 3, a similar trend was observed, but for concentrations up to 5 μmol/L. The superoxide anion-induced chemiluminescence intensities of probes 1 and 3, at all probe concentrations, were greater than those of MCLA and FCLA. In contrast, the intensity of probe 2 was lower than that of MCLA. In the case of probe 3, the γ-cyclodextrin moiety was effective for the greater luminescence intensity than that of probe 2. At probe concentrations less than 0.5 μmol/L, the luminescence intensity of probe 1 was superior to that of probe 3, whereas, at the concentrations more than 0.5 μmol/L, the intensity of probe 3 was higher than that of probe 3. Unfortunately, the photometer used in this study is not very sensitive toward red light (photo-sensitivity of ~18% for 460 nm, ~15% for 520 nm, and ~4% for 610 nm). The use of a photometer that is more sensitive toward red light would be able to detect the luminescence of probes 1-3 more effectively.

For probe 1, the ratio of the superoxide anion-induced chemiluminescence intensity (*S*) to background chemiluminescence intensity (*B*), over the probe concentration range of 0.0093 - 0.93 μmol/L, was similar to those of MCLA and FCLA. The *S/B* ratios of probes 2 and 3 were significantly lower than that of probe 1. Because the *S/B* ratios of probes 2 and 3 can be easily influenced by the probe concentrations under our test conditions, the use of probe 1 can offer improved performance as a highly useful probe.

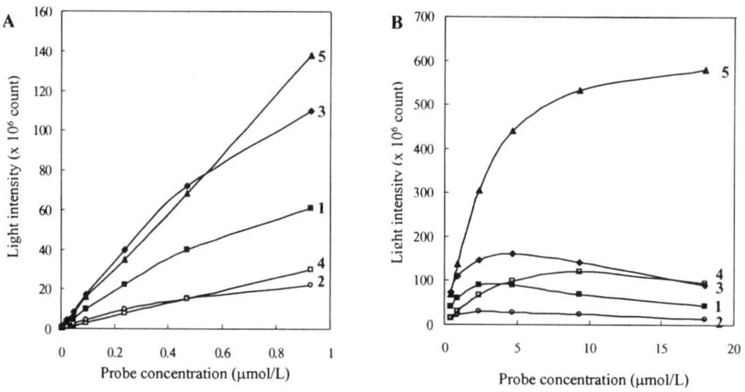

Fig. 2. Superoxide anion-induced chemiluminescence intensities of red-chemiluminescent probes 1-3, MCLA and FCLA for 1 min

Concentration effects of xanthine oxidase on the superoxide anion-induced chemiluminescence intensity, with the probe concentration at 0.093 μmol/L. Over the entire xanthine oxidase concentration range, the chemiluminescence intensities of probes 1 and 3 were higher than those of MCLA and FCLA. In contrast, the intensities of probe 2 were significantly lower over the concentration range.

Acknowledgments

I am grateful to the Grant-in-Aid for Scientific Research (C) (No. 17510176) from the Japanese Ministry of Education, Science, and Culture, and Research, and to the Development Program for New Bio-industry Initiatives from The Bio-oriented Technology Research Advancement Institution.

REFERENCES

1. Goto T, Takagi T. Chemiluminescence of a Cypridina luciferin analogue, 2-methyl-6-phenyl-3,7-dihydroimidazo[1,2-*a*]pyrazin-3-one, in the presence of the xanthine-xanthine oxidase system. Bull Chem Soc Jpn 1980;53:833-4.
2. Minakami H, Arai H, Nakano M, Sugioka K, Suzuki S, Sotomatsu A. A new and suitable reconstructed system for NADPH-dependent microsomal lipid peroxidation. Biochem Biophys Res Commun 1988;153:973-8.

AEQUORIN BIOLUMINESCENCE DISSECTED BY RANDOM MUTAGENESIS

LUDOVIC TRICOIRE,[1] KEISUKE TSUZUKI,[2] BERTRAND LAMBOLEZ[1]
[1] *Neurobiologie des Processus Adaptatifs, UMR 7102, CNRS, Université Pierre et Marie Curie-Paris, Paris, F-75005 France;* [2] *Department of Physiology, Gunma University School of Medicine, Maebashi, Gunma 371-8511, Japan*

INTRODUCTION

The photoprotein aequorin is a bioluminescent complex formed from the reaction of the luciferase apoaequorin with its luciferin coelenterazine, which emits light upon Ca^{2+} binding.[1] Apoaequorin is a 189 aminoacid protein containing three EF-hand Ca^{2+} binding sites[2] whose reaction with coelenterazine results in a stable intermediate containing a coelenterazine peroxide. The binding of Ca^{2+} destabilizes the peroxide and, in the continuous presence of Ca^{2+}, the bioluminescence reaction rapidly goes to completion with a quantum yield of ~ 0.2.[1] Aequorin luminescence is thus emitted as an exponentially decaying flash, whose peak amplitude increases with $[Ca^{2+}]$.[3] The decay rate of this response increases with $[Ca^{2+}]$, while the onset rate and the total light emitted remain roughly constant.[3] This suggests that the increase of peak amplitude with $[Ca^{2+}]$ is determined by an increase of the decay rate (i.e. rate of luminescence), but not of the reaction yield.

We used random mutagenesis and functional screening to investigate the links between peak amplitude, decay rate and yield of aequorin bioluminescence.[4,5] This method allows selecting mutants that efficiently affect a specific subfunction of a protein with minimal alteration of its overall structure-function relationships. Screening was aimed at isolating aequorin mutants exhibiting increased intensity (Bright mutants) or slow decay kinetics of luminescence (SloDK) in bacteria.

MATERIALS AND METHODS

Random mutagenesis and functional screening has been described elsewhere.[4] Mutants were next characterized in cell-free assays following *in vitro* expression using a transcription/translation system.[4] Aequorin was reconstituted in a final volume of 50 μL and its activity was measured in a PhL microplate luminometer (Mediator, Vienna) by injecting 100 μL of solutions containing variable $CaCl_2$ concentrations buffered with 10 μM EDTA.

RESULTS AND DISCUSSION

Bright and SloDK mutations affected residues conserved in other photoproteins,[4,5] either in EF-hands at Ca^{2+} binding positions or in their adjacent alpha-helices (Fig. 1). Four out of six SloDK mutations consisted in a D/E to G substitution at an EF-hand border. The luminescence intensity of Bright mutants exhibited at least a 10-fold increase as compared to wild-type (WT) aequorin when tested in bacteria, while their decay kinetics was similar to WT (Fig. 1). In contrast, luminescence decays of SloDK mutants were strikingly slow, with half-decay times ranging from 20- (F149S) to 57-fold (E35G) greater than that of WT aequorin (Fig. 1).

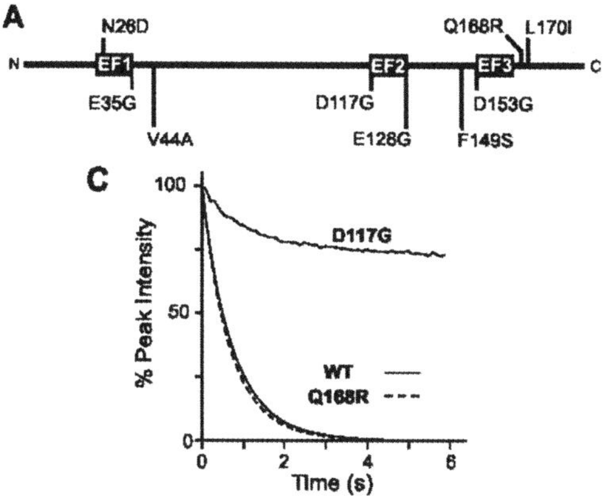

B

Bright mutation	Activity / WT
N26D	13 ± 2
Q168R	11 ± 3
L170I	16 ± 4

D

SloDK mutation	$t_{1/2}$(s)
WT	0.7 ± 0.01
E35G	39.9 ± 1.5
V44A	33.7 ± 0.8
D117G	28.6 ± 1.3
E128G	17.0 ± 0.6
F149S	13.7 ± 0.6
D153G	27.8 ± 0.5

Fig. 1. (A) Bright (above) and SloDK (below) mutations of apoaequorin. (B) Luminescence of Bright mutants vs. WT in bacteria. (C) Examples of Bright and SloDK luminescence decays as compared to WT. (D) half-decay times ($t_{1/2}$) obtained at saturating [Ca^{2+}] for WT and SloDK aequorins in cell-free assays

In vitro, none of the Bright mutants exhibited a higher maximal photon yield than WT aequorin, which would have explained their high luminescence intensity in bacteria. The N26D Bright mutant was presumably selected in intact bacteria because of a higher Ca^{2+} sensitivity, due to the replacement of a polar by a negatively charged residue at a Ca^{2+} binding position of EF1.[4,5] On the other hand, the Q168R and L170I Bright mutants exhibited an extended photoprotein lifetime at 37°C (6- and 2-fold that of WT, respectively). Consistent with this, Q168R and L170I mutants were resistant to higher temperatures than WT, retaining 50% of their activity after a 30 min heat shock at 53 and 48°C, respectively (versus 44°C for WT).[4] Q168 and L170 flank the H169 which interacts with the coelenterazine peroxide. Q168R and L170I mutations result in additional interactions with EF3 to stabilize the local protein structure.[4] These residues interact with both EF3 residues and coelenterazine and thus participate in the stabilization of the photoprotein and in the triggering of bioluminescence by this EF-hand.

As expected from their location in or near EF-hands, all SloDK mutations affected the Ca^{2+} sensitivity of aequorin response (Fig. 2). EF1 mutants exhibited a higher Ca^{2+} sensitivity, whereas EF2 and EF3 mutants exhibited a lower Ca^{2+} sensitivity than WT aequorin. This implies that EF1 has a lower Ca^{2+} affinity than EF2 and EF3. This was confirmed by analysis of double mutants where only one EF-hand is left intact. The EF1-only mutant exhibited a lower Ca^{2+} sensitivity than EF2-only and EF3-only mutants (Fig. 2). Furthermore, these results show that all individual EF-hands trigger luminescence and contribute to its [Ca^{2+}] dependence.

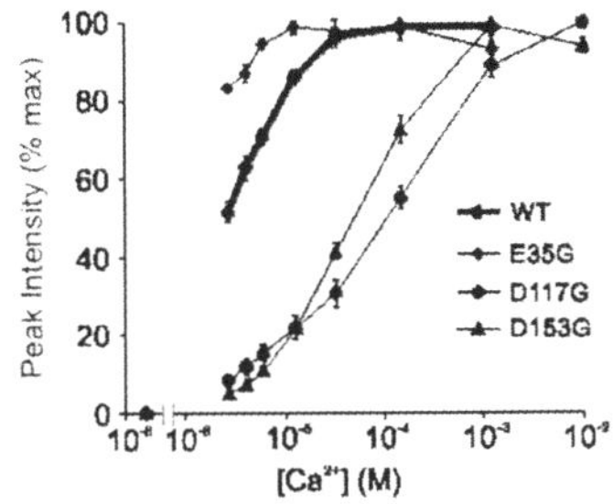

Mutation	Position	Ca^{2+} EC_{50} (μM)
WT	-	2.6 ± 0.3
E35G	EF1	< 2.6
V44A	EF1	< 2.6
D117G	EF2	109 ± 20
E128G	EF2	23 ± 2
F149S	EF3	61 ± 3
D153G	EF3	49 ± 4
E35G & E128G	EF1 & EF2	7.6 ± 0.6
E35G & D153G	EF1 & EF3	6.1 ± 0.1
E128G & D153G	EF2 & EF3	34 ± 0.9

Fig. 2. Ca^{2+} sensitivity of WT and SloDK aequorins

Analyses of WT aequorin responses showed that the increase of peak amplitude with $[Ca^{2+}]$ was accompanied by an increase of the decay rate (Fig. 3).[3] In the $[Ca^{2+}]$ range investigated, the total light emitted was already maximal at low $[Ca^{2+}]$ and only slightly decreased at higher $[Ca^{2+}]$. Hence, in WT aequorin, the increase of peak amplitude is determined by the increase of the decay rate: the luminescence rate increases with $[Ca^{2+}]$ but its yield remains roughly constant. All SloDK mutants exhibited a markedly different behavior,[5] exemplified in Fig. 3. The increase of peak amplitude with $[Ca^{2+}]$ was not accompanied by an increase of the decay rate, but was essentially determined by an increase of total light emitted. Hence, in WT aequorin, all three EF-hands contribute to the increase of decay rate with $[Ca^{2+}]$. Luminescence decays of WT and mutant aequorins are described by the sum of a fast and a slow exponential[5] defined by their time constants and integrals (expressed as their contribution to the total light emitted). The fast time constant varied little with $[Ca^{2+}]$ and was similar in WT and SloDK aequorins (Fig. 3). The slow time constant decreased with increasing $[Ca^{2+}]$, exhibiting much larger values in SloDK than in WT aequorin. In WT aequorin, the contribution of the fast component to total light increased with $[Ca^{2+}]$, but remained minimal at all $[Ca^{2+}]$ in SloDK mutants (Fig. 3). Hence, the increase of the decay rate of WT aequorin luminescence relies on variations of a fast and a slow component involving all three EF-hands. Our results further suggest the coexistence of a fast and a slow light emitting form of aequorin whose equilibrium varies with $[Ca^{2+}]$.[5]

Aequorin is specialized in the stabilization of a luciferase-luciferin intermediate, the binding of Ca^{2+} and the rapid emission of light. Random mutagenesis and functional screening yielded mutants affecting these three properties. Analyses of these mutants showed that all EF-hands contribute to Ca^{2+} sensitivity with EF1 showing lower affinity than EF2 and 3, and that each individual EF-hand triggers luminescence. Decay kinetics of WT aequorin consisted of a slow and a fast component whose variations determine those of peak amplitude in a large $[Ca^{2+}]$ range where total light emission is roughly constant. All EF-hands contribute to these variations. These findings may apply to other photoproteins, which share extensive structural and functional similarities with aequorin.[6]

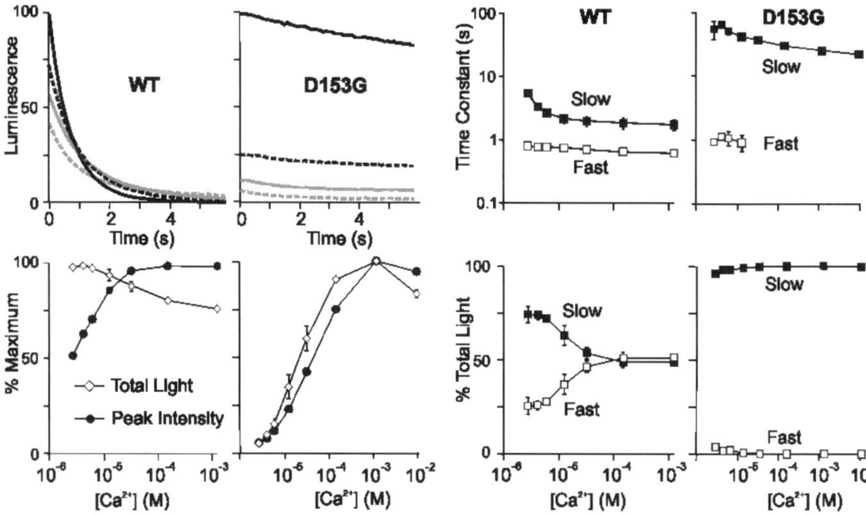

Fig. 3. Luminescence decays of WT aequorin and of the D153G SloDK mutant. Left, Row 1: Decays at low (dotted gray), medium low (gray), medium high (dotted black), and high (black) [Ca^{2+}]. Right, Row 1: For D153G, the fast time constant could not be determined above 12.5 μM Ca^{2+} because of breakdown of the fast exponential. Right, Row 2: Proportion of the light integrals of each exponential to total light emitted vs. [Ca^{2+}]

REFERENCES

1. Shimomura O, Johnson FH. Calcium binding, quantum yield, and emitting molecule in aequorin bioluminescence. Nature 1970; 227: 1356-7.
2. Inouye S, Noguchi M, Sakaki Y, Takagi Y, Miyata T, Iwanaga S, et al. Cloning and sequence analysis of cDNA for the luminescent protein aequorin. Proc Natl Acad Sci USA 1985;82:3154-8.
3. Hastings JW, Mitchell G, Mattingly PH, Blinks JR, Van Leeuwen M. Response of aequorin bioluminescence to rapid changes in calcium concentration. Nature 1969;222:1047-50.
4. Tsuzuki K, Tricoire L, Courjean O, Gibelin N, Rossier J, Lambolez B. Thermostable mutants of the photoprotein aequorin obtained by in vitro evolution. J Biol Chem 2005; 280:34324-31.
5. Tricoire L, Tsuzuki K, Courjean O, Gibelin N, Bourout G, Rossier J. Calcium dependence of aequorin bioluminescence dissected by random mutagenesis. Proc Natl Acad Sci USA 2006;103:9500-5.
6. Liu ZJ, Stepanyuk GA, Vysotski ES, Lee J, Markova SV, Malikova NP. Crystal structure of obelin after Ca^{2+}-triggered bioluminescence suggests neutral coelenteramide as the primary excited state. Proc Natl Acad Sci USA 2006;103:2570-5.

INFLUENCE OF INORGANIC SULFUR COMPOUNDS ON BIOLUMINESCENT SYSTEM NAD(P)H:FMN-OXIDOREDUCTASE – LUCIFERASE

EV VETROVA, EV PROKHOROVA
Institute of Biophysics SB RAS, 660036 Krasnoyarsk, Russia

INTRODUCTION

Recently, bioluminescent organisms and their fermentative systems which transform energy of conjugated redox reactions into light have become widely used for ecological monitoring of aquatic ecosystems. This has stimulated studies of interaction rules for the effect of different chemical compounds on bioluminescence and enzyme dynamics of the bacterial bioluminescence system in different mediums.[1,2]

The aim of the current study was investigation of the influence of inorganic sulphur containing reducing agents on bioluminescent coupled system NADH:FMN-oxidoreductase - luciferase. Inorganic sulfur containing compounds are widespread in the environment and may be a natural component or pollutant, so they have a great impact on the ecological state of aquatic ecosystems. For example, sodium salts Na_2S, Na_2SO_3, Na_2SO_4, $Na_2S_2O_4$, $Na_2S_2O_5$ and $Na_2S_2O_3$ containing sulfur in different oxidation rates and characterized by different redox properties have been chosen for study. Also it should be noticed that Na^+ cation influence, as has been shown earlier,[3] leads to weak activation of bioluminescence of this coupled enzyme system. Sulfur compounds have strong reduction properties and their effect on bioluminescence can be caused by their influence on redox processes in coupled enzyme systems.

METHODS

Reagents: NADH, FMN (Sigma, USA), inorganic sulfur compounds ("REACHIM", Russia). The lyophilized preparation of luciferase from *Photobacterium leiognathi* (0.5 mg) and NAD(P)H:FMN-oxidoreductase from *Vibrio fischeri* (0.15 unit activity) was produced in the Bacterial Bioluminescence Laboratory of the Institute of Biophysics, SB RAS (Krasnoyarsk, Russia).

The bioluminescence of the coupled enzyme system was studied as described earlier.[2] Bioluminometer BLM 8801 (scientific design office "Nauka", Krasnoyarsk, IBP SB RAS) has been used for bioluminescence signal measurement.

The effect of a sulfur compound on bioluminescence was estimated by the luciferase index (LI) using the following formula:

$$LI = I/I_c, \quad (1)$$

where I is the intensity of a bioluminescent signal in the presence of a redox compound of concentration C, M; I_C is the intensity of a bioluminescent signal in the absence of the compound. The efficiency of compound impact on bioluminescence has been defined by the inhibition coefficient (K_i, M^{-1}) calculated from

$$LI = \exp(-K_i \cdot C), \qquad \text{for} \quad C > C^* \tag{2}$$

and activation coefficient (K_a, M^{-1}) calculated from

$$LI = \exp(-K_a \cdot C), \qquad \text{for} \quad C < C^* \tag{3}$$

where C^* is the concentration of the sulfur compound referring to maximum bioluminescent activation signal (LI_{max}).

RESULTS

Influence of sulfur compounds on bioluminescence intensity of the coupled enzyme system NADH:FMN-oxidoreductase – luciferase was studied. Fig. 1 shows the bioluminescence maximum intensity dependencies on sulfur containing compound concentration in the coupled enzyme system. With increase of sulfur compound concentration in the bioluminescent system, activation and then inhibition was observed. Maximum intensity activation was observed with Na_2S and $Na_2S_2O_5$, with values for LI_{max} of 28 and 10, respectively.

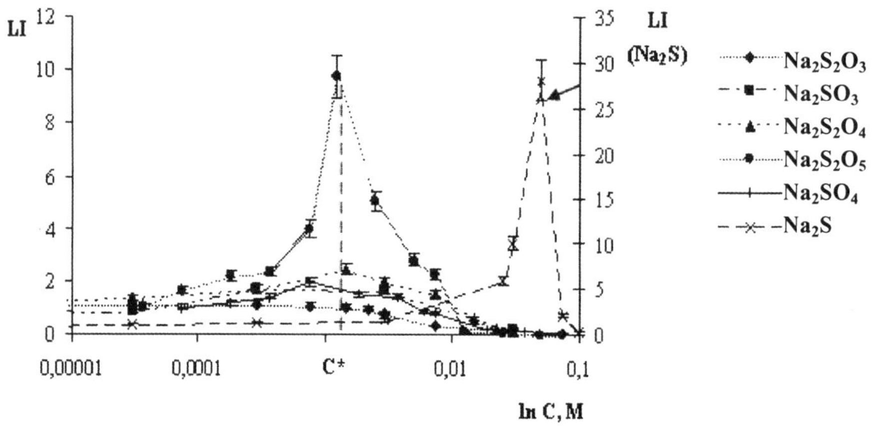

Fig. 1. Bioluminescence intensity dependence (LI) on sulfur compounds concentration (lgC, M)

For description of LI(C) dependencies, activation (K_a) and inhibition (K_i) coefficients, calculated according to formula2 and 3, have been used. The values of K_i and K_a versus the standard redox potential E^0 of a sulfur compound are presented on Fig. 2A and Fig. 2B, respectively.

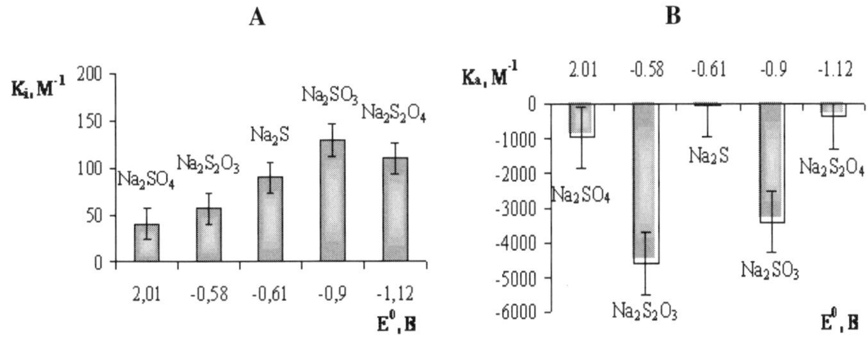

Fig. 2. Dependence of inhibition coefficients (K_i, M^{-1}). Panel A: and activation coefficients (K_a, M^{-1}). Panel B: bioluminescence on standard redox potential (E^0, B) of inorganic sulfur compounds

It was established that a decrease in the sulfur compound E^0 leads to K_i value increase, i.e. with increase of sulfur compound redox properties the efficiency of bioluminescence inhibition also increases. The largest values of K_i were found with $Na_2S_2O_4$ and Na_2SO_3, compounds with lowest redox potential, i.e. with highest reduction abilities. It is known that sulfur compounds are easily oxidized by molecular oxygen dissolved in water, so bioluminescence inhibition processes are probably affected by decrease of oxygen concentration (oxygen is a substrate of the reaction catalyzed by bacterial luciferase).

For K_a values, clear dependence on redox potential of sulfur compounds has not been established. Maximum K_a values have been established for $Na_2S_2O_3$ and Na_2SO_3. Thus, Na_2SO_4 has the highest E^0 value, and it's influence on bioluminescence is characterized by minimum K_i and K_a values. Bioluminescence activation processes intensity could be related to sulfur compound participation in reduction reactions of FMN in the coupled enzyme system.

It is also should be assumed that bioluminescence activation and inhibition processes occur simultaneously and appear to be competitive. Thus, influence of inorganic sulphur containing compounds (Na_2S, Na_2SO_3, Na_2SO_4, $Na_2S_2O_4$, $Na_2S_2O_5$ and $Na_2S_2O_3$) on the coupled enzyme system leads to activation as well as inhibition of bioluminescence. Impact is defined by concentration and redox potential of the sulfur compound on the bioluminescent system. On the basis of established rules it is shown that increasing the reduction cabilities of sulfur compounds is accompanied by an increase in the bioluminescence inhibition coefficients (K_i). The dependence of bioluminescence kinetic parameters on redox characteristics of sulfur compounds is an essential feature of the NAD(P)H:FMN-oxidoreductase – luciferase coupled enzyme system and this makes the system a good candidate for bioassays.

ACKNOWLEDGEMENTS

The work was financially supported by Award № REC-002 and Y1-B-02-12 of the U.S. Civilian Research & Development Foundation (CRDF) and RF Ministry of Education", BRHE Programs; Russian Foundation for Basic Research, "Molecular and Cellular Biology" program of the Russian Academy of Science; grant of President of Russian Federation N MK-1950.2005.4.

REFERENCES

1. Vetrova EV, Kratasyuk VA, Kudryasheva NS. Bioluminescent characteristics map of Shira lake water. Aquat Ecol 2002;36:309-15.
2. Kudryasheva N, Vetrova E, Kuznetsov A, Kratasyuk V, Stom D. Bioluminescence assays: effects of quinones and phenols. Ecotoxicol Environ Saf 2002;53:221-5.
3. Kudryasheva NS, Zuzikova EV, Gutnyk TV. Mechanism of influence of metal salts on bacterial bioluminescence system *in vitro*. Biophys 1999;44:244-50.

DEVELOPMENT OF THE *CYPRIDINA* BIOLUMINESCENT SYSTEM FOR HIGH THROUGHPUT SCREENING ASSAYS

CHUN WU,[1] KOSEI KAWASAKI,[2] SATROU OHGIYA,[2]
YOSHIHIRO OHMIYA[1]

[1]Research Institute for Cell Engineering, National Institute of Advanced Industrial Science and Technology, (AIST) Osaka, 563-8577, Japan; [2]Research Institute for Genome-based Biofactory, National Institute of Advanced Industrial Science and Technology, Sapporo, 062-8517, Japan, E-mail: chun-wu@aist.go.jp

INTRODUCTION

The ostracod bioluminescent system is a typical luciferin luciferase reaction. The gene of this luciferase has been cloned and successfully expressed in various mammalian cells as a bioluminescent reporter.[1,2] The first total synthesis of Cypridina *(Vargula)* luciferin was achieved in 1966, but the yield was very low.[3] Recent synthetic studies including our report on the synthesis of (*S*)-*Cypridina* luciferin have provided a method to obtain this luciferin in large quantities.[4,5] Here we describe a method to improve the stability of Cypridina luciferin in buffer and to introduce a blue shift in this bioluminescent system. Our experimental results may contribute to a novel high throughput promoter assay, BRET assay and immunoassay by using the *Cypridina* bioluminescent system.

MATERIALS AND METHODS

(*S*)-Cypridina luciferin (1, Fig. 1) was prepared according to our previous report. Recombinant Cypridina luciferase (CLuc) was expressed in yeast and purified by sequential chromatographies. Anti-oxidants were purchased from Sigma. Cypridina luciferin analogues were prepared via the route shown in scheme 1.

Analogue 5a: ^{1}H NMR (500 MHz, CD_3OD) 0.92 (3H, t, *J* =7 Hz), 1.36 (3H, t, *J* =7 Hz), 1.73-1.89 (2H, m), 2.25 (2H, quintet, *J* =7 Hz), 3.09 (1H, m), 3.26 (2H, t, *J* =7 Hz), 3.36 (2H, t, *J* =7 Hz), 7.55 (4H, m), 7.87 (1H, s), 7.88 (1H, s).

Analogue 5b: ^{1}H NMR (500 MHz, CD_3OD) 0.96 (3H, t, *J* =7 Hz), 1.47 (3H, t, *J* =7 Hz), 1.86 (2H, m), 2.35 (2H, quintet, *J* =7 Hz), 3.19 (1H, m), 3.43 (2H, t, *J* =7 Hz), 3.48 (2H, t, *J* =7 Hz), 7.55 (2H, d, *J* =7 Hz), 7.90 (1H, d, *J* =7 Hz), 8.01 (2H, d, *J* =7 Hz), 8.14 (1H, d, *J* =7 Hz), 8.60 (1H, s), 8.78 (1H, s).

Analogue 7a: ^{1}H NMR (500 MHz, CD_3OD) 2.24 (2H, quintet, *J*=7 Hz), 3.25 (2H, t, *J*=7 Hz), 3.39 (2H, t, *J*=7 Hz), 4.27 (2H, s), 7.10 (2H, dd, *J*=7 Hz), 7.24 (1H, t, *J*=7 Hz), 7.30–7.45 (5H, m), 7.93(1H, dd, *J*=7 Hz), 7.96 (1H, s), 8.27 (1H, s).

Analogue 7b: ^{1}H NMR (500 MHz, CD_3OD) 1.84 (3H, d, *J* =7 Hz), 2.29 (2H, quintet, *J*=7 Hz), 3.38 (2H, t, *J*=7 Hz), 3.42 (2H, t, *J*=7 Hz), 4.60 (1H, q, *J* =7 Hz), 7.16 (2H, dd, *J*=7 Hz), 7.24 (1H, t, *J*=7 Hz), 7.32 (2H, t, *J*=7 Hz), 7.40 (2H, d, *J* =7 Hz), 7.43(1H, dd, *J*=7 Hz) 7.97(1II, dd, *J*=7 IIz), 7.98 (1H, s), 8.27 (1H, s).

RESULTS AND DISCUSSION

Cypridina luciferin solution with antioxidants. Since the optimal pH for Cypridina luciferase activity is approximately 7.4, Cypridina luciferin was first dissolved in 0.1 M Tris-HCl buffer (pH 7.4). However, the half-life of luciferin in Tris-HCl buffer was observed to be approximately 2 h. Meanwhile a weak but apparent luminescence of the Cypridina luciferin in Tris-HCl buffer was observed. The short half-life of Cypridina luciferin was thought to be due to deprotonation of the imidazopyrazinone ring resulting in an auto-oxidation of luciferin.[6] Therefore, we decided to investigate the effect of adding anti-oxidants into Tris-HCl buffer on the stability of Cypridina luciferin. We found that the half-life of Cypridina luciferin was extended to 60 h in the presence of sodium ascorbate and sodium sulfite (Fig. 1). Interestingly, the luminescence of Cypridina luciferin in Tris-HCl buffer (pH 7.4) was reduced by approximately 10-fold by adding anti-oxidants. This stable luciferin solution enables us to obtain a steady and sensitive signal produced by the bioluminescence for batch processing reporter assay and immunoassay.

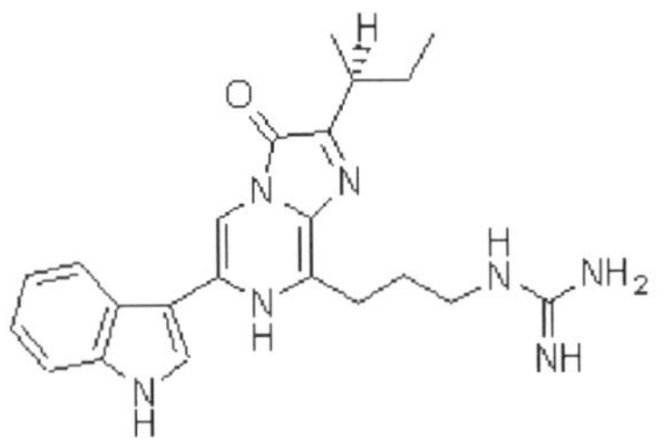

1

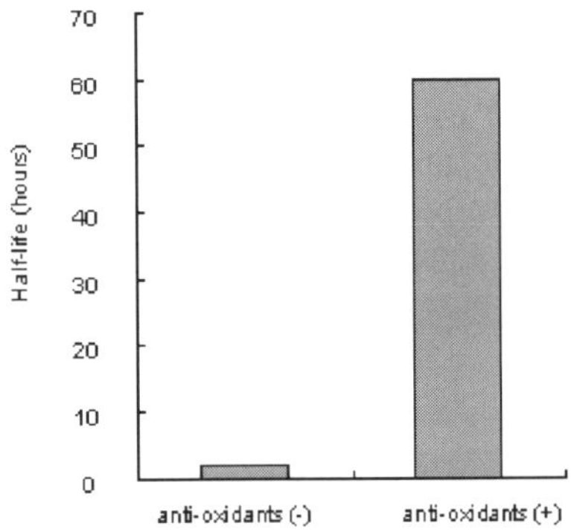

Fig. 1. Half-life of 1 in Tris-HCl buffer in the presence or absence of anti-oxidants

Cypridina luciferin analogues. Syntheses of 5a and 5b were started with the known compound 2 (Scheme 1). The cross-coupling reaction of 2 with boronic reagents gave 3a and 3b in good yields. Deprotection of 3a and 3b with TFA was followed by a guanylation to give 4a and 4b. Condensation of 4a and 4b with ketoacetal afford compound 5a and 5b in moderate yields. Analogue 7a and 7b were prepared by condensation of the known compound 6 with ketoacetal compounds prepared by our previously described method.[5]

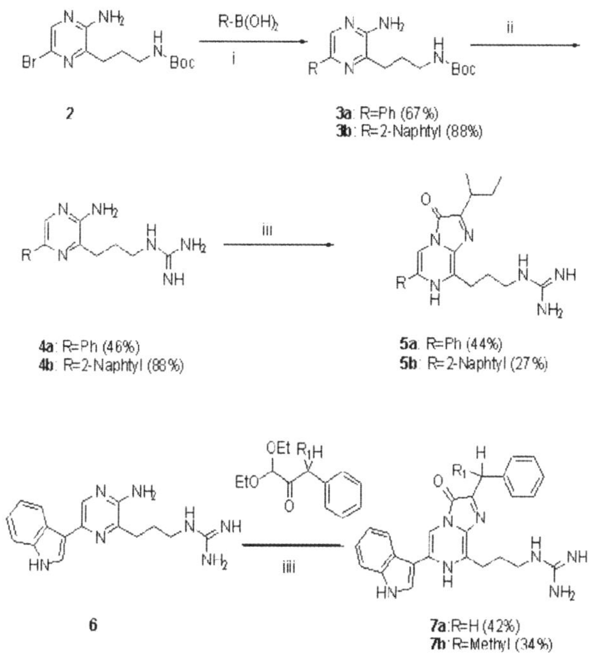

Scheme 1. Reagents and conditions: i)$Pd_3(PPh_3)_4$, Na_2CO_3, reflux, 1h; ii)TFA, 0 °C, 1h, then Pyrazole-1-carboxamidine, DIEA, r.t., o.n.; iii) 1,1-diethyoxy-3-methyl-2-pentanone, HBr, reflux, 1h; iiii) HBr, reflux, 1h

Table 1. Luminescent properties of *S*-Cypridina luciferin (1) and its analogs. Bioluminescence was performed at 1×10^{-6} M of luciferins in 50 mM Tris-HCl pH 7.4 with recombinant Cypridina luciferase (50 pg)

	1	**5a**	**5b**	**7a**	**7b**
Relative light yield	100	<1	2	25	10
λmax/nm	460	440 and 380	395	450	470

Compound 5a showed a bimodal spectrum with a large decrease in the light yield. Analogs 5b showed 2% of the light yield of (*S*)-1 with a large blue shift with an emission peak at 395 nm (Table 1). Analogue 7a showed a small blue shift with an emission peak at 450, whereas 7b showed a red shift with an emission peak at 470 nm. It is known that the analogue having a 3-benzofuranyl or a 3-benzothienyl group in the place of a 3-indolyl group has exhibited a slight blue shift of emission peak. The reason for the lager blue shift of the analogues with naphthyl described here is unknown. However this blue shift analogue may be used in bioluminescence resonance energy transfer study (BRET) as an energy donor, because its bioluminescence may allow a good separation from fluorescence generated by fluorescent protein.

REFERENCES

1. Thompson EM, Nagata S, Tsuji FI. Cloning and expression of cDNA for the luciferase from the marine ostracod *Vargula hilgendorifii*. Proc Natl Acad Sci USA 1989;86:6567-71.
2. Nakajima Y, Kobayashi K, Yamagishi K, Enomoto T, Ohmiya Y. cDNA cloning and characterization of a secreted luciferase from the luminous Japanese ostracod *Cypridina noctiluca*. Biosci Biotechnol Biochem 2004;68:565-70.
3. Kishi Y, Goto T, Inoue S, Sugiura S, Kishimoto H. Cypridina bioluminescence III: total synthesis of Cypridina luciferin. Tetrahedron Lett 1966;7:3445–50
4. Nakamura H, Aizawa M, Takeuchi D, Murai A, Shimomura O. Convergent and short-step syntheses of dl-Cypridina luciferin and its analogues based on Pd-mediated cross couplings. Tetrahedron Lett 2000;41:2185–8.
5. Wu C, Kawasaki K, Ohgiya S, Ohmiya Y. Syntheses and evaluation of the bioluminescent activity of (S)-Cypridina luciferin and its analogs. Tetrahedron Lett 2006;47:753-6.
6. Shimomura O, Johnson FH. Mechanism of the luminescent oxidation of Cypridina luciferin. Biochem Biophys Res Commun 1971;16:340-6.

QUANTITATIVE WESTERN BLOTTING USING A GZ-11 BASED CHEMILUMINOGENIC SIGNAL REAGENT

G ZOMER,[1] C SMITSMAN,[2] R ARTS,[2] M HAMZINK,[1] M KOOIJMAN[1]
[1] Unit Research and Development; [2] Dept of Operational Quality, Netherlands Vaccine Institute, P.O. Box 457 3720 AL Bilthoven, The Netherlands
Email: Bert.Zomer@nvi-vaccin.nl

INTRODUCTION

Acridan esters as part of a horseradish peroxidase detection reagent were introduced in 1994 by Akhavan-Tafti *et al.*[1] In 1998 we proposed other types of acridan esters (e.g., GZ-11) useful for ELISA purposes.[2] In this communication we present data on Western blotting using a GZ-11 derived signal reagent. A tentative mechanism for the light emission producing steps for the horseradish peroxidase signal reagent based on acridan ester GZ-11 is shown in Fig. 1.

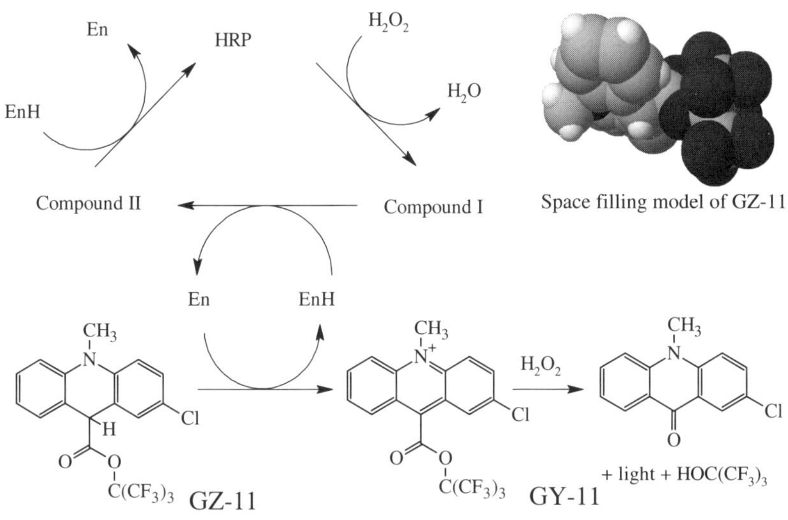

Fig. 1. Light generating pathway using GZ-11 signal reagent

Acridan ester GZ-11 is converted to acridinium ester (GY-11) through enhancer radicals which are formed by the HRP catalyzed oxidation reaction of enhancer by hydrogen peroxide. From the work of McCapra it is known that the pKa of the leaving group of acridinium ester is directly proportional to the efficiency of light emission.[3] Perfluoro-*t*-butanol is a rather acidic compound with a pKa of about 5.4 yielding a very efficient emitter. Furthermore, due to its bulkiness the perfluoro-*t*-butyl group protects the ester carbonyl of GZ-11 from hydrolysis (Fig. 1).

In Western European countries Meningococcal disease is mainly caused by serogroup B. At our institute a vaccine for this disease is under development.[4] With recombinant DNA-technology three trivalent MenB strains are constructed, each strain contains three Class 1 Outer Membrane Proteins (PorA's) of different subtypes. With these strains, trivalent outer membrane vesicle (OMV) bulk products are produced and combined in one nonavalent OMV vaccine.

MATERIALS AND METHODS

Separation of proteins on SDS-Page was done using PAGEr Duramide precast gels 12% Tris-Glycine Buffer (Cambrex),in the presence of molecular weight marker (Magic Marker 20, 30, 40, 50, 60, 80, 100, 120, and 220kD, Invitrogen). Blotting on PVDF and nitrocellulose was performed wet and semi-dry using standard procedures. Immunological detection was performed using MoAb's and HRP-labeled second Ab. After washing the blot, the bound antigens were detected using a GZ-11 containing signal reagent (GZ-11 0.025 mM, 4-phenylphenol 0.06 mM, H_2O_2 3.5 mM in PBST 10 mM, pH 7.4, 0.05% Tween20, 0.05 mM EDTA). The glowing blot was exposed to a CCD camera (Fluorchem 8900) for 12 minutes at medium sensitivity. The signals on the blots were quantified using Phoretix 1D Quantifier.

RESULTS AND DISCUSSION

In order to assess the stability of the porA protein in the vaccine a quantitative Western blotting method was developed using GZ-11 as part of the HRP signal reagent. Proteins were separated on gels and blotted on PVDF or nitrocellulose. The blots were incubated with different monoclonal antibodies specific for one PorA subtype, followed by incubation with detecting antibody (HRP-labeled goat-anti-mouse). The blots were visualized after addition of GZ-11 containing signal reagent using a CCD-camer0061 (Fig. 2), and quantified (Fig. 3).

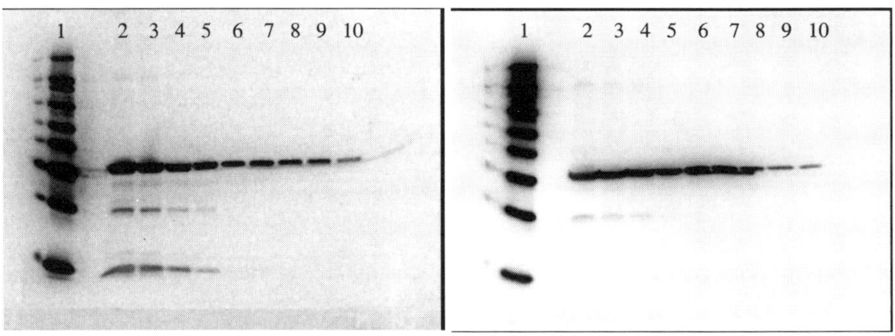

Fig. 2. Lane 1 mw markers (20, 30, 40, 50, 60, 80, 100, 120, 220kD), lanes 2-10 are dilutions (2-0.015 μg total protein of a MenB vesicle product) blotted on PVDF (left panel) or nitrocellulose (right panel)

The major band close to 40 kD is due to intact PorA. Especially on PVDF a number of bands derived from PorA fragments (at about 35, 28, and 19kD, respectively) are visible. Possibly due to other blotting characteristics of nitrocellulose these low molecular weight bands are less well visible on NC than on PVDF. Apparently, blots on nitrocellulose favor high molecular weight (this is also clear from the molecular weight markers). Fig. 3 shows that the signal for PorA generated by GZ11 is linear in the range in which the main protein band was added. Of the incomplete proteins not enough data are present to show linearity, although it is obvious, that the points are not part of a line with the same slope as that of the intact protein. This might be due to the difference in affinity of the monoclonals for the different proteins. Therefore, at this stage it is only possible to do a relative quantification of each individual PorA subtype. This is adequate to show if a product is unchanged in time. The signal of the proteins in the marker was used to calculate the precision of the test. The variation in the percentage of each relevant protein is <20%. The quantification limit is the volume of the band corresponding with the most diluted sample of the intact PorA.

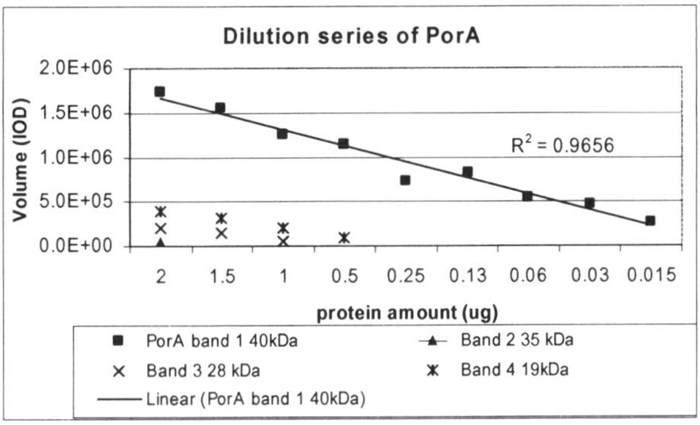

Fig. 3. Graph showing linearity of response for PorA band, together with data points for incomplete PorA derived bands

Diphtheria is an acute infectious disease caused by *Corynebacterium diphtheriae,* and characterized by the production of a systemic toxin and the formation of a false membrane on the lining of the mucous membrane of the throat and other respiratory passages, causing difficulty in breathing, high fever, and weakness. The toxin is particularly harmful to the tissues of the heart and central nervous system. A vaccine is produced based on the toxoid, an inactivated version of the toxin. In order to assess the quality of the vaccine it is important to know what part of the protein is recognized by the different monoclonal antibodies (epitope mapping). This was studied by point mutating the epitope, and measuring the binding of monoclonal antibodies with the mutant protein. Mutants were expressed in *E. coli* and isolated from the periplasm.

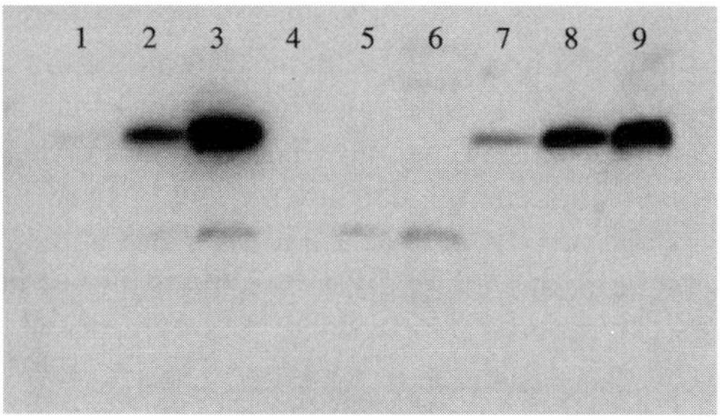

Fig. 4. Lanes 1-3 Diphtheria Toxin wild type, lanes 4-6 *Diphtheria* Toxin mutant 09, lanes 7-9 Diphtheria Toxin mutant 21. Lanes 1, 4, and 7 contain 10 ng, lanes 2, 5, and 8 contain 100 ng, and lanes 3, 6, and 9 contain 500 ng of protein, respectively

Samples of the different mutants and the native protein were separated on SDS-page and blotted on nitrocellulose membranes. Blots were incubated with the specific monoclonal antibody followed by incubation with detecting antibody, and visualization using the GZ-11 signal reagent (Fig.4). Fig. 4 shows that mutant DTA33 (lanes 4-6) in which one lysine is replaced by alanine is not recognized by the monoclonal antibody DIM09. The mutant DTA21 (lanes 7-9) in which another lysine is replaced by alanine is recognized by DIM33. Using a different MoAB (DIM09) DTA33 is recognized while DTA09 is not (results not shown).

REFERENCES

1. Akhavan-Tafti H, DeSilva R, Arghavani Z, Eickholt RA, Handley RS, Schaap AP. Lumigen PS: New chemiluminescent substrates for the detection of horseradish peroxidase. In: Campbell AK, Kricka LJ, Stanley PE. eds. Bioluminescence and Chemiluminescence: Fundamentals and Applied Aspects. Chichester: Wiley 1994:199-202.
2. Zomer G, Jacquemijns M. On site screening: Application of novel acridan esters as HRP detection reagent in immunoassay for the detection of beta-agonists in bovine urine. In: Roda A, Pazzagli M, Kricka LJ, Stanley PE. eds. Bioluminescence and Chemiluminescence: Perspectives for the 21st Century. Chichester: Wiley 1999:87-90.
3. McCapra F. Chemical mechanisms in bioluminescence. Acc Chem Res 1976;9:201-8.
4. Vermont CL, van den Dobbelsteen GPJM. Meningococcal serogroup B infections. A search for a broadly protective vaccine. Expert Rev Vaccines 2003;2:673-81.

PART 4

OPTICAL IMAGING AND MULTICOLOR ASSAYS

DEMONSTRATION OF TRI-COLORED REPORTER ASSAY USING A FILTER LUMINOMETER "SIRIUS C"

TOMOMI ASAI,[1] ANSELM BERTHOLD,[2] YOSHIHIRO OHMIYA[3,4]

[1] Toyobo Co., Ltd., Tsuruga, Fukui, 914-0047 Japan; [2] Berthold Detection Systems GmbH, Pforzheim 75173 Germany; [3] Research Institute for Cell Engineering, National Institute of Advanced Industrial Science and Technology, (AIST) Osaka, 563-8577, Japan; [4] Graduate School of Medicine, Hokkaido University, Sapporo 011-8638, Japan, Email: Tomomi_asai@bio.toyobo.co.jp

INTRODUCTION

A reporter assay system to monitor the expression of several genes simultaneously is required to investigate a cascade of transcriptional regulations, which is the molecular basis of various biological functions. Recently, we have developed a novel reporter assay system, a tricolor reporter *in vitro* assay system, in which the expression of three genes can be monitored simultaneously by splitting the emissions from green-, orange-, and red-emitting luciferases with an optical filter.[1] In this study, we demonstrated this novel system to analyze a transcription-translation feedback loop of c-fos promoter using a luminometer "Sirius C" fitted with optical filters.

MATERIALS AND METHOD

Plasmids. pSLR-cfos(1) construct was generated by cloning the -446 to +38bp 5'-flanking region of c-Fos upstream of the reporter red-emitting luciferase gene (SLR). pSLO-cfos(4) construct was generated by inserting c-fos promoter deletion region (-235 to +38bp) upstream of the reporter orange-emitting luciferase gene (SLO).[2] pHSVtk-SLG construct was generated by inserting tk promoter of the reporter green-emitting luciferase gene (SLG).

Reporter assays. CHO cells were maintained in Ham's F12 medium supplemented with 10% fetal calf serum (FCS). The cells were seeded into a 12-well plate 24 h before transfection. 0.3 μg of pHSVtk-SLG, 0.3 μg of pSLR-cfos(1), and 0.4 μg of pSLO-cfos(4) were transfected using GeneJuice Transfection Reagent (Novagen) in Ham's F12 medium supplemented with 0.1% FCS. After 24 h at 37°C, medium was replaced and cells were treated with 10% FCS or 0.1 μM TPA for another 5 h. Luciferase assays were conducted using the Multi Reporter Assay System – Tripluc® –Detection Reagents (TOYOBO) using by 'Sirius C' (Berthold Detection Systems GmbH).

RESULTS AND DISCUSSION

Concept of tri-colored reporter assay. The beetle luciferases, which have pH-insensitive spectra, have been developed for expression in a mammalian cell. Table 1 summarizes the basic character of green-, orange-, and red-emitting luciferases. In a tricolor reporter *in vitro* assay system, the expression of three genes can be monitored simultaneously by splitting the emissions from green-, orange-, and red-emitting luciferases with optical filters. This system allows us to simply and rapidly monitor the expression of three genes simultaneously (two are test reporters and one is an internal control) by using a single luminescent substrate in one tube (Fig. 1).

Table 1. Basic character of green, orange, and red luciferases

Luciferase	Symbol	λmax	Origin	Amino acid sequence
Green-emitting luciferase	SLG	550 nm	*Rhagophthalmus ohbai*	Wild type
Orange-emitting luciferase	SLO	580 nm	*Rhagophthalmus ohbai*	Mutant T226N
Red-emitting luciferase	SLR	630 nm	*Phrixothrix hirtus*	Wild type

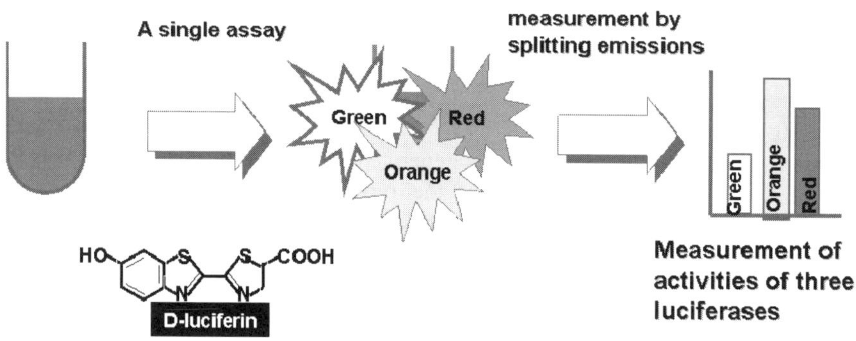

Fig. 1. Concept of tri-colored reporter assay in a single tube

Quantitative relationships of tri-color luciferase using a filter luminometer. The activity of tri-color luciferase mixtures was measured for 2s in the absence (F0) or presence of 575 nm-long path or 610 nm-long path filter (F575, F610) after 20 s incubation.

$$\begin{pmatrix} F0 \\ F1 \\ F2 \end{pmatrix} = \begin{pmatrix} 1 & 1 & 1 \\ \kappa GO575 & \kappa OO575 & \kappa RO575 \\ \kappa GR615 & \kappa OR615 & \kappa RR615 \end{pmatrix} \begin{pmatrix} G \\ O \\ R \end{pmatrix}$$

Each activity was calculated using the simultaneous equation where G, O, and O are the green, orange, and red luciferase activities, respectively; F0 is the total relative light units (RLU) measured in the absence of the optical filter; F1 and F2 are the RLU that pass through the O575 and R615 filters, respectively; κG_{O575}, κO_{O575}, and κR_{O575} are the transmission coefficients of the green, orange and red luciferase of the O575 filter, respectively; κG_{R615}, κO_{R615}, and κR_{R615} are the transmission coefficients of the green, orange and red luciferase of the R615 filter, respectively.

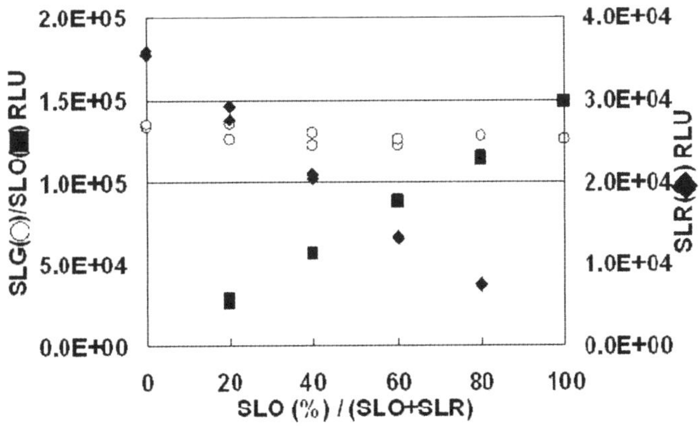

Fig. 2. Quantitative relationships among green- (SLG), orange- (SLO) and red-emitting (SLR) luciferase activities in a mixture

The activities in the mixtures with various volume ratios were measured to examine whether the respective activities of the three luciferases in a mixture can be measured simultaneously. The measured activities of the orange and red luciferases were found in increase in proportion to the amount added and showed good linearity, whereas the activity of the green luciferase was constant. The results indicate that tri-color reporter assay using by a filter luminometer "Sirius C" can accurately measure three luciferase activities simultaneously.

Monitoring of a transcription-translation feedback loop of c-fos promoter. We analyzed in detail the role of the serum response element (SRE) and TPA response element (TRE) in the trans-activation of c-fos promoter. Reporter plasmid pSLR-cfos(1), pSLO-cfos(4), and pHSVtk-SLG (an internal control) were co-transfected into CHO cells, in which the pSLR-cfos(1) contains the both response elements but the pSLO-cfos(4) does not contain them. After 24 h at 37°C, the medium was replaced and cells were treated with 10% FCS or 0.1 μM TPA for another 5 h. Both induced-factors enhanced the activity of cfos(1) promoter, respectively, whereas the activities of cfos (4) promoter was not changed, indicating in tri-color reporter assay, different promoter regions can be compared directly and easily.

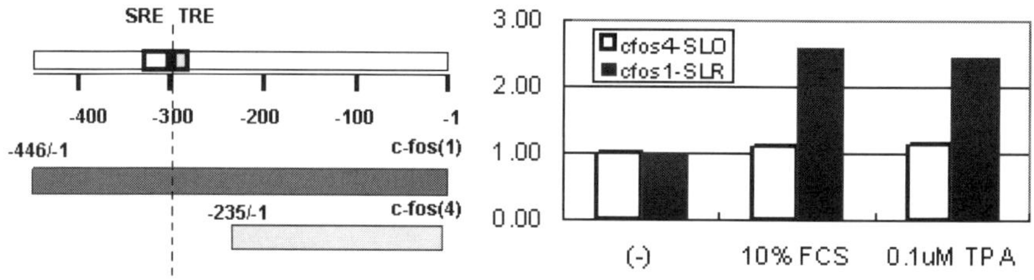

Fig. 3. Simultaneous monitoring of SRE- and TRE-dependent induction of cfos(1) and cfos(4) promoter transcription activities

Splitting emissions using long-pass filters (cut-off filters) is advantageous because emission loss is less than that when interference filters are used, thereby improving the signal-to-noise ratio. As a result of our knowledge of the chemical reaction and the structure of luciferin and luciferase, the firefly bioluminescence system is now used widely as a reporter for gene expression and as a cellular marker. The different color luciferases are able to provide a large amount of information about processes in living cells, and will result in new applications.

REFERENCES

1. Nakajima Y, Kimura T, Sugata K, Enomoto T, Asakawa A, Kubota H, et al. Multicolor luciferase assay system: one-step monitoring of multiple gene expressions with a single substrate. Biotechniques 2005;38:891-4.
2. Chaudhary J, Skinner MK. Basic Helix-Loop-Helix proteins can act at the E-Box within the serum response element of the c-fos promoter to influence hormone-induced promoter activation in sertoli cells. Mol Endocrinol 1999; 13:774-86.

REAL-TIME MONITORING OF VACCINIA VIRUS INFECTION IN CULTURED CELLS AND IN LIVING MICE USING LIGHT-EMITTING PROTEINS

NANHAI CHEN,[1] YONG A. YU,[1] QIAN ZHANG,[1]
VIKTORIA RAAB,[2] ALADAR A. SZALAY [1,2]

[1] *Genelux Corporation, San Diego Science Center, San Diego, CA 92109, USA;*
[2] *Virchow Center for Experimental Biomedicine, Institute for Biochemistry and Institute for Molecular Infection Biology, University of Würzburg, Am Hubland, D97074 Würzburg, Germany*

INTRODUCTION

Vaccinia virus (VV) is the prototypic member of the genus *Orthopoxvirus* of the family *Poxviridae,* which was used worldwide as a live vaccine against smallpox virus that was declared eradicated by the World Health Organization in 1980. Since the early 1980s when the first genetically engineered VVs were created, VV has been widely used for high-level expression of foreign genes in mammalian cells and as a vector for vaccines against other infectious diseases and cancer.[1] Recently, VV has also been explored in tumor-directed oncolytic virotherapy and imaging of tumors and metastases.[2,3] As an oncolytic agent, the capability of a virus to replicate *in vitro* in cultured tumor cells is a must. To facilitate evaluation of virus replication *in vitro* and monitoring virus therapeutic effects and spread *in vivo* we have created two recombinant vaccinia viruses that express click beetle luciferase-mRFP1 (*CBG99-mRFP1*) and *Renilla* luciferase-GFP (*Ruc-GFP*) fusion genes, respectively. We have shown that vaccinia virus replication in cultured cells and in living mice can be monitored by both fluorescence and bioluminescence imaging.

RESULTS AND DISCUSSION

Generation of two recombinant VVs expressing luciferase and fluorescent protein fusion genes. GLV-1h84 and GLV-1h86 express *CBG99-mRFP1* and *Ruc-GFP* fusion genes, respectively (Fig. 1). *CBG99* and *mRFP1* are connected through a picornavirus 2A element. During translation these two proteins will be cleaved into two individual proteins at picornavirus 2A element, whereas *Ruc-GFP* will be expressed as a fusion protein.

Real-time monitoring of infection of CV-1 and GI-101A cells with GLV-1h84 and GLV-1h86, respectively. Infection of CV-1 (green monkey kidney cells) and GI-101A (human breast adenocarcinoma) cells with GLV-1h84 at an m.o.i. of 0.01 was monitored in real-time using fluorescence microscopy. In both cell types, individual red plaques were seen 24 hours post infection (hpi). By 48 hpi, more than 80% of cells were infected; at 72 hpi, almost all cells were infected (Fig. 2). GLV-1h86 showed similar spreading patterns in both CV1 and GI-101A cells compared to GLV-1h84 (Fig. 3). Infection of both cell types with either GLV-1h84 or GLV-1h86 was also imaged by measuring luciferase activities (data not shown).

Real-time monitoring of co-infection of CV-1 and GI-101A cells with GLV-1h84 and GLV-1h86. When CV-1 cells were co-infected with GLV-1h84 and GLV-1h86 at an m.o.i of 0.01 for each virus, individual red or green plaques were seen at 24 hpi under a fluorescence microscope, confirming a general belief that a virus plaque is derived from a single virus particle. At 48 hpi, most cells were infected. Interestingly, some of infected cells were yellow, indicating that cells infected with one virus can also be infected later with another virus. By 72 hpi, most cells were co-infected. In contrast, when GI-101A cells were co-infected with both viruses, there were not many yellow cells even by the time of 72 hpi (Fig. 4), suggesting that the virus can't spread very well from one infected cell to another in GI-101A cells.

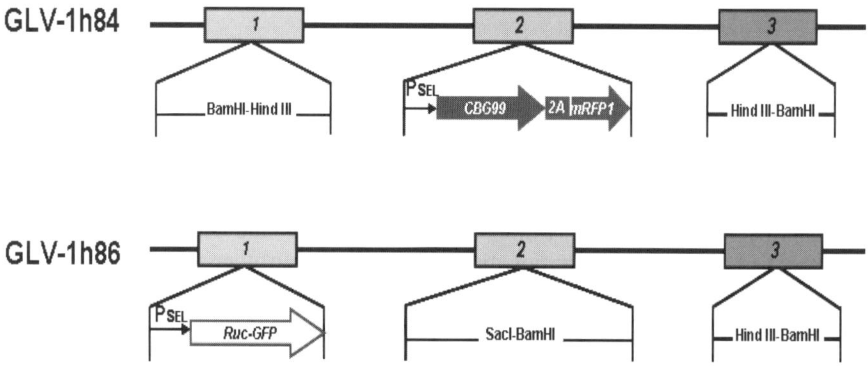

Fig. 1. Schematic representation of genotypes of GLV-1h84 and GLV-1h86. PSEL, vaccinia virus synthetic early/late promoter

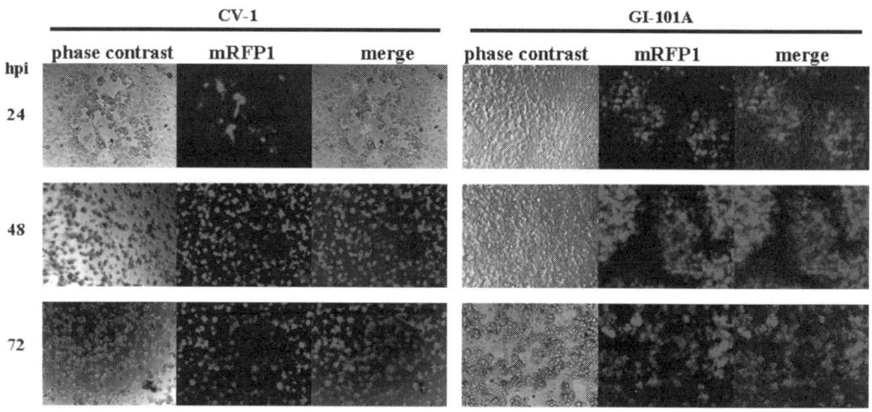

Fig. 2. Spread of GLV-1h84 in CV-1 and GI-101A cells monitored by fluorescence imaging

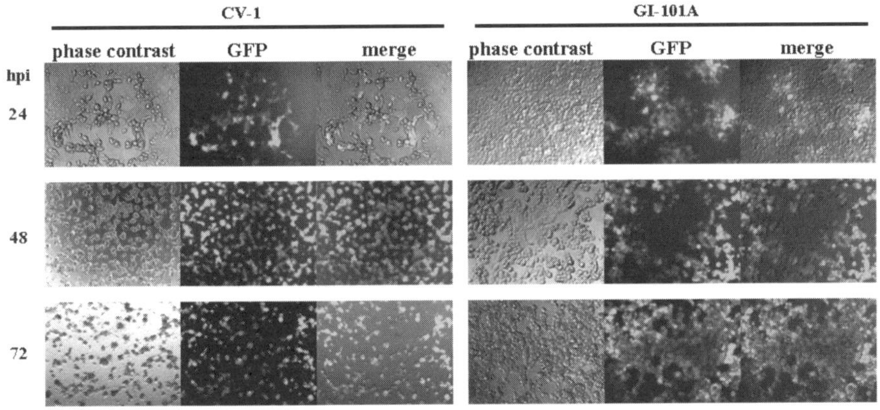

Fig. 3. Spread of GLV-1h86 in CV-1 and GI-101A cells monitored by fluorescence imaging

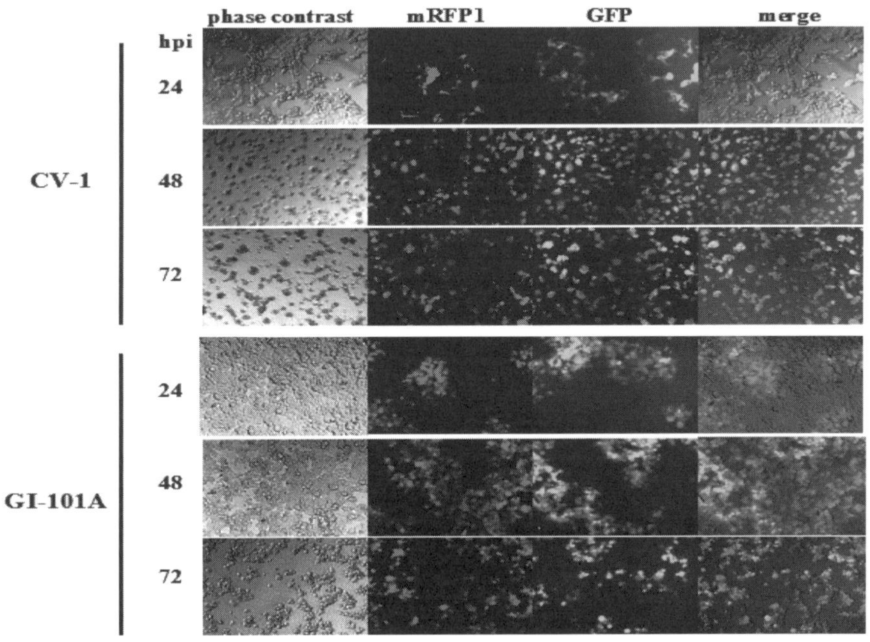

Fig. 4. Co-infection of CV-1 and GI-101A cells with GLV-1h84 and GLV-1h86 monitored by fluorescence imaging

***In vivo* imaging**. C6(pLEIN) tumor-bearing nude mice were injected i.v. with 5 x 10^6 pfu of GLV-1h84 and imaged at 7 days post infection. Tumor and poxes on the tail lit up by both mRFP1 and CBG99 imaging (Fig. 5).

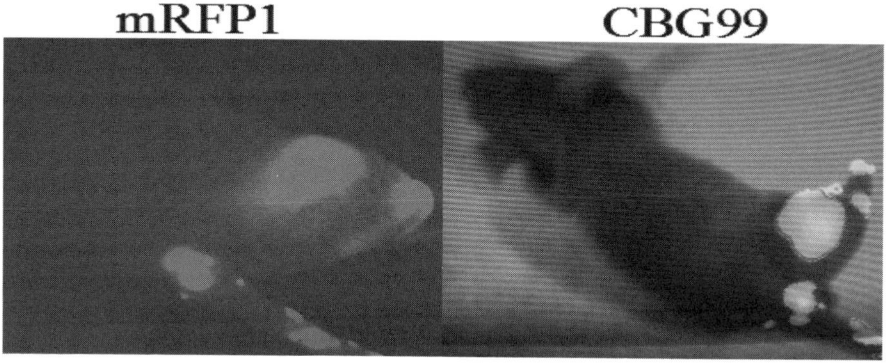

Fig. 5. Fluorescence (left) and bioluminescence (right) imaging of C6(pLEIN) tumor-bearing mice

CONCLUSION

We have shown that *in vitro* and *in vivo* infection with our two recombinant VV, GLV-1h84 and GLV-1h86, separately or together can be monitored by both fluorescence and bioluminescence imaging. GLV-1h84 and GLV-1h86 showed similar spreading patterns in both CV-1 and GI-101A cells. The separate red and green plaques were seen in CV-1 and GI-101A cells 24 hpi by fluorescence imaging when these cells were co-infected with GLV-1h84 and GLV-1h86 at low m.o.i., confirming a general belief that individual virus plaques are derived from single virus particles. The fact that some yellow CV-1 and GI-101A cells were seen 48 hours post co-infection by fluorescence imaging indicated that cells infected with one virus can also be infected later with another virus. The virus seemed to spread much better from one infected cell to another in CV-1 cells than in GI-101A cells. GLV-1h84 and GLV-1h86 are great tools for the preclinical evaluation of VV as an oncolytic agent.

REFERENCES

1. Moss B. Genetically engineered poxviruses for recombinant gene expression, vaccination, and safety. Proc Natl Acad Sci USA 1996;93:11341-8.
2. McCart JA, Ward JM, Lee J, Hu Y, Alexander HR, Libutti SK, et al. Systemic cancer therapy with a tumor-selective vaccinia virus mutant lacking thymidine kinase and vaccinia growth factor genes. Cancer Res 2001;6:8751-7.
3. Yu YA, Shabahang S, Timiryasova TM, Zhang Q, Beltz R, Gentschev I, et al. Visualization of tumors and metastases in live animals with bacteria and vaccinia virus encoding light-emitting proteins. Nat Biotechnol 2004;22:313-20.

USE OF BIOLUMINESCENT REPORTER BACTERIA TO STUDY INVASION AND SURVIVAL WITHIN MAMMALIAN CELL LINES

KL CUTTER, SM NELSON, J GREENMAN, GM ROBINSON, VC SALISBURY
Faculty of Applied Sciences, University of the West of England, Coldharbour Lane, Bristol, BS16 1QY, UK

INTRODUCTION

Neisseria meningitidis is the major causative agent of bacterial meningitis, with over 500,000 cases occurring globally each year. The lack of any animal models for this human specific disease has hampered our knowledge of exactly how this pathogen enters endothelial cells and breaches the blood-brain barrier to cause fulminating meningitis.

Bacterial bioluminescence has been shown to be an accurate real-time reporter of metabolic activity of bacteria. A strain of *N. meningitidis* C751 has been genetically modified with *luxCDABE* from *Photorhabdus luminescens* on the plasmid pGLITE to give a self-bioluminescent reporter construct. This self-bioluminescent strain of *Neisseria meningitidis* was used in the development of an assay to study invasion and survival within the endothelial cell line ECV-304. In addition to this, systems where internalisation is known to occur, such as that of *Escherichia coli* DH5a pLITE27 within the monocytic cell line THP-1 were used as positive controls.

Once optimum conditions for this assay have been identified, this method could be utilised to study real-time *in situ* bacterial passage across the blood-brain barrier, with the potential to influence antibiotic studies and also reduce animal testing in this area of research.

METHODS

Internalisation of *E. coli DH5a* pLITE27 within THP-1 cells.

Preparation of inoculum. *E. coli DH5a* pLITE27 was inoculated in 10 mL RPMI-1640 supplemented with 50 mg/L ampicillin and grown on a shaking incubator at 37°C overnight. The bacteria were washed twice in pre-warmed phosphate buffered saline (PBS) and resuspended in pre-warmed RPMI-1640.

Maintenance and propagation of cell lines. Human acute monocytic leukemia cell line THP-1 was cultured in RPMI-1640 (10% foetal bovine serum, 2 mM glutamine) at 37°C in an atmosphere enriched with 5% CO_2. THP-1 cells were subcultured 1:4 into fresh RPMI-1640 every 3-4 d.

Experimental setup. A flask of confluent THP-1 cells was centrifuged at 1033 x g for 5 min, washed once in pre-warmed PBS and resuspended in 10 mL fresh pre-warmed RPMI-1640. The cells were then incubated at 37°C, 5% CO_2 for 40 min. Bacteria were added to cells at a multiplicity of infection of 400:1 and incubated at 37°C, 5% CO_2 for 2 h. Cells and bacteria were centrifuged at 1033 x g for 5 min, washed twice with pre-warmed PBS to remove the majority of non-internalised bacteria and resuspended in fresh RPMI-1640. Samples were added to a 96-well microtitre plate and 150 mg/L colistin was added to inactivate all remaining non-internalised bacteria. Bioluminescence was measured in a luminometer (Fluostar Optima) every 15 min for 2 h at 37°C, 5% CO_2. 0.1%w/v saponin was added to permeabilize cells allowing inactivation of internalised bacteria by colistin and bioluminescence was measured as before for a further 22 h.

Internalisation of *N. meningitidis* C751 pGLITE within ECV-304 human endothelial cells.

Preparation of inoculum. *N. meningitidis* C751 pGLITE was streaked onto BHI agar supplemented with 10% heat-inactivated foetal bovine serum (FBS) and 10 mg/L kanamycin and grown at 37°C, 5% CO_2 overnight. The bacteria were washed twice in pre-warmed PBS and resuspended in pre-warmed Medium 199 (M199).

Maintenance and propagation of cell lines. Human endothelial cell line ECV-304 was cultured in M199 (25mM HEPES buffer, 10% foetal bovine serum, 2mM glutamine) at 37°C in an atmosphere enriched with 5% CO_2. ECV-304 cells were trypsinised on a weekly basis and subcultured 1:10 into fresh M199. 24 h prior to experimental use, a flask of confluent ECV-304 cells was trypsinised and the cells resuspended in fresh M199. A 96-well microtitre plate was seeded with approximately 5 x 10^4 ECV-304 cells per well in M199 and incubated at 37°C, 5% CO_2 overnight.

Experimental setup. Once ECV-304 cells in microplates were confirmed confluent, they were washed twice with pre-warmed PBS, fresh M199 was added to each well and the cells were incubated at 37°C, 5% CO_2 for 40 min. Bacteria were added to the cells at a multiplicity of infection of 500:1 and incubated at 37°C, 5% CO_2 for 2 h. Two washes were done using pre-warmed PBS to remove the majority of extracellular bacteria and fresh M199 was added. Ampicillin (128mg/L) was added to inactivate any remaining non-internalised bacteria and bioluminescence was measured in a luminometer (Fluostar Optima) every 15 min for 21 h at 37°C, 5% CO_2. 0.1%w/v saponin was added to permeabilize cells and allow inactivation of internalised bacteria by ampicillin after either 2 h or 4 h from the start of bioluminescence measurement.

RESULTS AND DISCUSSION

The results shown in Fig. 1 suggest low-level internalisation of *E. coli* DH5a pLITE27 within the monocytic cell line THP-1 which persists for the duration of the experiment. This positive outcome serves to validate the use of bioluminescence as a direct, sensitive, real-time means for assessing intracellular bacterial survival as the bacteria continued to emit light when internalised within the cells.[1]

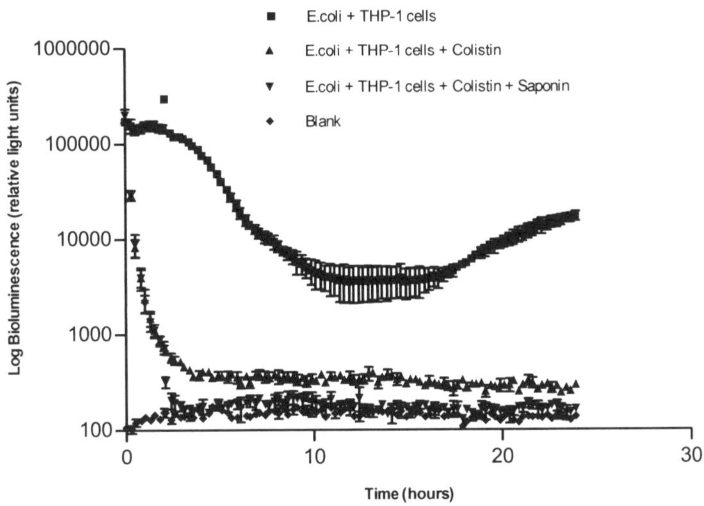

Fig. 1. Internalisation of *Escherichia coli DH5a* within the monocytic cell line THP-1. Colistin (150 mg/L) was added at time 0 to remove extracellular bacteria and saponin (0.1%) was added after 2 h to permeabilize cells, allowing inactivation of intracellular bacteria

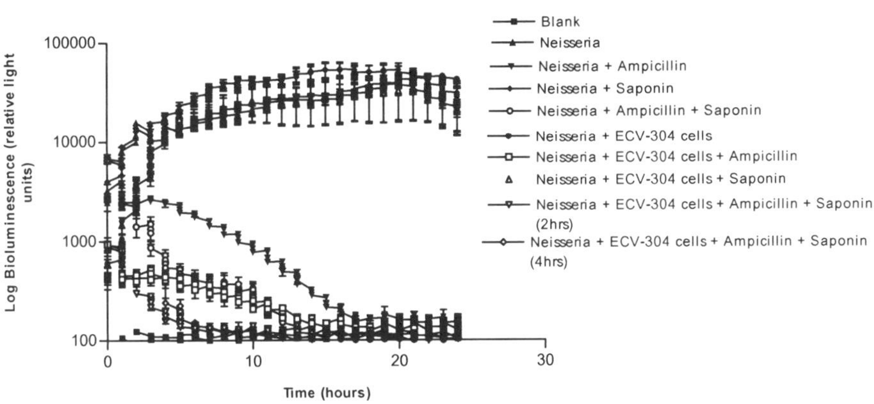

Fig. 2. An assay to determine whether *Neisseria meningitidis* C751 pGLITE is able to internalise within ECV-304 human endothelial cells. Ampicillin (128 mg/L) was added at time 0 to remove extracellular bacteria and saponin (0.1%) was added after either 2 h or 4 h to permeabilize cells, allowing inactivation of intracellular bacteria.

The results shown in Fig. 2 do not show any evidence of internalisation of *N. meningitidis* C751 pGLITE within ECV-304 endothelial cells after 2 h and 4 h as a synergistic relationship between ampicillin and saponin is seen in the treatment of *N. meningitidis* C751 pGLITE without cells. Therefore, it is necessary to identify alternative agents for removal of extracellular bacteria and permeabilizing cells where a synergistic relationship is not seen. There is no evidence of any intracellular bacteria following treatment of bacteria and cells with ampicillin alone, as bioluminescence falls to zero by 17 h.

Previous studies using viable counts have found that only very few bacteria were actually internalised, with the majority remaining as extracellular adherent pathogens.[2,3] However, it has been suggested that even though bacteria have not as yet been seen between cells, the possibility that this route is used should not be excluded.[4] Further investigation into this possibility will be carried out using Caco-2 cells which, when cultured as a monolayer differentiate to form tight junctions between cells and serve as a model of paracellular movement. Also, they are known to express many cell adhesion molecules which mediate entry into the cell.

REFERENCES

1. Angell JE, Mccleod JD, Lewis RJ, Salisbury VC. Use of bioluminescent *Escherichia coli* within a monocytic cell line to assess antimicrobial efficacy. Abstr Intersci Conf Antimicrob Agents Chemother 2002; abstract no. B-1110.
2. Pujol C, Eugene E, de Saint Martin L, Nassif X. Interaction of *Neisseria meningitidis* with a polarized monolayer of epithelial cells. Infect Immun 1997;65:4836-42.
3. Virji M, Kayhty H, Ferguson DJP, Alexandrescu C, Heckels JE, Moxon ER. The role of pili in the interactions of pathogenic *Neisseria* with cultured human endothelial cells. Mol Microbiol 1991;5:1831-41.
4. Nassif X, Bourdoulous S, Eugene E, Couraud OP. How do extracellular pathogens cross the blood-brain barrier? Trends Microbiol 2002;10:227-32.

NETWORK ACTIVITIES OF THE CEREBRAL CORTEX MONITORED BY BIOLUMINESCENT AEQUORIN IMAGING

ESTELLE DROBAC, LUDOVIC TRICOIRE, BERTRAND LAMBOLEZ
Neurobiologie des Processus Adaptatifs, UMR 7102, CNRS, Université Pierre et Marie Curie-Paris, Paris, F-75005 France

INTRODUCTION

The cerebral cortex is involved in sensory-motor processing and cognitive and emotional brain functions. This structure integrates afferent signals through a neuronal network comprising excitatory pyramidal neurons, which use glutamate as a neurotransmitter, and inhibitory interneurons, which use GABA as a neurotransmitter.[1] Pyramidal neurons form the majority of cortical neurons (80%) and project to other brain regions. Interneurons project locally and have more highly diverse morphological, functional and molecular properties than pyramidal neurons.[2,3] Information processing in the cortical network involves large neuronal assemblies whose activity patterns have been little explored. Here, we show that the Ca^{2+} dependent bioluminescence of aequorin[4] allows real-time imaging of cortical network activity patterns with a high signal-to-background ratio.

MATERIALS AND METHODS

A GFP-aequorin (GA) fusion protein allows both fluorescence labeling of expressing neurons and single cell luminescence recording.[5] We expressed this bifunctional reporter in cortical slices using a recombinant neurotropic sindbis virus (SIN-GA). Acute 300 µm-thick brain slices obtained from 15-day-old rats were prepared in an O_2/CO_2 95:5 bubbled standard artificial cerebro-spinal fluid supplemented with 1 mM kynurenate. Slices were then rinsed in culture medium (50% MEM, 50% HBSS, glucose adjusted to 6.5 g/L, Penicillin-Streptomycin) and transferred onto a millicell-CM membrane pre-equilibrated with culture medium in a Petri dish. Infection with SIN-GA was performed by adding the viral solution (titer: 10^7 i.p./mL) dropwise on the top of the slice. Slices were incubated overnight at 35°C in a humidified 5% CO_2 atmosphere. Next, aequorin was reconstituted by adding 10 µM coelenterazine for 3 h. After transfer into the recording chamber, bioluminescence was imaged at video rate using an ICCD125 intensified CCD camera (Photek, UK) coupled to a 10X or 60X objective (Olympus). Electrical stimulation was performed using a tungsten bipolar electrode. Data analysis was performed with Visilog 6.3 (Noesis, France).

RESULTS AND DISCUSSION

As visualized with GFP fluorescence, most of the GA-expressing cells displayed the typical morphology of pyramidal neurons with a prominent apical dendrite (Fig. 1), localized in layers II/III and V. GA was distributed in the cytosol and nucleus, throughout the soma and dendrites.

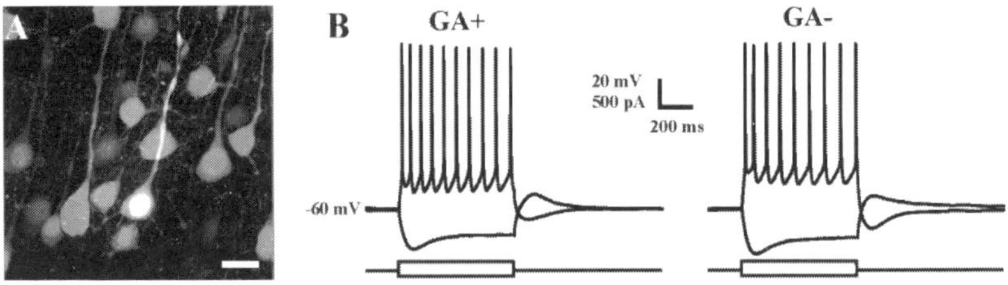

Fig. 1. Expression of GFP-Aequorin in cortical slices. A. Morphology of layer V pyramidal neurons expressing GA. Scale bar: 20 μm. B. Patch-clamp recordings of GA-expressing (GA+) and control (GA-) layer V pyramids in current-clamp mode

Patch-clamp recordings showed that GA-expressing layer V pyramidal neurons exhibited action potential (AP) firing and membrane properties similar to control neurons from non-infected slices (Fig. 1). This demonstrates the good viability of GA-expressing neurons after overnight incubation of cortical slices.

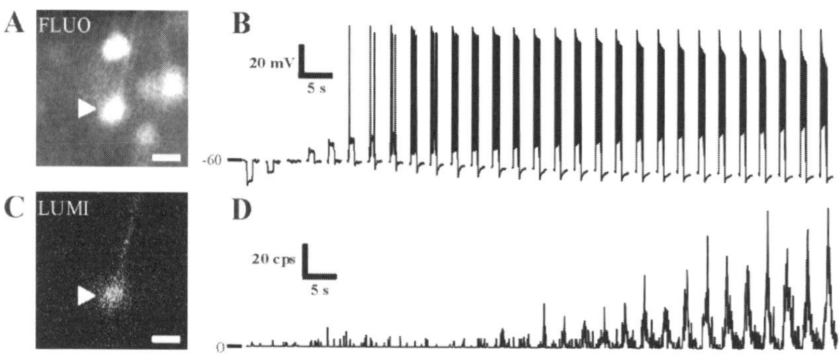

Fig. 2. Electrical and luminescent responses of a neuron expressing GA. A. A fluorescent layer V pyramidal neuron was patched (arrowhead). Scale bar: 10 μm. B. Voltage responses of this neuron to a sequence of increasing current steps. C. Image of luminescence emitted during the sequence. D. Luminescence recording shows synchronization with electrical activity

We next examined the ability of GA to report bioelectrical activities with single cell resolution using combined patch-clamp recording and bioluminescent responses of GA-expressing layer V pyramidal neurons (Fig. 2). Depolarizing currents steps of increasing amplitude generated AP discharges of increasing frequency. Above an AP frequency of around 30 Hz, a bioluminescence response was recorded in stimulated neurons at each AP burst (Fig. 2). This indicates that GA has a relatively low sensitivity to Ca^{2+} transients associated with AP firing, as observed with fluorescent protein Ca^{2+} sensors.[6] Very little luminescence background was observed in resting neurons. Hence, recombinantly-expressed GA is a suitable reporter of bioelectrical activities when those are of sufficient intensity.

Bioluminescence imaging of cortical network responses to local electrical stimulations was also performed (Fig. 3). Electrical stimulations performed in layer I or in white matter yielded comparable results (not shown). Aequorin luminescence was recorded in layers I to VI GA expressing neurons. Whole field recordings showed that robust light responses were obtained at each electrical stimulus with a very high signal-to-background ratio. The luminescence signal was observed in discrete regions distributed throughout layers I to VI. Evoked light responses were blocked in the presence of tetrodotoxin (TTX, Fig. 3) or the calcium channel blocker Ni^{2+} (not shown). This shows that bioluminescent signals resulted from bioelectrical activity of the neurons and subsequent calcium entry.

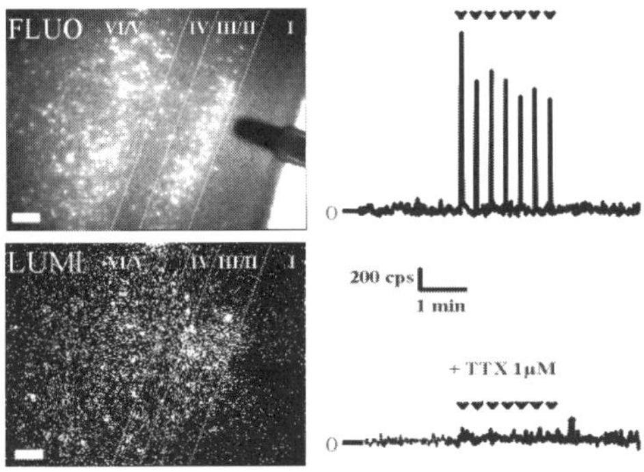

Fig. 3. Luminescent responses to electrical stimulation. Left: Superimposed fluorescence and bright field images show the GA-expressing slice and the position of the stimulation electrode in layer I. Image of luminescence emitted during a sequence of 7 stimuli (0.5 mA, 100 μs, 50 Hz for 500 ms). Right: Whole field recording shows light responses, obtained for each stimulus in control conditions, were blocked by TTX. Note the high signal to noise ratio. Scale bar: 100 μm.

Finally, we tested the effects of acetylcholine on responses of the cortical network evoked by electrical stimulation. Acetylcholine is a key neuromodulator of cortical functions involved in memory and attention. Experiments were done with the non-degradable cholinergic agonist carbachol (CCH, 20 μM). In control conditions, electrical stimulation elicited peak responses which returned to baseline immediately after the end of the stimulus (Fig. 4). In absence of electrical stimulation, application of CCH induced a smooth plateau response of small amplitude. In the presence of CCH, the amplitude of electrically evoked light responses was increased by 400 ± 100 % as compared to control conditions (n=6, see example in Fig. 4). This effect of CCH was due to a large increase in the spatial extent of the response, which spread much further away from the stimulus electrode than in control conditions. Furthermore, the response to electrical stimulation persisted for more than 20 s after the end of electrical stimulation (Fig. 4). Previous electrophysiological recordings of single neurons have shown that in the presence of cholinergic muscarinic agonists, electrical stimulation induces long lasting responses of cortical pyramidal neurons.[7] Our results obtained using aequorin bioluminescence imaging show that these cholinergic agonists increase the spatial extent and the duration of the response of cortical network to transient stimuli. These effects may participate in the memory and attention processes where acetylcholine is involved.

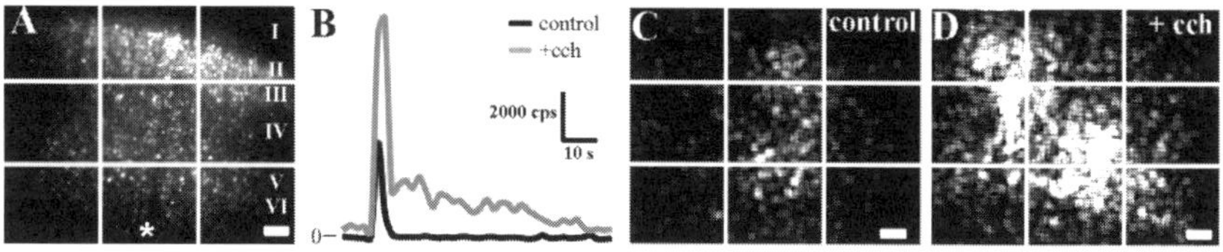

Fig. 4. Cholinergic modulation of electrically evoked cortical responses. A. Fluorescence image of a GA-expressing slice. *Stimulating electrode in white matter. B. Whole field recording shows luminescent response to a single stimulus in control conditions (black) and in 20 μM CCH (gray). C, D. Spatial distribution of local mean light response (binning:18x18) integrated every 5 s. Note: CCH increases the spatial extent of evoked response. Scale bar: 100 μm

CONCLUSIONS

We used a bifunctional GFP-aequorin fusion protein to image cortical network activities in brain slices. GFP fluorescence allowed visualizing the probe, which was preferentially expressed in layers II/III and V pyramidal neurons. The Ca^{2+}-dependent aequorin luminescence allowed real-time imaging of neuronal activities at single cell and multicellular network levels. Acetylcholine increases the extent and duration of cortical network responses to focal electrical stimulation.

REFERENCES

1. Peters A, Jones EG. eds. Cerebral cortex, Vol 1, Cellular components of the cerebral cortex. New York: Plenum, 1984:107-21.
2. Cauli B, Audinat E, Lambolez B, Angulo MC, Ropert N, Tsuzuki K, et al. Molecular and physiological diversity of cortical nonpyramidal cells. J Neurosci 1997;17:3894-906.
3. Cauli B, Porter JT, Tsuzuki K, Lambolez B, Rossier J, Quenet B, et al. Classification of fusiform neocortical interneurons based on unsupervised clustering. Proc Natl Acad Sci USA 2000;97:6144-9.
4. Shimomura O, Johnson FH. Calcium binding, quantum yield, and emitting molecule in aequorin bioluminescence. Nature 1970;227:1356-7.
5. Baubet V, Le Mouellic H, Campbell AK, Lucas-Meunier E, Fossier P, Brulet P. Chimeric green fluorescent protein-aequorin as bioluminescent Ca^{2+} reporters at the single-cell level. Proc Natl Acad Sci USA 2000;97:7260-5.
6. Pologruto TA, Yasuda R, Svoboda K. Monitoring neural activity and $[Ca^{2+}]$ with genetically encoded Ca^{2+} indicators. J Neurosci 2004; 24: 9572-9.
7. Egorov AV, Hamam BN, Fransen E, Hasselmo ME, Alonso AA. Graded persistent activity in entorhinal cortex neurons. Nature 2002; 420:173-8.

BIVALENT FLUORESCENT REPORTERS FOR GENE EXPRESSION STUDIES IN GRAM-POSITIVE BACTERIA

SR GADDIPATI,[1] TM PEREHINEC,[1] SNA QAZI,[2] CED REES,[1] AND PJ HILL[1,2]
[1] *School of Biosciences, University of Nottingham, Sutton Bonington Campus, Loughborough, Leics, LE12 5RD, UK;* [2] *Institute of Infection, Immunity and Inflammation, University of Nottingham, Nottingham, NG7 2UH, UK*

INTRODUCTION

Listeria monocytogenes is a Gram-positive intracellular pathogen that has developed numerous complex regulatory mechanisms in order to establish successful infection. A central transcriptional activator (PrfA) regulates the expression of the virulence genes required for the intracellular growth and cell-to cell spread. The role of this gene has been studied extensively and, in addition to deletion mutants, mutations with increased levels of PrfA activity (PrfA* mutants) have been identified. PrfA protein binds to a palindromic sequence (the PrfA box) in the promoter region of genes controlled by this protein. Haemolysin (*hly*) and phosphatidyl phospholipase C (*plcA*) are the first two virulence genes involved in the escape from the vacuole to initiate intracellular growth.[1] The promoters of these genes are located in a 243 bp intragenic region and are transcribed divergently from promoters that have overlapping -35 sequences, containing one PrfA-box (see Fig. 1). Despite this shared binding site for the PrfA regulatory protein, different patters of expression of the two genes is seen, with *hly* being expressed more strongly and slightly earlier than *plcA*.[2] This differential pattern of expression of promoter expression may be of relevance for the temporal control of gene expression during cell infection.

This region has previously been used to create *lux*-based reporter gene constructs for the *hly* and *plcA* genes[3] and we have also successfully created *Egfp* single reporter gene plasmids and shown that the promoter fragments are correctly regulated by PrfA and that introducing multiple copies of the PrfA binding site into *Listeria* cells did not affect expression of the other chromosomally-encoded virulence genes due to protein titration (Hill, unpublished). These *gfp* reporter plasmids can be used to follow patterns of expression of individual virulence genes during cell infection,[2] but only one gene at a time can be monitored. However construction of bacterial strains carrying two compatible plasmids with equivalent copy numbers to monitor both promoters during infection is problematic. To try and further understand the differential regulation of the *hly* and *plcA* genes we have created promoter fusions to two different fluorescent reporter genes in a non-FRET application to facilitate microscopic visualisation of cells during cell infection. This work shows that such constructs are stable and the signals from the two different genes can be distinguished allowing the expression of two different genes to be simultaneously monitored.

MATERIALS AND METHODS

E. coli cells were cultivated at in Luria-Bertani (LB) broth or agar supplemented with appropriate antibiotics. *L. monocytogenes* strains were grown in brain–heart infusion broth (BHI, Oxoid) at 37°C with aeration. Antibiotics Erythromycin (Erm; 5 μg/mL) and Kanamycin (Km; 50 μg/mL) were used to select for plasmid maintenance.

PCR reactions contained 10 μM of each primer, template DNA at 0.2 ng/μL and 2.5u of AccuPrime *Taq* DNA Polymerase (Invitrogen) and the buffer supplied by the manufacturer. *Lm EGD* DNA was used as template DNA to amplify the 243 bp P_{hly}-P_{plcA} promoter fragment. Plasmids pSG1192 and pSG1193[4] were used as templates for amplification of CFP and YFP, respectively. Generally PCR reactions consisted of denaturation at 95°C for 2 min, followed by 30 cycles of 95°C for 30 s, 64°C for 30 s, and 68°C for 1 min, with a final extension period at 68°C for 7 min. The bivalent fluorescent reporter plasmid was constructed by overlap extension PCR and Invitrogen Gateway Cloning. Plasmid constructs were transformed into *E. coli* competent cells (Invitrogen) or *L. monocytogenes* by electroporation.[5]

RESULTS

To construct the bivalent reporter plasmid, the *hlyA-plcA* intragenic region of LmEGD was amplified with primers containing homology to the required att site and reporter gene (*att*B1 with *Ecfp* and attB2 with *Eyfp*; see Fig. 1).

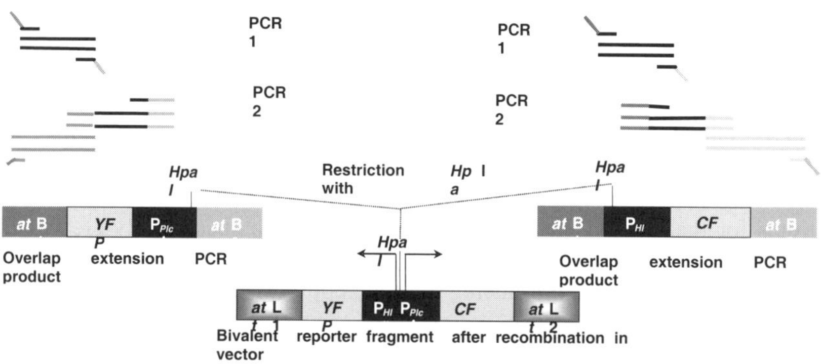

Fig. 1. Construction of the bivalent reporter plasmid. PCR1 indicates the amplification of the *hlyA-plcA* intragenic region (black) with primers with *att*B1 (grey) for *cfp* (light grey) and *att*B2 (mid grey) for *yfp* (light grey). PCR2 indicates the overlap extension PCR strategy to fuse reporters to the promoter fragment with *att*B2 or *att*B1. Restriction with *Hpa*I was used to fuse the two reporter constructs together.

The product from this reaction was used in overlap extension PCR with primers containing homology to *cfp* or *yfp* and the cognate att site to fuse the two promoters to a reporter gene with flanking *att*B1 and *att*B2 sites. These were cloned separately by recombination into *E. coli* Gateway vectors (pDONOR-221; Invitrogen). A unique *Hpa*I site located in the middle of the promoter fragment allowed the two separate reporter plasmids to be fused creating a construct with the structure attL1::*yfp*::P_{plcA}-P_{hly}::*cfp*::attL2. This construct was introduced into a Gram-positive Gateway shuttle vector based on plasmid pUNK1.[6] In order to demonstrate correct PrfA regulation of the reporter genes, this bivalent plasmid was transformed into three *L. monocytogenes* strains with different PrfA activity levels; EGD (*prfA* wt), NCTC 7973 (*prfA**) and NCTC 10357 (Δ*prfA*).

To investigate the stability and functionality of both reporter genes the bacterial cultures were first grown in BHI broth. Wet mounts were prepared and Cfp and Yfp fluorescence visualized using a confocal microscope. Clear signals were observed in both the wild-type EGDe and the PrfA* strain Lm7973, with a higher signal being detected in Lm7973 as expected. No signal was observed from the Δ*prfA* strain Lm10357 confirming the PrfA-dependent expression of fluorescent reporter proteins from both promoters in the bivalent expression vector. Superimposition of microscopic images indicated that all cells were expressing both reporter genes and no plasmid instability or plasmid loss was seen (data not shown).

The PrfA wt strain EGD containing the bivalent reporter plasmid was used to infect Caco2 cells invasion experiments[7] and visualised by confocal microscopy. To image the eukaryotic cells, and allow the location of bacteria to be determined, the red-fluorescent lipophilic styryl dye FM4-64 was used to stain membranes. Both CFP and YFP signals were detectable in intracellular bacteria showing that the plasmid copies of P_{plcA} and P_{hly} were being activated during cell infection (Fig. 2).

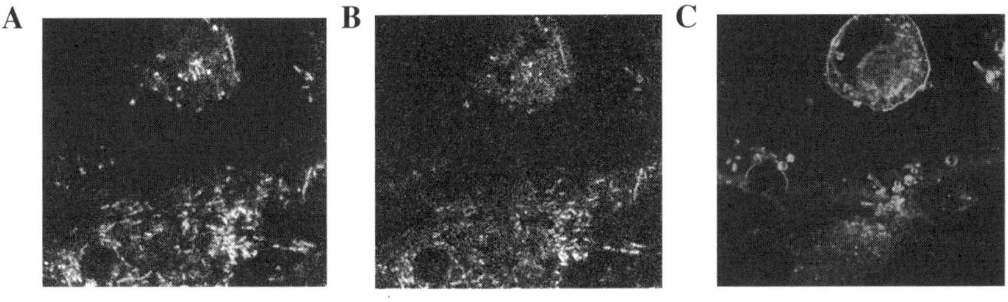

Fig. 2. Gene expression by *L. monocytogenes* EGDe during infection of Caco2 cells in a costar cover slip-base dish. Panel A, CFP expression from P_{hly} activity (excitation at 405nm); Panel B, YFP expression from P_{plcA} (excitation at 514 nm); Panel C, FM4-64 staining of cell membranes (excitation at 520 nm)

As expected no CFP and YFP signals obtained were obtained during infection with the Δ*prfA* strain Lm10357, confirming that fluorescence detected was due to

PrfA-directed expression of the plasmid-based reporter genes and not due to cellular autofluorescence (data not shown).

DISCUSSION

Although single GFP reporter gene plasmids have been successfully used to monitor gene expression during bacterial cell infection experiments, the difficulties maintaining two plasmids in the same cell at the same copy number mean that only one gene can be monitored at a time. By constructing a reporter construct to measure the activity of two divergent promoters using the YFP and CFP pair in a non-FRET application we have shown that, despite of a high degree of sequence homology of both *cfp* and *yfp*, they are both stable and functional when present on the same plasmid molecule. These constructs have enormous potential to give us clear insight into the regulatory mechanisms of the virulence genes in the pathogenic islands under *in vivo* conditions. This approach should provide, faster, more detail of temporal gene control events with less variation in the conditions which inevitably occurs when comparing results of parallel experiments when using separate reporter gene plasmids.

REFERENCES

1. Vazquez-Boland J, Kuhn M, Berche P, Chakraborty T, Dominguez-Bernal G, Goebel W, et al. *Listeria* pathogenesis and molecular virulence determinants. Clin Microbiol Revs 2001;14:584–640.
2. Bubert A, Sokolovic Z, Chun S-K, Papatheodorou L, Simm A, Goebel W. Differential expression of *Listeria monocytogenes* virulence genes in mammalian host cells. Mol Gen Genet 1999;261:323–36.
3. Park S, Stewart G, Kroll R. The use of bacterial luciferase for monitoring the environmental regulation of expression of genes encoding virulence factors in *Listeria monocytogenes.* J Gen Microbiol 1992;138:1-8.
4. Feucht A, Lewis P. Improved plasmid vectors for the production of multiple fluorescent protein fusions in *Bacillus subtilis.* Gene 2001;264:289-97.3.
5. Park S, Stewart G. High efficiency transformation of *Listeria monocytogenes* by electroporation of penicillin treated cells. Gene 1990;94:129-32.
6. Pilgrim S, Stritzker J, Schoen C, Kolb-Maurer A, Geginat G, Loessner J, et al. Bactofection of mammalian cells by *Listeria monocytogenes*: improvement and mechanism of DNA delivery. Gene Ther 2003;24:2036-45.
7. Qazi S, Harrison S, Self T, Williams P, Hill P. Real-time monitoring of intracellular *Staphylococcus aureus* replication. J Bacteriol 2004;186:1065-77.

FOUR-COLOR LABELING OF CELL CULTURE AND TUMORS OF LIVE MICE UPON INFECTION WITH: *GFP-RUC* AND *RFP-CBG99* EXPRESSING VACCINIA VIRUS STRAINS

V RAAB,[1*] C HORBASCHEK,[1*] N CHEN,[2] Q ZHANG,[2] YA YU,[2]
C SEUBERT,[1] U GEISSINGER,[1] A WORSCHECH,[1] CJ TIETZE,[1] E ZELLNER,[1]
TH LE,[1] F GRUMMT,[1] J STRITZKER,[2] AA SZALAY [1,2]

[1] *Virchow Center for Experimental Biomedicine, Institute for Biochemistry and Institute for Molecular Infection Biology, University of Würzburg, Am Hubland, D97074 Würzburg, Germany;*
[2] *Genelux Corporation, Science Center, San Diego, CA 92109, USA;*
[*]*Both of these authors contributed equally to the presentation*

INTRODUCTION

Among viruses, *Vaccinia* virus (VV) has the most potential for therapy and/or diagnosis of solid tumors in patients. The relatively small size of mice makes these animals a valuable tool for *in vivo* optical imaging that might not (yet) be applicable to larger animal models or to humans. Therefore, basic research of oncolytic viruses may allow labeling with multiple alternative fluorescent proteins and/or luciferases at the same time or different time intervals.

MATERIALS AND METHODS

Tumors. Five times 10^6 GI101A tumor cells were implanted subcutaneously on the lateral side of mice and tumors were developed for 30 days prior to injection of VV.
Vaccinia Virus. The 2 strains of VV were produced in CV-1 cells. Five times 10^5 pfu were injected into the lateral tail vein of GI101A tumor bearing mice.
Luciferase assay. For each assay 0.5 μg cell extract were used. VV *Ruc-GFP* infected cells were assayed with 0.0375 μg coelenterazine and VV- *CBG99-RFP* infected cells with 0.5 μg beetle luciferin.
Flow cytometry. At different times post infection GI101A cells were analyzed for the expression of GFP and/or RFP by flow cytometry using a Beckam Coulter Epics XL-MCL and the WinMDI 2.8 software.

RESULTS AND DISCUSSION

Two strains of VV, one expressing Renilla luciferase and GFP, the other RFP and Click Beetle luciferase (CBG99) were constructed and luciferase activity was measured *in vitro* and *in vivo* after addition of coelenterazine and click beetle luciferin respectively as shown in Fig. 1 and Fig. 2.

Ten days after injection of VV into tumor-bearing mice, the expression of the fluorescent proteins was detectable. Fig. 3 depicts the RFP and GFP expression from live mice and from isolated tumors of the same mice.

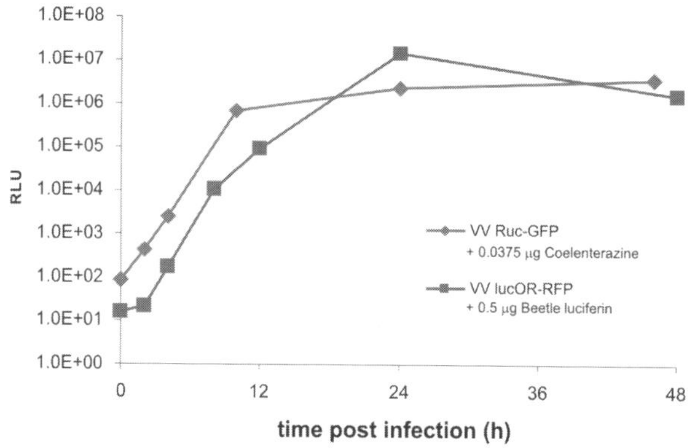

Fig. 1. Luciferase assay of 0.5 μg cell extract at different times post infection. Even though the click beetle luciferase was incubated with much more luciferin, the photon emission during the assay is comparable for both luciferases

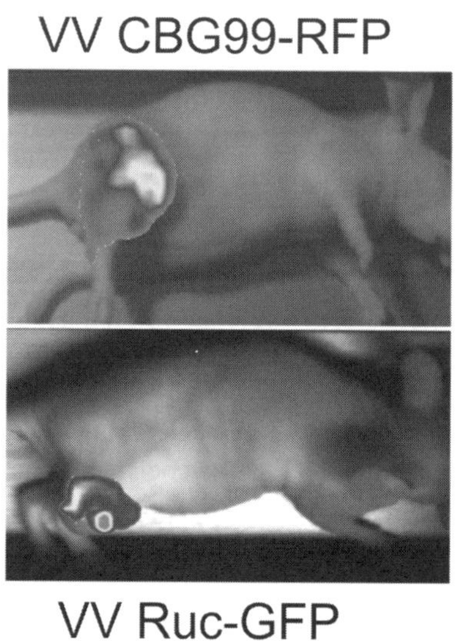

Fig. 2. Low light imaging of VV injected GI101A tumor bearing nude mice at 7 days *p.i.* Lower panel: VV Ruc-GFP and coelenterazine injected mouse. Upper panel: VV CBG99-RFP and click beetle luciferin injected mouse

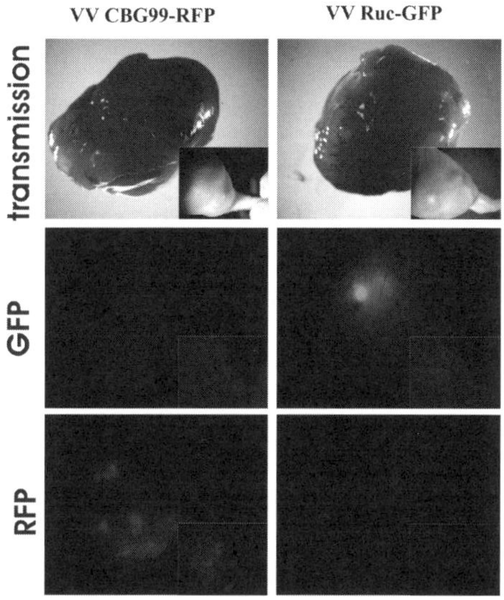

Fig. 3. Fluorescent images of excised tumors and live tumor-bearing mice. Mice were injected with VV *CBG99-RFP* (left panels) or VV *Ruc-GFP* (right panels) 10 days prior fluorescence imaging. Inlays represent the live mouse image while the large image represents the excised tumor from the same mouse

In order to assess the VV properties, GI101A cells were infected with a multiplicity of infection (MOI) of 0.5 and flow cytometry data were obtained at different times post injection as shown in Fig. 4 and Fig. 5.

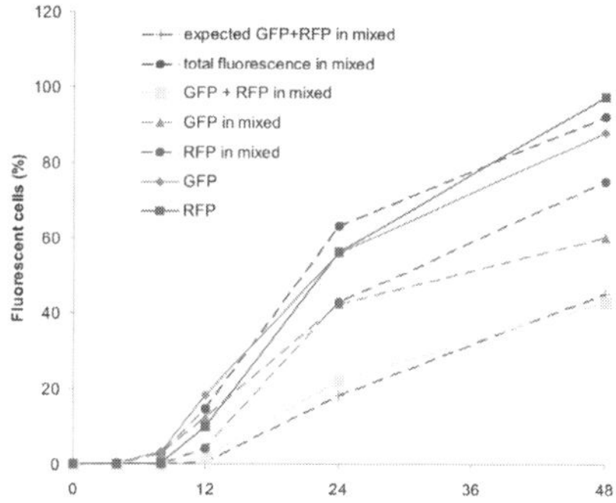

Fig. 4. Analyses of the flow cytometry data. No big differences are seen between the VV strains regarding replication. GFP is detectable earlier. VV strains do not influence each other as can be seen from the comparison of expected and GFP and RFP expressing cells in the mixed infection

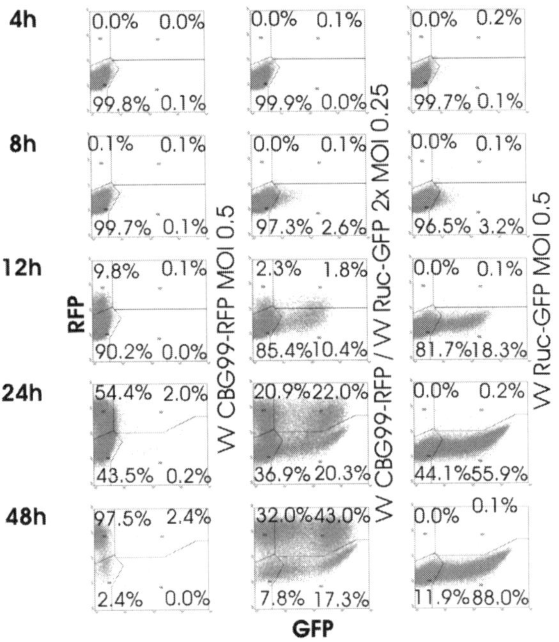

Fig. 5. Flow cytometric analysis of VV *CBG99-RFP* and VV *Ruc-GFP* infected GI101A cells at different times post infection

The replication of the two VV strains is independent of each other and already infected cells can still be infected by other VV.

In summary, the described VV strains seem to be ideal tools to investigate the influence of one Vaccinia infection followed by a second since they did not influence each other in their replication but were clearly distinguishable from each other due to their multi color labeling. One virus expresses GFP (max. emission wavelength: 509 nm) and photon emission (482 nm) occurs after adding coelenterazine – the other VV emits light (537 nm) when beetle luciferin is added and expresses RFP (583 nm). Thus both viruses can be used independently in the same mouse for quantitative low light imaging and high resolution in *in* and *ex vivo* histology analysis.

ACKNOWLEDGEMENTS

Genelux Corporation for funding. C.H. is a recipient of a Graduate fellowship of Genelux Corporation to the University of Würzburg.

QUANTITATIVE IMMUNOHISTOCHEMISTRY AND *IN SITU* HYBRIDIZATION WITH LUMINESCENCE DETECTION: A NEW POTENTIAL DIAGNOSTIC TOOL

A RODA,[1] M GUARDIGLI,[1] M MIRASOLI,[1] S AMBRETTI,[2]
S VENTUROLI,[2] M MUSIANI,[2] S CASANOVA,[3] E RODA[4]
[1] *Dept. of Pharmaceutical Sciences;* [2] *Dept. of Clinical and Experimental Medicine, Division of Microbiology;* [3] *"F. Addarii" Institute of Oncology;*
[4] *Dept. of Internal Medicine and Gastroenterology, University of Bologna, Bologna, Italy, Email: aldo.roda@unibo.it*

INTRODUCTION

The ultrasensitive localization and quantification of analytes inside tissues and cells is fundamental both for the understanding of physiopathological events and for the early diagnosis of pathologies. Luminescence imaging techniques allow one to perform multiplexed detection via the combination of different detection principles and/or labels, while chemiluminescence (CL) imaging of enzyme-labeled biospecific probes combines high sensitivity with easy and reliable evaluation of the amount of the probe, thus enabling the quantitative analysis of the target molecules. Here we report applications of luminescence immunohistochemistry (IHC) and *in situ* hybridization (ISH) procedures for the detection and quantification of proteins and nucleic acids.

MATERIALS AND METHODS

Measurements of the luminescence emission from the target surface were performed using a modified epifluorescence microscope (BX 60, Olympus Optical), which is enclosed in a dark box to prevent interference from ambient light and connected to a slow-scan ultrasensitive CCD camera (LN/CCD, Princeton Instruments) cooled to −100°C with liquid nitrogen. Fluorescence measurements were performed by employing an USH-102D 100-Watt mercury excitation lamp (Ushio Inc.) and fluorescence cubes with suitable excitation and emission filters.

RESULTS AND DISCUSSION

Quantitative detection of MRP2 in liver biopsies. The MRP2 transport protein (a member of the human Multidrug Resistance–Associated protein family) is the most important hepatocellular transporter involved in the excretion of organic anions into bile. Alteration of its expression and/or function is thought to play an important role in the pathogenesis of cholestasis. Moreover, reduced MRP2 levels have been observed in patients affected by primary biliary cirrhosis (PBC), thus MRP2 levels could represent an indicator of PBC progression. A direct, simple and rapid CL-IHC method has been developed for the quantitative evaluation of MRP2 expression in liver biopsy sections.[1] The protein was localized by means of an anti-MRP2 monoclonal antibody, followed by a biotin-labeled anti-mouse antibody and a streptavidin-horseradish peroxidase (HRP) conjugate, which was then detected by means of a CL HRP substrate (ECL, Amersham Bioscience) and CL imaging (Fig. 1a). The quantitative analysis of CL images allowed the reliable and reproducible evaluation of the protein content of the tissue and, when applied to the analysis of clinical samples from PBC patients under therapy with ursodeoxycholic acid, the technique showed an increase of the MRP2 levels (Fig. 1b), in line with results obtained with conventional protein expression analysis techniques. Most interestingly, the changes in the mean CL signals showed a good correlation with the histological data: most of the subjects that gave increased CL signals (thus incremented MRP2 levels) also showed regression of the PBC-associated fibrosis. Therefore, this method could represent an alternative technique to Western blotting and realtime PCR for the quantitative evaluation of protein expression in tissue sections, with the advantages of not requiring pre-analytical extraction steps, or large amounts of tissue.

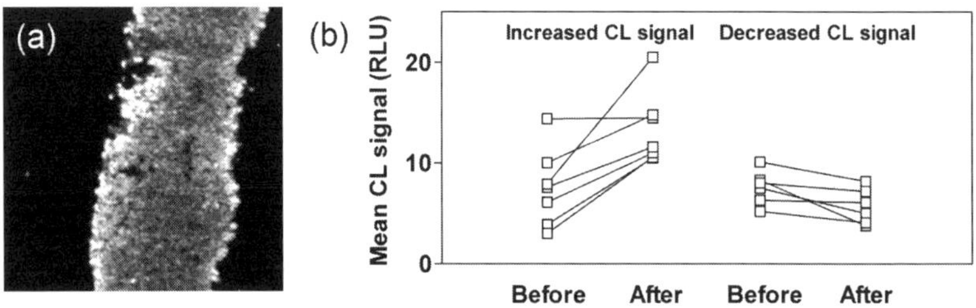

Fig. 1. (a) CL image showing the localization of MRP2 in a liver biopsy section. (b) Mean CL signals measured in liver biopsies obtained before and after therapy with ursodeoxycholic acid

Analysis of the p16^{INK4a} concentration profile. The p16^{INK4a} neoplastic marker protein is overexpressed in cervical intraepithelial neoplasias (CINs) caused by human papillomavirus (HPV). Since the positivity of p16^{INK4a} in the epithelium varies from low-grade CIN1 lesions (only the lower third of epithelium is positive) to high-grade CIN3 lesions (the whole epithelium is stained), the oncogenic risk of CINs could be evaluated by measuring the protein expression profile.

We have developed a CL-IHC technique for the quantitative detection of the p16^{INK4a} protein in cervical biopsy sections. The protein was localized by means of a monoclonal anti-p16^{INK4a} antibody, followed by a biotinylated anti-mouse antibody, streptavidin and a biotin-alkaline phosphatase (AP) conjugate, which enzymatic activity was measured by means of the Lumiphos Plus AP CL enzyme substrate (Lumigen). The quantitative analysis of the CL signal showed a p16^{INK4a} concentration profile across the epithelium which changed depending on the degree of CIN lesion, according to data obtained by IHC with colorimetric detection (Fig. 2a). In order to discriminate between low-risk (CIN1) and high-risk (CIN2 and

CIN3) lesions we defined a score value useful for CIN classification. In particular, once the background CL signal in the connectival portion of the tissue was measured, the "positive" epithelium area - in which the CL signal was higher than a threshold value proportional to the background CL signal - was selected. Then, the score value was calculated for each sample as the product between the area fraction of "positive" epithelium and its mean CL signal, expressed as the ratio between the CL signal and the background CL. Statistical analysis of data obtained from 39 clinical samples (classified by histological analysis as 9 negatives, 11 low-risk and 19 high-risk lesions) demonstrated that the score value can distinguish between low-risk and high-risk lesions with statistical significance (Fig. 2b).

The newly developed method thus allows objective evaluation of the risk of progression of CIN lesions, thus representing an advancement with respect to the subjective evaluation obtained by means of histological analysis or p16^{INK4a} detection performed by IHC with colorimetric detection.

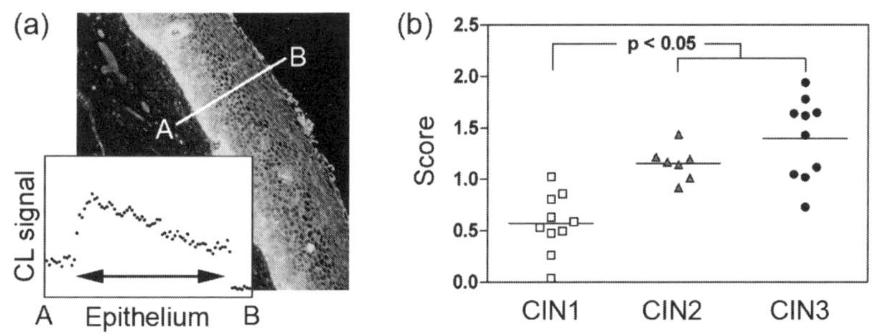

Fig. 2. (a) Localization of p16^{INK4a} by CL imaging in a bioptic section of a CIN2 lesion and protein concentration profile across the epithelium. (b) Discrimination between low-risk (CIN1) and high-risk (CIN2 and CIN3) lesions on the basis of their score values

Colocalization of HPV DNA and HMB-45. The correlation between high-risk mucosal HPV and basal cell carcinoma and squamous cell carcinoma has been frequently reported but not conclusively demonstrated. In order to assess such relationship, we have developed a sensitive method that combines an enzyme-amplified fluorescent ISH (FISH) for the localization of high-risk HPV DNA with a CL-IHC method for the localization of the tumoral melanocytic marker HMB-45.[2] The two localization procedures were performed sequentially on the same primary melanoma biopsy section. The HPV DNA was detected by means of a digoxigenin-labeled oligonucleotide probe, followed by an AP-labeled anti-digoxigenin monoclonal antibody and the ELF97 AP substrate (Molecular Probes), which produces a fluorescent precipitate at the site of the enzymatic activity, then HMB-45 was localized by using a biotin-labeled anti-HMB-45 monoclonal antibody, an HRP-labeled anti-mouse antibody and the ECL HRP substrate. Digital images of

FISH and CL-IHC were separately recorded and merged using the specific software for image analysis. The method was applied to eight biopsy specimens of skin melanomas, previously found positive by PCR for high-risk mucosal HPVs. Positive signals for the presence of both HPV DNA and HMB-45 were detected in the same sections of melanoma biopsies, showing a sharp colocalization (Fig. 3) and thus stressing the association between HPV and melanoma cells and excluding the possibility that positivity for HPV DNA could be merely ascribed to its presence only in few cells of the tumor harboring the virus.

The assay thus represents a rapid, specific and sensitive method suitable to perform an extensive study of the interesting and not yet clear relationship between high risk mucosal HPV infection and skin melanoma carcinogenesis.

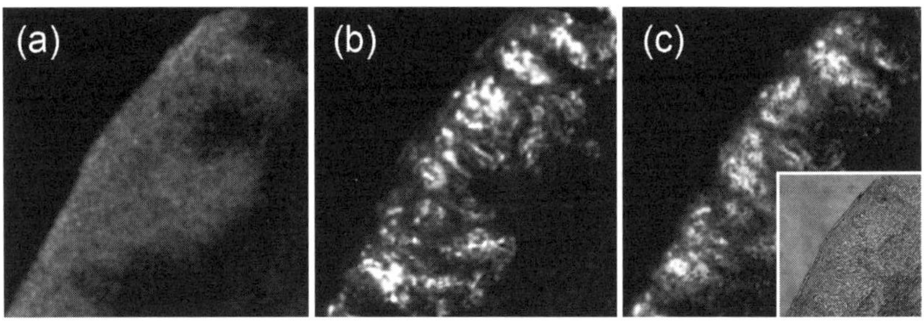

Fig. 3. Colocalization of HPV and HMB-45 in a primary melanoma biopsy section. (a) Fluorescence image showing the localization of HPV DNA. (b) CL image showing the localization of HMB-45. (c) Colocalization of HPV DNA and HMB-45 assessed by multiplying pixel-by-pixel images (a) and (b). The inset shows the transmitted light image of the section

CONCLUSIONS

Immunohistochemical and *in situ* hybridization techniques with luminescence detection represent powerful bioanalytical tools which could be used for early and reliable diagnosis. They allow highly sensitive and quantitative detection of target analytes (potentially single-molecule level) and the development of multiple assays by combining different labels and/or detection techniques.

REFERENCES

1. Guardigli M, Marangi M, Casanova S, Grigioni WF, Roda E, Roda A. Chemiluminescence quantitative immunohistochemical determination of MRP2 in liver biopsies. J Histochem Cytochem 2005;53:1451-7.
2. Ambretti S, Venturoli S, Mirasoli M, La Placa M, Bonvicini F, Cricca M, et al. Assessment of the presence of mucosal human papillomaviruses in malignant melanomas using combined fluorescent in situ hybridization and chemiluminescent immunohistochemistry. Br J Dermatol (doi:10.1111/j.1365-2133.2006.07541.x).

EVALUATION OF AN *IN VIVO* GENE INDUCTION SYSTEM IN INFECTED TUMOR-BEARING MICE

JOCHEN STRITZKER,[1*] STEPHANIE WEIBEL,[2*] PHILIP J. HILL,[2,3] ALADAR A. SZALAY[1,2]
[1] *Genelux Corporation, Science Center, San Diego, CA 92109, USA;*
[2] *Virchow Center for Experimental Biomedicine, Institute for Biochemistry and Institute for Molecular Infection Biology, University of Würzburg, Am Hubland, D97074 Würzburg, Germany;* [3] *On sabbatical leave from: University of Nottingham, School of Biosciences, Sutton Bonington Campus, Leics, LE12 5RD, UK;*
[*] *Both of these authors contributed equally to the presentation*

INTRODUCTION

An increasing number of bacterial strains are known to colonize solid tumors in mice. Here we investigated the histologic distribution of a tumor-targeting form of *E. coli*. We also investigated whether we can introduce a gene induction system that is only turned on when a non-toxic compound is exogenously administered.

Tumor-specific targeting of bacteria can in theory be used to diagnose and treat certain types of cancer by expression of suitable genes.[1,2] Therefore, a tight regulation of—possibly toxic—protein production is desired. Here we describe the evaluation of a gene induction system *in vivo* monitored through *lux* expression and low light imaging.

MATERIALS AND METHODS

Tumors. Thirty-three thousand 4T1 tumor cells were implanted subcutaneously on the lateral side of mice and tumors were developed for 14 days prior to injection of bacteria.

Bacteria. *E. coli* was grown in LB medium containing ampicillin with or without addition of the inducer (as indicated). For injection into tumor-bearing mice, bacteria were grown until they reached an OD of 0.4 at 600 nm, washed twice in phosphate-buffered saline (PBS) and then injected into the lateral tail vein.

Induction of gene expression. The inducer compound was either added to the growth medium (for *in vitro* experiments) or intravenously injected into tumor-bearing mice that were previously injected with the engineered *E. coli*.

RESULTS AND DISCUSSION

Histology. DsRed expressing *E. coli* were injected into 4T1 tumor-bearing mice and histology was performed on days 2 and 3 post-injection, as shown in Fig. 1.

Interestingly, the necrotic area within the tumors seemed to enlarge due to the injected *E. coli*. Control tumors that are not colonized by *E. coli* do not show a visible necrotic center.

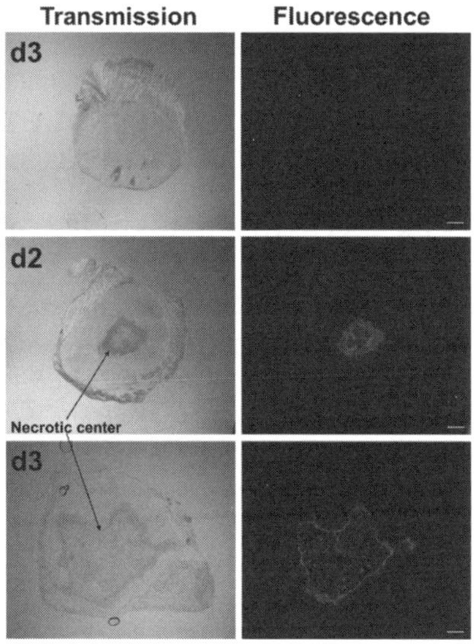

Fig. 1. Histologic distribution of DsRed expressing *E. coli* in 4T1 tumors. Tumors were succesfully colonized by *E. coli*, which localized in the necrotic center of the tumor (DsRed expression in the right column). This necrotic region enlarged from day 2 to day 3 post-injection. Control tumors (upper row) did not have a clearly visible necrotic center

***In vitro* gene regulation.** Next we tested our inducible promoter system *in vitro* by growing *E. coli* in LB broth with or without a non-toxic inducer supplementation (see Fig. 2 for details).

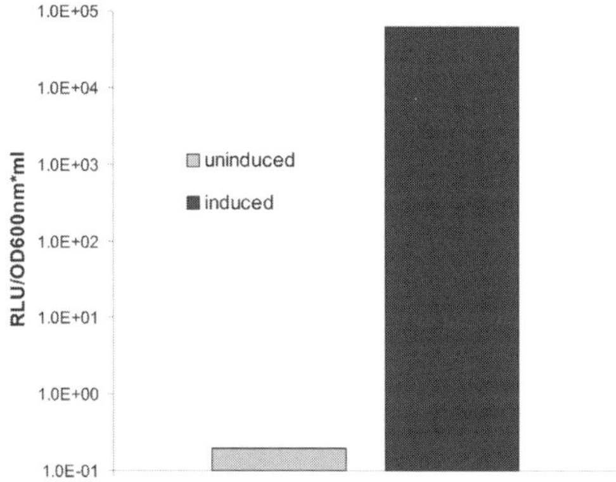

Fig. 2. Specific *luxA-E* gene activation by non-toxic inducer

Induction of the promoter system resulted in specific expression of the *luxABCDE* operon, causing a 10^5-fold increase in photon emission from the bacteria. **Gene induction in tumors of live mice.** Knowing that our *E. coli* strain specifically replicates in solid tumors and that gene induction is possible, at least in culture, we wanted to go one step further and test the system in live mice. Therefore, 4T1 tumor-bearing mice were injected with the *E. coli* strain transformed with the inducible *luxABCDE* cassette. Two and four days later we administered the inducer compound and performed low light imaging on live mice, as shown in Fig. 3.

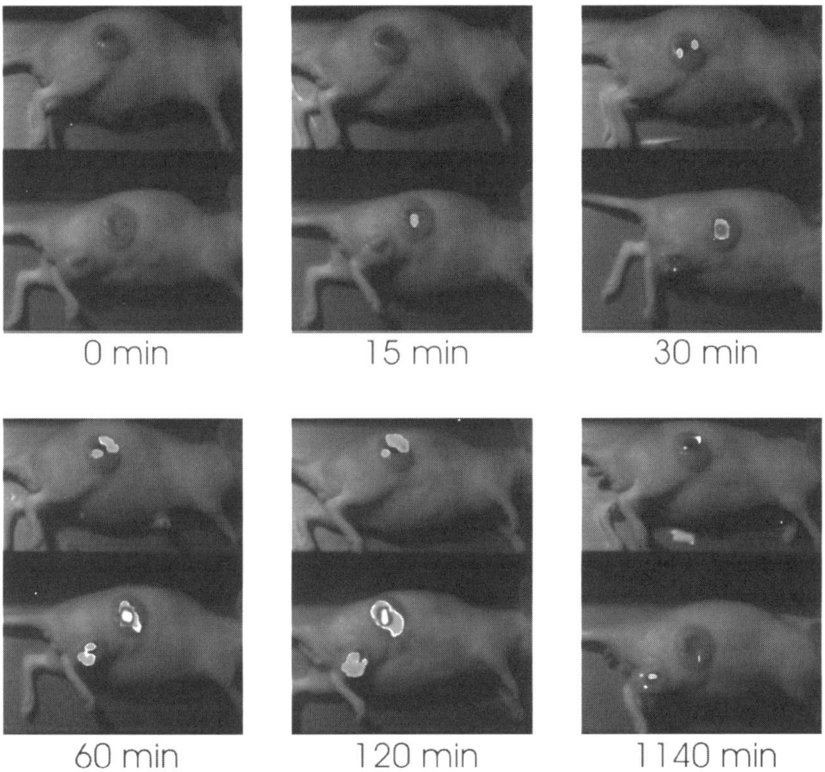

Fig. 3. Inducer-dependent activation of *luxABCDE* gene expression in *E. coli* colonized 4T1 tumors of live mice. Mice were injected with inducer 4 days after systemic administration of *luxABCDE* encoding *E. coli*. Pictures were taken at the indicated time points after inducer administration

Just 15 min after *i.v.* injection. photon emission from the tumors was detectable. The signal strength increased over the next 2-3 h and then started to decline if no more inducer was injected. When administered orally, we could also detect light emission from the tumors but the signal was delayed compared to the intravenous inducer injection.

Moreover, we could see that photon emission on day 4 was about twice as strong as compared to day 2 (see Fig. 4).

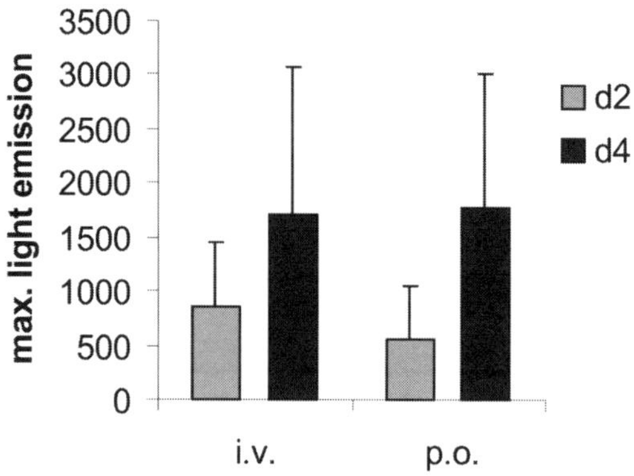

Fig. 4. Maximal light emission from mice injected with inducer on days 2 and 4 post injection of *luxABCDE* encoding *E. coli*. Inducer was given intravenously or orally

The reason for this is probably the depth from which the signal originates since the amount of bacteria per tumor does not change during the same time (data not shown).

In summary, we found a new tumor-targeting bacterial strain that can be used for the expression of therapeutic or diagnostic genes and also be strictly controlled by an exogenously applied trigger signal. The presented inducer system is applicable in many bacteria and might become be a valuable tool in basic research as well as in applied medicine.

REFERENCES

1. Yu YA, Shabahang S, Timiryasova TM, Zhang Q, Beltz R, Gentschev I, et al. Visualization of tumors and metastases in live animals with bacteria and vaccinia virus encoding light-emitting proteins. Nat Biotechnol 2004;22:313-20.
2. Zhao M, Yang M, Li XM, Jiang P, Baranov E, Li S, et al. Tumor-targeting bacterial therapy with amino acid auxotrophs of GFP-expressing *Salmonella typhimurium*. Proc Natl Acad Sci USA 2005;102:755-60.

EXAMINATIONS OF BACTERIUM-MEDIATED DETECTION OF TUMORS IN MICE MODELS

YONG A. YU,[1] ALADAR A. SZALAY[1,2]

[1] *Genelux Corporation, San Diego Science Center, San Diego, CA, USA;*
[2] *Virchow Center for Experimental Biomedicine, Institute for Biochemistry and Institute for Molecular Infection Biology, University of Würzburg, Am Hubland, D97074 Würzburg, Germany; Correspondence to: Aladar A. Szalay, Genelux Corporation, San Diego Science Center, 3030 Bunker Hill St., Ste. 310, San Diego, CA 92109, USA. Email: aaszalay@genelux.com*

INTRODUCTION

In our previous study, we demonstrated that after intravenous injection bacteria gained entry and replicated in both solid tumors and their metastases in live animals.[1] The tumor-specific amplification process was visualized in real time based on luciferase-catalyzed luminescence, which served to reveal the locations of the tumors and metastases. *E. coli* DH5α and attenuated pathogens such as *V. cholerae* and *S. typhimurium* all gained entry into tumors and replicated. We found that, while the luminescence signal in *Vibrio*-colonized tumors increased exponentially overtime, the signal in *Salmonella*-infected tumors decreased from a peak time of 3-4 days post intravenous delivery. In this study, we investigated the possible reason for such signal decrease. Furthermore, we have also examined whether systemically injected bacteria could differentially colonize benign vs. malignant tumors.

MATERIALS AND METHODS

Cell Lines. C6 Rat glioma cell line was obtained from the American Type Culture Collection (ATCC, Manassas, VA, USA). Cells were cultured in Dulbecco's modified Eagle's medium (DMEM) supplemented with antibiotic-antimycotic solution (100 U/mL penicillin G, 250 ng/ml amphotericin B, 100 U/mL streptomycin; Sigma, St. Louis, MO, USA) and 10% fetal bovine serum (Invitrogen Corporation, Carlsbad, CA, USA) at 37°C under 5% CO_2 (all reagents except noted were supplied by Mediatech, Inc., Herndon, VA, USA).

Plasmid and bacteria. Plasmid pLITE201 carrying the *luxCDABE* expression cassette was transformed into *Salmonella typhimurium* or *Vibrio cholerae*, as described previously.[1]

Animal Studies. All mice were cared for and maintained in accordance with animal welfare regulations under an approved protocol by the Institutional Animal Care and Use Committee of LAB Research International Inc. (San Diego, CA, USA). Nude mice with subcutaneous C6 glioma tumors were developed (NCI:Hsd:Athymic Nude-***Foxn1***^{nu}, Harlan) to assess the biodistribution of bacteria. Tumors were generated by implanting C6 tumor cells (5×10^5 in 100 µL of PBS) subcutaneously on the right hind leg of 6-8 weeks old nude mice. C57BL/6-*Min* (*Min*/+) mice (The

Jackson Laboratory, Bar Harbor, ME, USA) develop benign tumors spontaneously throughout the intestinal tract.[2] *Min* mice have an average life span of 119 days.

Subcutaneous C6 tumor growth was recorded twice weekly in three dimensions using a digital caliper. Tumor volume was calculated as [(length × width × height)/2], and reported in mm^3. On day 14 after cell implantation (tumor volume ~ 300-500 mm^3), a single *i.v.* inoculation (1×10^8 cfu in 100 µL PBS) was delivered. At each designated time post-injection, 4 animals from each group were sacrificed for analysis of bacteria titer. The tumors and organs were excised and homogenized using MagNA Lyser (Roche Diagnostics, Indianapolis, IN, USA) at speed of 6,500 for 30 s. The bacterial titers were determined in duplicate on LB-amp agar plates.

Luminescence imaging. Immediately before imaging, mice were anesthetized with nembutal (60 mg/kg body weight). The animals were placed inside the dark box for photon counting and recording superimposed images (ARGUS 100, Hamamatsu, Japan). To image luminescent bacteria on agar plates and in excised intestines, the agar plates and intestinal tissues are placed directly inside the dark box of the imager. Photon collection was carried out for one minute. An image was then recorded and the low light image was then superimposed over the light image to record the location of luminescent activity.

RESULTS

Decreased luminescence signal in tumors is due to plasmid loss in *Salmonella*. In our previous study, we showed that, following systemic delivery, both *Vibrio cholerae* and *Salmonella typhimurium* were able to enter and replicate specifically in tumors.[1] The difference between the two types of bacteria, however, is that while the luminescence signal increase exponentially in tumors colonized by *Vibrio* within a week, the signal from *Salmonella*-colonized tumors decreased in a week from the peak time of 3-4 days. We went further to analyze the reason for such signal decrease. Since both bacteria carry the pLITE201 plasmid with ampicillin resistance, we analyzed ampicillin-resistant bacteria titer as well as total bacterial counts from grounded s.c. C6 glioma tumors on replica plates with or without drug selection, respectively. We found that, while the total numbers of bacteria found in tumors are identical, much lower drug-resistant colonies are found in *Salmonella*-infected tumors than in *Vibrio*-infected tumors (Fig. 1). We therefore concluded that pLITE201 plasmid is less stable in *Salmonella* than in *Vibrio*, and the decrease in luminescence signal in *Salmonella*-infected tumors is due to loss of plasmids and not due to clearance of bacteria from the tumors.

Bacterial colonization is not observed in benign tumors in *Min* mice. In our previous studies, we demonstrated the colonized of subcutaneous malignant adenocarcinoma tumors by bacteria and vaccinia virus.[1] In this study, we used mice with benign tumors and analyze whether these tumors could be equally colonized by systemically delivered bacteria. *Min* mice are known to carry autosomal dominant germline mutation in the adenomatous polyposis coli (*Apc*) gene and develop benign adenomas throughout the intestinal tract.[2]

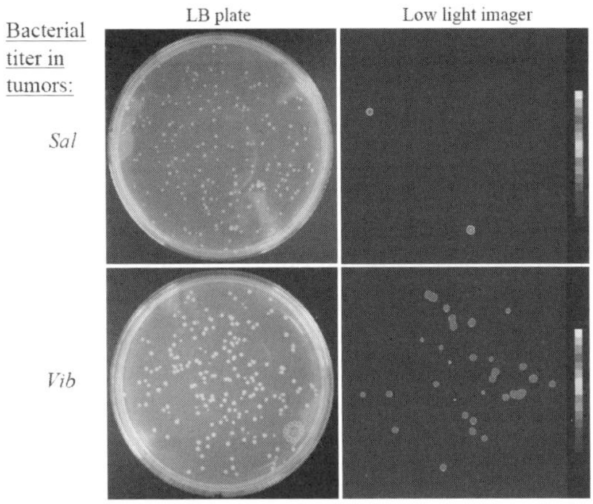

Fig. 1. Decreased luminescence signal in tumors is due to plasmid loss in *Salmonella. Sal: Salmonella typhimurium; Vib: Vibrio cholerae.*

We first i.v. injected 1×10^8 cfu of *Vibrio cholerae* in 8-week-old *Min* mice. One week after bacteria injection, mice are killed and the intestines are analyzed for the presence of tumors and bacteria colonization. A small number of tumors were found on the intestinal tract. However, when examined under the ARGUS100 low light imager, no luminescent bacteria were found in the tumors (Fig.2). This was confirmed by plating grounded tumor tissues on LB+amp agar plates. Therefore, we concluded that the systemically delivered bacteria were not able to colonize benign intestinal tumors in *Min* mice.

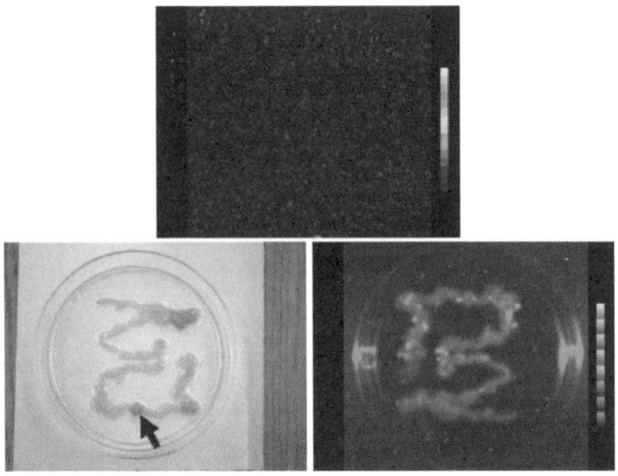

Fig. 2. Bacterial colonization is not observed in benign tumors of intestine in *Min* mice. Arrow is pointed at a tumor. Low light photon counting (top panel), bright field (lower left), and overlay (lower right) images are shown

DISCUSSION

In this study, we showed that systemically delivered *Salmonella* could colonize subcutaneous tumors as efficient as that of *Vibrio*. However, the reason for decreased luminescence signal overtime in *Salmonella*-infected tumors is due to loss of plasmids by *Salmonella*. The pLITE201 plasmid is more stable in *Vibrio* and therefore the bacterial luciferase was increasingly expressed and luminescence signal continue to increase along with replicating bacteria. We also showed that systemically delivered bacteria were not able to colonize benign tumors of intestinal tract in *Min* mice. One possible explanation is that tumor-colonization by bacteria is impaired in *Min* mice due to their immunocompetent status.[3] However, this is not likely to be the case because intravenously delivered bacteria have been shown to be able to enter and replicate specifically in spontaneous breast tumors in immunocompetent mice (unpublished data). A more likely explanation is that the benign tumors do not provide the same immune protection as the malignant tumors for the bacteria to permanently colonize. Further examination of the *Min* mice with spontaneous benign tumors could help to shed the light on the possible mechanisms of tumor colonization by bacteria.

REFERENCES

1. Yu YA, Shabahang S, Timiryasova TM, Zhang Q, Beltz R, Gentschev I, et al. Visualization of tumors and metastases in live animals with bacteria and vaccinia virus encoding light-emitting proteins. Nat Biotechnol 2004;22:313-20.
2. Moser AR, Pitot HC, Dove WF. A dominant mutation that predisposes to multiple intestinal neoplasia in the mouse. Science 1990;247:322–4.
3. Pierre F, Perrin P, Bassonga E, Bornet F, Meflah K, Menanteau J. T cell status influences colon tumor occurrence in min mice fed short chain fructo-oligosaccharides as a diet supplement. Carcinogenesis 1999;20:1953-6.

PART 5

BIOCHEMISTRY/LUMINESCENCE

DIRECT OBSERVATION OF A RADICAL-ION INTERMEDIATE IN THE CHEMIEXCITATION STEP OF PEROXYOXALATE CHEMILUMINESCENCE

EL BASTOS,[1] LFML CISCATO,[2] FH BARTOLONI,[2] WJ BAADER[2]

[1] Centro de Ciências Naturais e Humanas, Fundação Universidade Federal do ABC, Rua Santa Adélia, 166, Santo André, SP 09210-170, Brazil; [2] Instituto de Química, Universidade de São Paulo - Av. Lineu Prestes, 748 Bloco 12S, São Paulo – SP 05508-900, Brazil; E-mail: wjbaader@iq.usp.br

INTRODUCTION

The chemically initiated electron exchange luminescence (CIEEL) mechanism was proposed by Schuster in the late 1970's and represents the most general explanation for efficient bio- and chemiluminescence phenomena. It consists in a combination of the triplet chemiexcitation, observed in unimolecular decomposition or rearrangement of high-energy compounds (e.g. organic peroxides and hydroperoxides), and electrochemiluminescence, the light emission observed by annihilation of radical-ions generated electrochemically.[1]

The *intermolecular* variation of the CIEEL mechanism is widely associated with the chemiluminescent reaction of cyclic and linear peroxides with highly fluorescent polycondensate aromatic hydrocarbons, the so-called activators (ACTs). As light emission intensities and peroxide decomposition rate constants are both dependent on the oxidation potential of the activator used, an electron transfer process should take place in the rate-limiting step. This mechanism proposes the formation of a charge transfer complex between a low oxidation potential activator and the peroxide, followed by an irreversible electron transfer (ET) and loss of a neutral fragment or rearrangement. The resulting radical ion pair is annihilated, and the back electron transfer (BET) is able to release enough energy to generate the ACT in its first singlet-excited state. Despite the fact that this proposition is based on activated diphenoylperoxide decomposition, which is known to be an inefficient system,[2] it was also used to interpret the results obtained with the peroxyoxalate system, which shows chemiexcitation quantum yields (Φ_S) of *ca.* 0.3. In the latter reaction, a peroxidic high energy intermediate (HEI) is produced by the reaction between an oxalic ester and H_2O_2 in the presence of a base. The interaction between the HEI and the ACT, which is supposed to involve the CIEEL mechanism, results in chemiexcitation of the ACT.

Our group has pursued experimental evidence on the occurrence of the CIEEL mechanism, especially concerning the chemiexcitation step, and on the nature of the HEI in the peroxyoxalate system.[3,4] This work shows direct experimental evidence for the occurrence of the CIEEL in the reaction of bis(2,4,6-trichlorophenyl)oxalate (TCPO) with hydrogen peroxide in the presence of imidazole (IMI-H) and an ACT, as it presents direct experimental evidence on the formation of the carbon dioxide radical anion in this system. A radical adduct could be detected by electron paramagnetic resonance (EPR) spectroscopy in the reaction mixture in the presence of the spin trap α-phenyl-N-*t*-butylnitrone (PBN) and is attributed to the PBN-CO_2 radical anion adduct (Fig. 1).

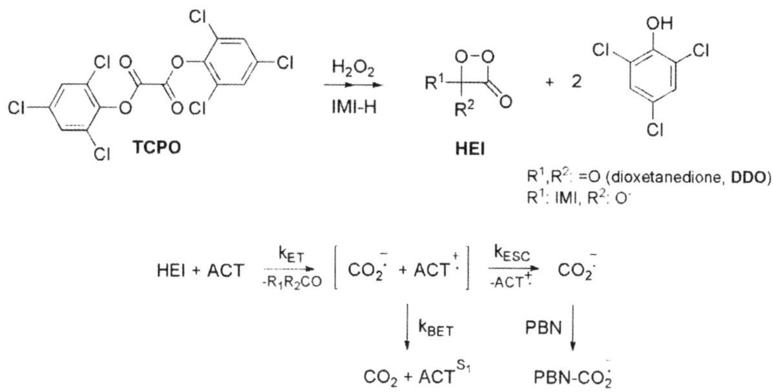

Fig. 1. Peroxyoxalate reaction and formation of the PBN-$CO_2^{\cdot -}$ adduct

RESULTS AND DISCUSSION

The reaction between TCPO and H_2O_2 in the presence of imidazole, 9,10-diphenylanthracene (DPA) and PBN was initially performed in anhydrous ethyl acetate as solvent. In addition to the complete system, the following control experiments were performed: absence of TCPO, absence of DPA, absence of TCPO and DPA, absence of DPA and H_2O_2, and absence of IMI-H. The complete system resulted in a characteristic signal of a PBN adduct with six peaks (Fig. 2A). No signal was observed in absence of TCPO or H_2O_2. However, also in absence of DPA a lower intensity signal was observed. This intriguing result (since radical formation depends on the presence of an ACT) can be explained considering the oxidation potential of PBN (1.47 V *vs.* SCE), which is within the range of conventional ACTs. In this way, PBN may act as an ACT and consequently the PBN-$CO_2^{\cdot -}$ adduct can also be observed in the absence of an ACT.

The experiments were also performed in acetonitrile in order to obtain EPR signals with better resolution. Unfortunately, the peaks observed for the PBN adduct in acetonitrile were broad, similar to that obtained in ethyl acetate (Fig. 2B). This broadening is probably due to the presence of dissolved residual oxygen. Furthermore, since the hyperfine coupling constants in ethyl acetate were never described, we decided to acquire the EPR spectra in water. As activated oxalic esters easily hydrolyze in water, the reaction was carried out in ethyl acetate, the solvent was removed, and deionized water was added to the residue. The resolution of the signal obtained by this methodology was much improved (Fig. 2C). Hyperfine coupling constants are listed in Table 1, along with that of literature.

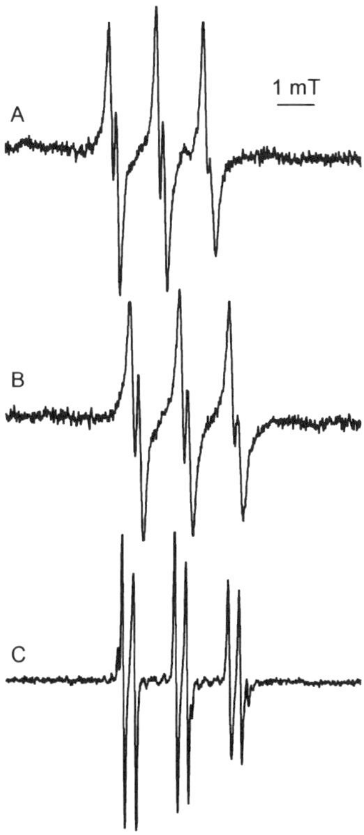

Fig. 2. EPR spectra of the observed PBN adduct in: (A) ethyl acetate, (B) acetonitrile, and (C) water. [TCPO]: 0,1 mmol/L; [H_2O_2]: 10 mmol/L; [IMI-H]: 1.0 mmol/L; [DPA]: 1.0 mmol/L; [PBN]: 50 mmol/L

Table 1. Hyperfine coupling constants for the PBN adduct in several solvents

	a^N (mT)	a^H (mT)	a^N/a^H	Ref
Ethyl Acetate	1.33	0.19	6.81	
Acetonitrile	1.40	0.23	5.97	
Water	1.49	0.32	4.65	
PBN-$CO_2^{\cdot -}$ (Acetonitrile)	**1.44**	**0.23**	**6.26**	4
PBN-$CO_2^{\cdot -}$ (Benzene)	**1.39**	**0.23**	**6.04**	4
PBN-$CO_2^{\cdot -}$ (Water)	**1.59**	**0.46**	**3.46**	5

The observed discrepancy between the hyperfine constants reported for the PBN-$CO_2^{\bullet -}$ adduct in water can be explained by the presence of residual ethyl acetate. However, the values measured in acetonitrile are in better agreement with that reported in literature.[5,6] One should notice that $CO_2^{\bullet -}$ is usually generated by much cleaner systems (e.g., formate/formic acid) and that the components present in peroxyoxalate reaction media would interfere with coupling constants.[7]

CONCLUSIONS

We report here, for the first time, experimental evidence on the occurrence of the carbon dioxide anion radical in the chemiexcitation step of the peroxyoxalate reaction. Therefore this work constitutes the first experimental report on a reaction step in the CIEEL sequence which occurs after the initial electron transfer from the ACT to the peroxide intermediate, adding important evidence on the validity of the widely quoted but still not proven CIEEL mechanism.

ACKNOWLEDGEMENTS

Financial support for this work has been provided by Fundação de Amparo à Pesquisa do Estado de São Paulo – FAPESP, Coordenação de Aperfeiçoamento de Pessoal de Nível Superior – CAPES and Deutscher Akademischer Austauschdienst – DAAD. We thank Prof. Dr. Ohara Augusto and Dr. Edlaine Linares from IQ-USP, for the use of the EPR facilities.

REFERENCES

1. Schuster GB. Chemiluminescence of organic peroxides - Conversion of ground state reactants to excited state products by the chemically-initiated electron exchange luminescence mechanism. Acc Chem Res 1979; 12: 366.
2. Baader WJ, Stevani CV, Bastos EL. Chemiluminescence of Organic Peroxides. In: Rappoport Z (ed.) The Chemistry of Peroxides. Chichester: Wiley, 2006: 1211-79.
3. Stevani CV, Silva SM, Baader WJ. Studies on the mechanism of the excitation step in the peroxyoxalate chemiluminescence. Eur J Org Chem 2000; 4037-46.
4. Stevani CV, Lima DF, Toscano VG, Baader WJ. A kinetic study on the peroxyoxalate reaction: imidazole as nucleophilic catalyst. J Chem Soc Perkin Trans.2 1996; 5; 989-95.
5. Rehorek D, Benedix M, Thomas P. Photocatalytic systems.13. Detection of oxalate free-radicals by EPR spin trapping in the photolysis of tris(oxalato)iron(III). Inorg Chim Acta 1977; 25: L100.
6. Connor HD, Thurman RG, Galizi MD, Mason, RP. The formation of a novel free-radical metabolite from CCl4 in the perfused-rat-liver and in vivo. J Biol Chem 1986, 261: 4542.
7. Vasquez-Vivar J, Denicola A, Radi R, Augusto O. Peroxynitrite-mediated decarboxylation of pyruvate to both carbon dioxide and carbon dioxide radical anion". Chem Res Toxicol 1997;10:786.

KINETIC OBSERVATION OF THE CHEMIEXCITATION STEP IN PEROXYOXALATE CHEMILUMINESCENCE

LFML CISCATO, FH BARTOLONI, WJ BAADER
Instituto de Química da Universidade de São Paulo - Av. Lineu Prestes, 748 Bloco 12S, São Paulo – SP 05508-900, Brazil, E-mail: wjbaader@iq.usp.br

INTRODUCTION

The developments of the application of the peroxyoxalate reaction in the last 25 years in analytical chemistry have made possible the detection of traces of H_2O_2 and derivatized fluorophores.[1] Mechanistic studies and the elucidation of the reaction steps leading to light emission can bring more knowledge which may lead to new developments and improved detection methods (Fig. 1).[2,3]

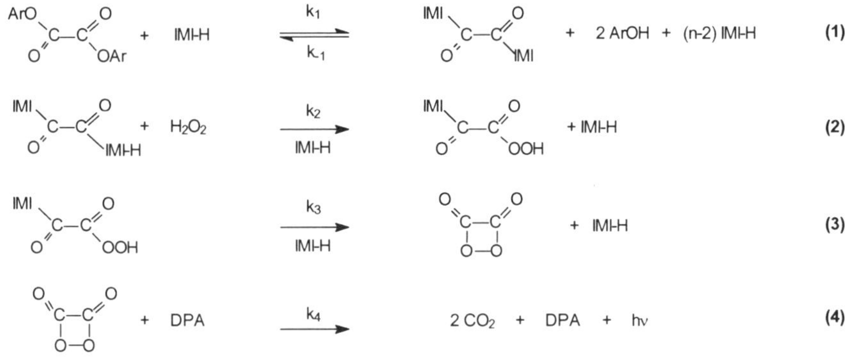

Fig. 1. Mechanism of the imidazole-catalyzed peroxyoxalate reaction

In this work, a new approach has been made to study the chemiexcitation step in peroxyoxalate chemiluminescence, where the fluorescent activator (ACT) is added to the reaction mixture after a delay time. The high-energy intermediate (HEI) is then accumulated in the reaction mixture and interacts with a measurable kinetic rate constant with the ACT, presumably by the CIEEL mechanism (Chemically Initiated Electron Exchange Luminescence, Fig. 2).[3,4]

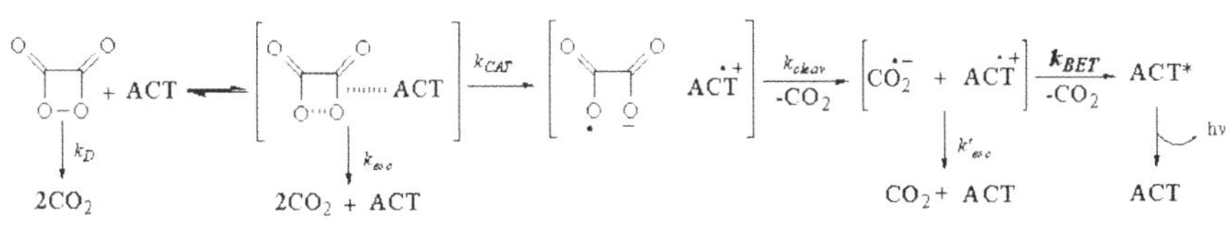

Fig. 2. Chemiexcitation in the peroxyoxalate reaction according to the CIEEL mechanism

The rate constant of the interaction of the high-energy intermediate and the ACT (k_{CAT}) and the rate of dark unimolecular decomposition of the HEI (k_D) could be determined directly for the first time, using the method described here. The activation parameters of the reaction could also be determined by measuring the reaction rates at different temperatures.

MATERIALS AND METHODS

The kinetic studies were performed by measuring the chemiluminescence emission, using a spectrofluorimeter equipped with a thermostated cell holder, using oxalyl chloride,[5] hydrogen peroxide and different ACT as reactants. This complete system reacts emitting light immediately, but if the ACT is added after a delay time, a burst of light can be observed and detected, with a rate at least 10 times higher than that of the complete system (Fig. 3).

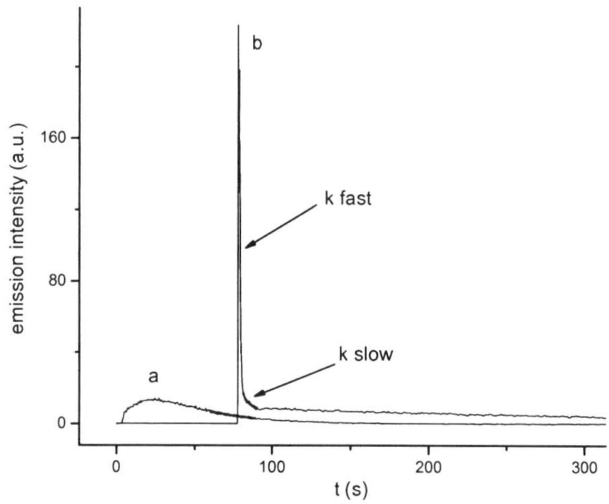

Fig. 3. Light emission kinetics of the reaction between oxalyl chloride and hydrogen peroxide in the presence of DPA. a: complete system; b: delayed addition of DPA

The kinetic data can be fitted with a monoexponential function to obtain an observed rate constant (k_{obs}), and the intensity of the light burst depends on the injection time of the ACT. Performing the experiments with different ACT concentrations injected after the same delay time and determining k_{obs} of the emission decay, the values of k_D and k_{CAT} can be determined according to the equation $k_{obs} = k_D + k_{CAT}[ACT]$. For the determination of activation parameters, k_{obs} was determined in the temperature range between 20 and 30°C, with 9,10-diphenylanthracene (DPA) as ACT.

RESULTS AND DISCUSSION

Values of k_{CAT} between 160000 and 60 $M^{-1}s^{-1}$ and an average value for k_D of 0.04 ± 0.02 s^{-1} were obtained for the different ACTs studied. Relating the k_{CAT} values with the half-wave oxidation potential (E_{ox}) of the ACTs, obtained by means of cyclic voltammetry, by the plot $\ln(k_{CAT})$ *vs.* E_{ox}, a linear correlation is observed (Fig. 4). This correlation is the first direct experimental evidence for the occurrence of an electron transfer from the ACT to the HEI in the rate-limiting step of the chemiexcitation in the peroxyoxalate reaction. An electron transfer correlation coefficient of α = 0.20 ± 0.03 is obtained from the linear fit of this free-energy correlation.

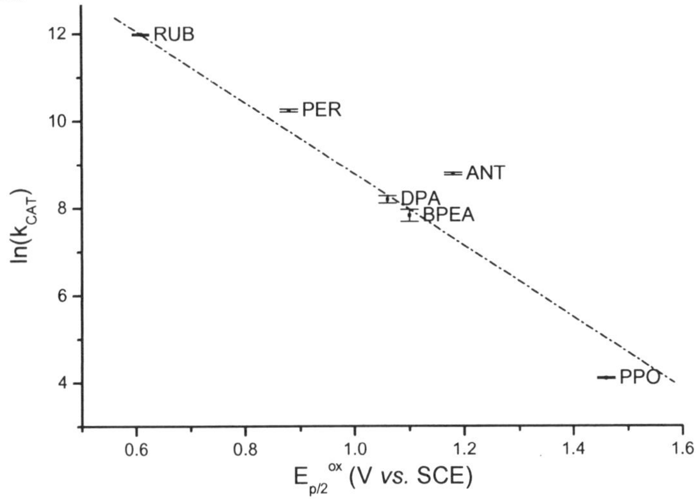

Fig. 4. Linear free-energy correlation between the catalytic rate constant (k_{CAT}) and the oxidation potential (E_{ox}) for various activators. RUB: rubrene, PER: perylene, ANT: anthracene, BPEA: 9,10-bis-(phenyl-ethynyl)anthracene and PPO: 2,5-diphenyloxazole

Using DPA as activator and performing the experiments with different DPA concentrations at different temperatures, a linear correlation between k_{obs} and DPA concentration has been observed for temperatures of 20, 25 and 30°C, and values of k_D and k_{CAT} could be determined. The Eyring plot of those values does not show systematic dependence of k_D with the temperature, indicating a very low activation energy for the unimolecular HEI decomposition. However, the k_{CAT} values correlates well with temperature and the following activation parameters have been obtained: $\Delta H^{\neq}$ = +10.5 ± 0.9 kcal.mol^{-1}, $\Delta S^{\neq}$ = -6.0 ± 3.0 cal.K^{-1}.mol^{-1}, $\Delta G^{\neq}$ = +12.3 ± 0.9 kcal.mol^{-1} (25 °C).

CONCLUSIONS

Direct experimental evidence has been obtained for the accumulation of an HEI in the peroxyoxalate reaction in certain experimental conditions, and the rate constants related to the chemiexcitation step, k_D and k_{CAT}, could be directly determined for the first time in this work. The value determined for k_D is independent of the ACT used, which is consistent with the proposed mechanism for the peroxyoxalate reaction. The value of $\alpha = 0.2$ for the electron transfer coefficient indicates an early transition state for this process. The activation parameters obtained from the dependency of k_{CAT} on the temperature support the existence of a bimolecular process in the chemiexcitation step.

ACKNOWLEDGEMENTS

Financial support for this work has been provided by Fundação de Amparo à Pesquisa do Estado de São Paulo – FAPESP, Coordenação de Aperfeiçoamento de Pessoal de Nível Superior – CAPES, Conselho Nacional de Pesquisa (CNPq) and Deutscher Akademischer Austauschdienst – DAAD.

REFERENCES

1. McCapra F. Chemical generation of excited states: The basis of chemiluminescence and bioluminescence. Methods Enzymol 2000; 305:3-47.
2. Stevani CV, Lima DF, Toscano VG, Baader WJ. A kinetic study on the peroxyoxalate reaction: imidazole as nucleophilic catalyst. J Chem SoC Perkin Trans 2 1996;5:989-95.
3. Stevani CV, Silva SM, Baader JW. Studies on the mechanism of the excitation step in the peroxyoxalate chemiluminescence. Eur J Org Chem 2000;4037-46.
4. Schuster GB. Chemi-luminescence of organic peroxides - Conversion of ground-state reactants to excited-state products by the chemically-initiated electron-exchange luminescence mechanism. Acc Chem Res 1979;12:366-73.
5. Rauhut MM, Roberts BG, Semsel AM. A Study of chemiluminescence from reactions of oxalyl chloride hydrogen peroxide and fluorescent compounds. J Am Chem Soc 1966;88:3604.

SYNTHESIS AND STUDIES OF FENCHYL-DERIVED 1,2-DIOXETANES

LFML CISCATO,[1] D WEISS,[2] HJ FLAMMERSHEIM,[2] WJ BAADER[1]
[1] *Instituto de Química da Universidade de São Paulo - Av. Lineu Prestes, 748 Bloco 12S, São Paulo SP 05508-900, Brazil;* [2] *Friedrich Schiller Universität, Humboldtstr. 10, Jena D-07743, Germany, E-mail: wjbaader@iq.usp.br*

INTRODUCTION

Studies on chemiluminescent reactions contributes for a better understanding of bioluminescence phenomena and improvement of analytical methods.[1] These two perspectives converge through mechanistic investigation of the chemical formation of excited states.

Most reactions that result in high chemiluminescence quantum yields involve four-ringed peroxides called 1,2-dioxetanes. Induced decomposition of 1,2-dioxetanes is unlikely to involve electron transfer, and the high quantum yields for excited states formation can be justified by the CIEEL (Chemically Induced Electron Exchange Luminescence) mechanism.[2] This mechanism involves two electron transfers and is a controversial mechanism that, however, was used in rationalization of several process of chemi- and bioluminescence.[3]

The investigation of the CIEEL mechanism resulted in the development of new compounds widely used in several analytical methods leading to a decrease of radioisotope use. The induced decomposition of 1,2-dioxetanes and acridinium esters by chemicals and enzymes,[4] and the use of peroxyoxalate system provides for a diversity of analytical applications.

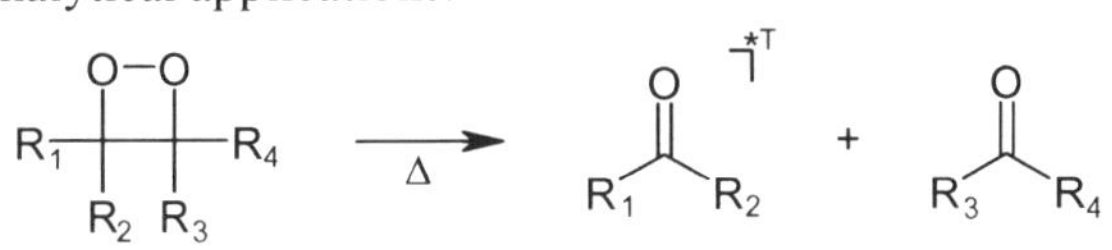

Fig. 1. Thermal decomposition of 1,2-dioxetanes

The thermal cleavage of 1,2-dioxetanes generates two carbonyl fragments (Fig. 1), one of which can be formed in an electronically excited state, predominantly in the triplet state, except for electron-rich substituents, for which excited singlet state formation may prevail.[5]

The thermal stability of 1,2-dioxetanes seems to be related to the steric hindrance of the substituent, so the fenchyl group[6] was chosen to develop a new class of stables 1,2-dioxetanes. Fenchone (1,3,3-trimethyl-bicyclo[2.2.1]heptan-2-one) is a very hindered ketone and has low reactivity toward nucleophilic attack. The yield of classical olefination reactions (Wittig, McMurry, Horner-Wadsworth-Emmons) with fenchone are very low, achieving maximally 15% in optimized conditions. However, a new synthetic route was developed to obtain the fenchyl-derived enol ethers, involving the Barton-Kellogg reaction,[7-10] which consists of the condensation of a diazocompound and a thioester generating the enol ether in high yields (Fig. 2) (steric hindrance is not a problem in this kind of reaction).

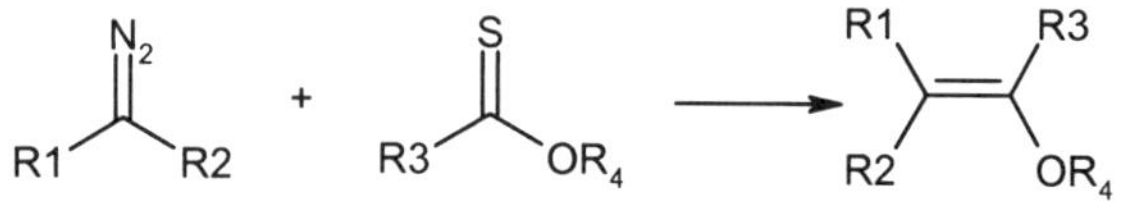

Fig. 2. The Barton-Kellog olefination reaction

RESULTS

Three 1,2-dioxetanes were synthesized, two of them not yet reported in the literature (Scheme 1). All three bear the fenchyl moiety and are relatively stable at room temperature. All the 1,2-dioxetanes were obtained by photooxygenation of the correponding enol ethers, which were prepared by Barton-Kellogg olefination for 1 and 2. Dioxetane 3 is reported in the literature;[11] however no activation parameters for its unimolecular decomposition have been determined.

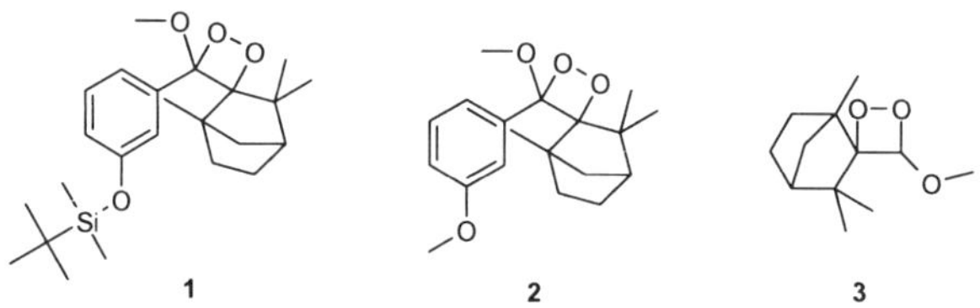

Scheme 1. 1,2-dioxetanes

The 1,2-dioxetane 1 produces a bright blue chemiluminescence when treated with tetrabutylammonium fluoride. The thermal stability of the dioxetanes was measured by two methods: one is the isothermal method (IM), measuring the kinetics of light emission at different temperatures; the other is differential scanning calorimetry (DSC), which measures the instantaneous heat flow, which can be converted to the rate of the dioxetane decomposition (Fig. 3). This method simultaneously allows the evaluations of both the activation parameters and the heat of reaction. The investigation of the reaction with a number of different heating rates supplies more reliable kinetic results than isothermal scans in a relatively narrow temperature window.

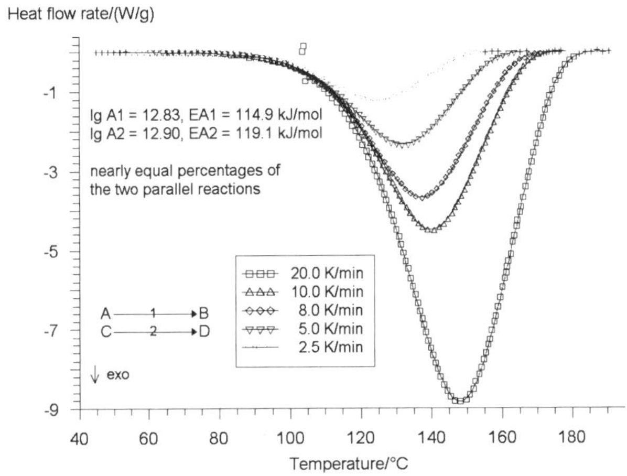

Fig. 3. DSC measurement of 3 with five heating rates (symbols) and the recalculated curves (lines)

The thermal decomposition of 1,2-dioxetanes is a classical first order process. The high quality of data produced by the DSC measurements allows the calculation of an ideal first order process, fitting perfectly in the case of adamantyl based dioxetanes.[12-14] In the case of the fenchyl-based dioxetanes, the recalculation fits only if we use an overlay of two nondependent first order reactions with slightly different activation parameters, but with nearly equivalent concentrations. This surprisingly fact we explain as a result of the synthesis and the geometry of the resulting 1,2-dioxetanes. The 1,2-dioxetanes 1 and 2 produced were obtained as a mixture of diastereisomers which could not be resolved by chromatography. The measure of the stability of the isomeric mixture showed the presence of compounds with two different stabilities (Table 1).

Table 1. Activation parameters for 1,2-dioxetanes 1-3

1,2-Dioxetane	Method	Ea (kcal.mol^{-1})	ΔH (total, kcal.mol^{-1})
1	DSC	26.3 24.6	86.2
2	DSC	28.1 26.9	81.6
3	IM	28.7	

The E_a of the most stable 1,2-dioxetane known, with two adamantyl groups, is ca. 34 kcal.mol^{-1}, and of the named "stable dioxetanes" is between 25 and 30 kcal.mol^{-1}. The data show that the fenchyl-derived 1,2-dioxetanes can be considered stable ones. The synthesis proved to be efficient, due to the Barton-Kellogg method, which generates very hindered enol-ethers, no matter how bulky the starting product.

ACKNOWLEDGEMENTS
Financial support for this work has been provided by Fundação de Amparo à Pesquisa do Estado de São Paulo – FAPESP, Coordenação de Aperfeiçoamento de Pessoal de Nível Superior – CAPES and Deutscher Akademischer Austauschdienst – DAAD, by means of a sandwich grant BEX1132/04-0.

REFERENCES

1. Adam W. Thermal generation of electronic excitation with hyperenergetic molecules. Pure Appl Chem 1980;52:2591-608.
2. Schuster GB. Chemiluminescence of organic peroxides: conversion of ground-state reactants to excited-state products by the chemically-initiated electron-exchange luminescence mechanism. Acc Chem Res 1979;12:366-73.
3. McCapra F. Chemical generation of excited states: the basis of chemiluminescnece and bioluminescence. Method Enzymol 2000;305:370-90.
4. Lee S, Singer LA. Structural effects on the intramolecular electron-transfer induced decomposition of a series of 1,2-dioxetanes derived from 9-alkylidene-10-methylacridans. J Am Chem Soc. 1980;102:3823-9.
5. Adam W, Reinhardt DJ. Indole dioxetanes and epoxides by oxidation of n-acylated indoles with singlet oxygen and dimethyldioxirane - kinetics and chemiluminescence yields of the thermal dioxetane decomposition and fluoride ion-induced CIEEL emission. Chem Soc Perkin Trans 2 1994:1503-7.
6. Kondakow JJ. Über fenchylderivate. J Prakt Chem 1908;79:271-80.
7. Barton DHR, Willis BJ. Olefin synthesis by 2-fold extrusion processes 1. Preliminary experiments. J Chem Soc Perkin I 1972; 305-10.
8. Barton DHR, Guziec FS, Shahak IJ. Olefin synthesis by two-fold extrusion processes 2. Synthesis of some very hindered olefins. Chem Soc Perkin I 1974:1794-9.
9. Barton DHR, Back TG, Britten-Kelly, Guziec FS. Olefin synthesis by 2-fold extrusion processes 3. Synthesis and properties of hindered selenoketones (selones). J Chem Soc Perkin I 1976:2079-89.
10. Kellogg RM, Wassenaar S, Buter J. Thiocarbonyl ylides: generation, properties and reactions. J Org Chem 1972;37:4045-60.
11. Meijer EW, Wynberg H. Singlet oxygen oxidation of enol ethers - synthesis of optically-active 1,2-dioxetanes-II. Tetrahedron Lett 1979;41:3997-4000.
12. Bastos EL, Ciscato LFML, Weiss D, Beckert R, Baader WJ. Comparison of convenient alternative synthetic approaches to 4-[(3-tert-butyldimethylsilyloxy) phenyl]-4-methoxyspiro[1,2-dioxetane-3,2'-adamantane]. Synthesis 2006;11:1781-6.
13. Adam W, Encarnacion LAA, Zinner K. Thermal stability of spiro[adamantane-[1,2]-dioxetanes]. Chem Ber 1983;116:839-46.
14. Flammersheim HJ, Weiß D, Opfermann JR. Investigations on the thermal decomposition of 1,2-dioxetanes via DSC - some peculiarities concerning the sample preparation. J Therm Anal Cal 2004;77:883-92.

ENZYME EMBEDDED ORGANIC MONOLITH SENSOR FOR FLOW INJECTION: CHEMILUMINESCENCE

OSAMU NOZAKI,[1] MOTONORI MUNESUE,[2] HIROKO KAWAMOTO[3]

[1] *Dept of Clinical Laboratory Medicine, Kinki University School of Med. Japan, 377-2 Ohno-Higashi, Osaka-Sayama, 589-8511, Japan;*
[2] *Chemco Scientific Co., Ltd., 2-8-14, Nakazaki, OSAKA, 530-0016, Japan;*
[3] *Dept of Biological Regulation, School of Health Science, Faculty of Medicine, Tottori University, 86 Nishimachi, Yonago, 683-8503, Japan*

INTRODUCTION

A novel method for making a fixed enzyme reactor for flow injection - chemiluminescence was developed. Fixed enzyme reactors have been used for flow injection - chemiluminescence (FI-CL), and are useful tools for continuous assay of enzyme substrates. However, there are some problems in fabricating fixed enzyme reactors.[1] These include: 1) The complicated and time consuming fabrication process. 2) Difficulty in fitting the reactor into a narrow space, 3) Optical properties for chemiluminescent detection and 4) Size of the connector for the reactor. This study aimed to solve the problems by developing an oxidase enzyme embedded monolith in a capillary glass tube that needs only one step photo-polymerization process at room temperature.[2] Two kinds of the reactors were prepared. One was an horseradish peroxidase (HRP) embedded reactor, and the other mixed HRP and glucose oxidase (GOD) embedded reactor. We employed the flow injection - HRP catalyzed imidazole chemiluminescence reaction because the method was suitable for assay of water-soluble specimens.[3] The analytical principles of the imidazole chemiluminescence for assay of hydrogen peroxide (H_2O_2) and glucose are as follows:

1.1 H_2O_2 + imidazole (alkaline pH)→ imidazole hydroperoxide
1.2 Imidazole hydroperoxide + HRP→ light emission

2.1 Glucose + GOD → H_2O_2 + alpha – gluconolactone
2.2 H_2O_2+ imidazole → imidazole hydroperoxide
2.3 Imidazole hydroperoxide + HRP → light emission

MATERIALS AND METHODS

Enzyme embedded organic monolith. HRP (14 mg), or a mixture of GOD (12 mg) and HRP (14 mg) were mixed with the solution consisted of 3-Methacryloxypropyltrimethoxysilane (MPTMS) 83 μL, hydrochloric acid (0.1 mol/L) 17 μL, 2-hydroxy-2-methyl-1-phenylpropan-1-one (DAROCUR 1173,

Ciba specialty chemicals, Tokyo, Japan) 20 μL, toluene (190 μL) and n-propanol (190 uL). The solution was introduced into the glass capillary tube (75 mm long, 1.5 mm i.d., Terumo, Tokyo, Japan) by aspiration. The glass capillary tube was then irradiated in a UV reactor (Chemco, Osaka, Japan) for 12 h at a room temperature.

A chemiluminescence system for image assay. The enzyme embedded organic monolith reactor (18 mm long, 0.85 mm i.d.) was set in the chamber of the CCD-based chemiluminometer (NightOWL NC100, Berthold, Germany), which was set in a flow injection system. A mobile phase solution (imidazole (0.1 mol/L)- Tricine (0.05 mol/L) solution, pH8.6) was delivered with a HPLC pump (PU-850, JASCO, Tokyo, Japan) at the flow rate of 0.1 mL/min. Specimen (H_2O_2 or glucose; 50 μL) was injected with a manual injector into the stream of the mobile phase.

A chemiluminescence system for quantitative assay. The enzyme embedded organic monolith reactor was set in a cell holder of the chemiluminometer with a photomultiplier (825-CL, JASCO), and placed close to the photomultiplier. The two kinds of mobile phase (S1: water and S2: imidazole (0.1 mol/L) -Tricine (0.05 mol/L) were delivered with two HPLC pumps (PU-850, JASCO) at the flow rate of 0.1 mL/min, respectively. Specimens (H_2O_2 or glucose solutions; 50 μL) were injected into the stream of S1 with an auto sampler (AS-950, JASCO).

RESULTS AND DISCUSSION

Enzyme monolith reactor. A picture of the outside appearance of the monolith was taken with a microscope (MIC - D, OLYMPUS, Tokyo, Japan), and revealed that the monolith layer was attached to the inside wall of the capillary glass tube (Fig. 1). The vertical section of the monolith was observed with a microscope. The view of the section revealed that the center of the reactor was hollow. This was the reason that the back pressure from the monolith in a flow injection - chemiluminescence system was zero and why the connector (a silicon tube) of the monolith did not leak fluid from the connection site. This simplified the connection system of the monolith.

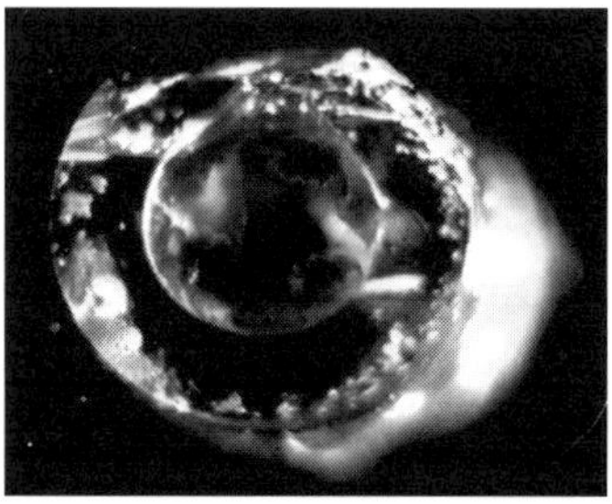

Fig. 1. Vertical section of the monolith reactor.
The center of the monolith revealed hollow

Imaging assay of the monolith reactor. Activity of the enzyme fixed in a photo polymerized monolith reactor was examined by imaging. An image of light emission with H_2O_2 (970 μmol/L, 20 μL) was taken qualitatively with the HRP embedded monolith (Fig. 2). The whole area of the monolith glowed when a H_2O_2 specimen reacted with the monolith. Light intensity (36999, 80950, 328618 RLU, respectively) increased in proportion to amount of H_2O_2 (97 μmol/L; 5, 10, 20 μL) injected. The GOD and HRP mixed monolith was used to detect light from glucose reactions. Light intensity of the image (11447, 14322, 15311 RLU, respectively) increased in proportion to amount of glucose (1.0 g/L; 5, 10, 20 μL) injected.

Fig. 2. A picture of light emission at the HRP embedded monolith reactor by injection of H_2O_2 specimen. The H_2O_2 specimen injected (20 μL) was 970 μmol/L

Quantitative assay of the monolith reactor. Light emitted from the enzyme monolith was assayed quantitatively with a chemiluminometer with a photomultiplier for consecutive and automatic assay of the enzyme substrates. H_2O_2 specimens (0, 122, 243, 485, 970, and 9700 μmol/L) were assayed with the HRP embedded monolith reactor. The chromatogram was obtained by consecutive injection of H_2O_2 specimens (0, 122, 243, 485, 970, 9700 μmol/L) (Fig. 3).

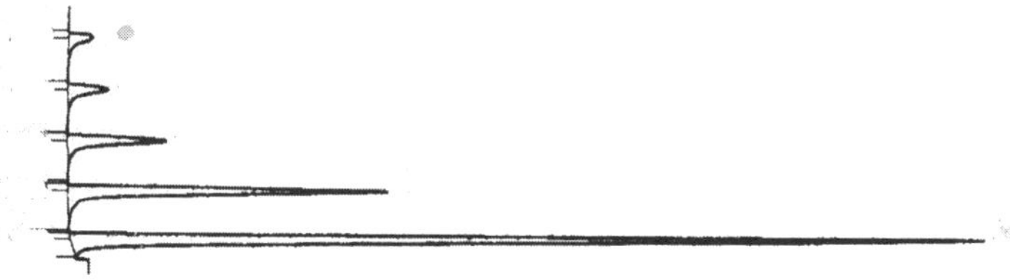

Fig. 3. Consecutive assay of H_2O_2 specimens for a calibration curve. The H_2O_2 specimens (50 μL) assayed were 0, 122, 243, 485, 970, 9700 μmol/L

Peak areas increased with the increase of H_2O_2 concentration [regression equation of the calibration curve was $Y = 0.698 X^2 + 442 X + 25650$ ($R^2 = 0.998$), where Y was light intensity (RLU) and X was concentrations of H_2O_2 (μmol/L)]. Lower detection limit for H_2O_2 (S/N = 2) was 21.4 μmol/L. Reproducibility for assaying H_2O_2 was CV=3.87 % (n = 20), 2.19 % (n = 7), and 14.3 % (n = 6) when H_2O_2 was 970 μmol/L, 485 μmol/L, and 122 μmol/L, respectively. When glucose specimens (0, 62.5, 125, 250, 500 mg/dL) were injected consecutively into the mixed GOD and HRP monolith reactor, the peak areas increased with the increase of glucose concentration [regression equation of a calibration curve was $Y = -0.173 X^2 + 177 X + 4238$ ($R^2 = 0.97$) where Y was light intensity (RLU) and X was concentration of glucose (mg/dL)] (Fig. 4).

Fig. 4. Consecutive assay of glucose specimens for a calibration curve.
The glucose specimens (50 μL) assayed were 0, 62.5, 125, 250, 500 mg/dL

The lower detection limit for glucose was 16.9 mg/dL (S/N = 2). Reproducibility of glucose assay was CV= 8.7 % (n = 5) ats 500 mg/dL. These results revealed that the enzyme embedded monolith was accessible for quantitative assay of the substrates. In conclusion, the novel enzyme embedded reactor was made by the photo polymerization method using mild conditions and one-step procedure. The enzyme reactor was effective in chemiluminescent assays to determine H_2O_2 or glucose using both imaging and quantitative assays coupled to imidazole chemiluminescence.

REFERENCES

1. Nozaki O, Kawamoto H. Determination of hydrogen peroxide by micro flow injection - chemiluminescence coupled the flow cell reactor in a chemiluminometer. Luminescence 2000;15: 137-42.
2. Svec F. Preparation and HPLC applications of rigid macroporous organic polymer monoliths. J Separ Sci 2004; 27: 747-66.
3. Nozaki O, Kawamoto H. Determination of hydrogen peroxide by micro-flow injection-horseradish peroxidase catalyzed "Imidazole chemiluminescence". In: Bioluminescecnce and Chemiluminescence: Progress and Current Applications. Stanley PE, Kricka LJ eds. Singapore:World Scientific 2002:335-8.

EFFICIENCY OF ELECTRON-TRANSFER INDUCED CHEMIEXCITATION: A COMPARISON OF INTER- AND INTRAMOLECULAR PROCESSES

MA OLIVEIRA AND WJ BAADER
Instituto de Química, Universidade de São Paulo - Av. Lineu Prestes, 748 Bloco 12S CEP 05508-900, São Paulo – SP, Brazil. E-mail: jbaader@iq.usp.br

INTRODUCTION

The CIEEL (Chemically Initiated Electron Exchange Luminescence) mechanism has been utilized for the rationalization of a wide variety of bioluminescence and chemiluminescence (CL) processes,[1,2] however, the efficiency for excited state formation of the originally studied CIEEL active systems has been questioned due to the re-determination of the quantum yields for the extensively studied diphenoyl peroxide system.[1] Although 1,2-dioxetanes containing simple alkyl and aryl substituents do not undergo catalyzed decomposition according to the CIEEL mechanism, 1,2-Dioxetane derivatives containing easily oxidizable substituents undergo fast decomposition, which is thought to be initiated by an intramolecular electron transfer from the electron-rich substituent to the peroxidic ring and leads to the efficient formation of singlet excited cleavage products.[2]

1,2-Dioxetanes containing protected electron-rich phenolate groups, which can by deprotected by the action of appropriate chemical or enzymatic trigger reagents, constitute the bases of several commercial immunoassay detection kits.[1]

The induced decomposition of phenoxy-substituted 1,2-dioxetanes is initiated by an intramolecular electron transfer from the phenolate – generated by the deprotecting reagent (fluoride in the case of silyl-protected phenolates) – to the dioxetane ring, simultaneously with O-O bond cleavage (Fig. 1).

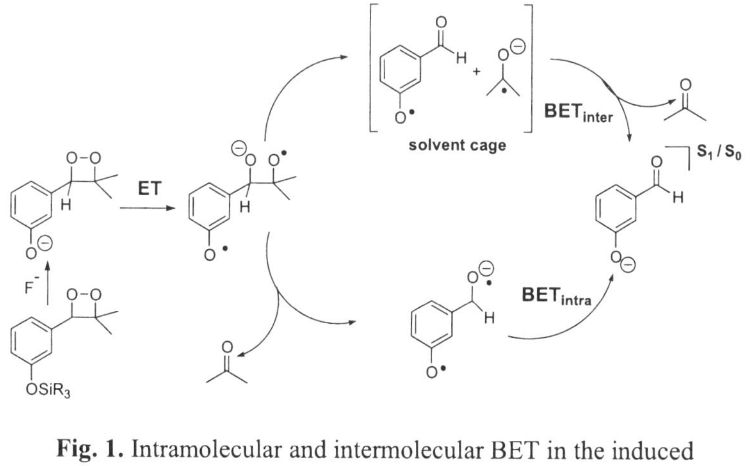

Fig. 1. Intramolecular and intermolecular BET in the induced decomposition of phenoxyl-substituted 1,2-dioxetanes

Cleavage of the C-C bond in the resulting diradical anion may occur in two different fashions: (i) leading to a neutral carbonyl fragment and a new diradical anion, which upon *intramolecular* back-electron transfer (BET) may lead to efficient phenolate singlet excited state formation (Fig. 1, BET_{intra}); (ii) a pair of radicals may form, which can undergo *intermolecular* BET within the solvent cage, resulting in formation of the phenolate in its ground or excited state (Fig. 1, BET_{inter}).

Recent work on the solvent viscosity dependence (solvent-cage effect) of the excitation efficiencies in the induced decomposition of CIEEL active spiro-adamanty-substituted 1,2-dioxetanes appear to demonstrate the occurrence of the intermolecular BET in this kind of transformation.[1] However, the occurrence of an intermolecular BET in the induced 1,2-dioxetane decomposition is in contrast to the quantum yields observed for intra and intermolecular CIEEL systems. With the notable exception of the peroxyoxalate system,[1] *intermolecular* CIEEL systems appear to show very low singlet quantum yields, whereas *intramolecular* systems posses high quantum efficiencies.[2]

RESULTS AND DISCUSSION

The recent determination of the singlet quantum yields for the catalyzed decomposition of the peroxides 1 and 2 (Table 1) clearly indicates that the initially reported values for two of the most investigated CIEEL active peroxides were overestimated in several orders of magnitude. The value obtained for diphenoyl peroxide in this work agrees well with the results of the re-determination reported by Catalani and Wilson.[3] Moreover, a comparison of the quantum yields obtained for inter and *intramolecular* CIEEL systems indicates that the *intramolecular* CIEEL can occur with very high singlet quantum yields, whereas the classical intermolecular CIEEL appears to show much lower excitation efficiencies.

Table 1. Compilation of singlet quantum yields for inter- and intramolecular CIEEL Systems

Process	Catalyzed decomposition of:	Φ_S (E/mol)[a]
Intermolecular	diphenoyl peroxide (**1**)	0.10 ± 0.05 (2 ± 1) 10^{-5} ***(4.9 ± 0.5) 10^{-5}***
	dimethyl-1,2-dioxetanone (**2**)	0.10 ± 0.05 ***(7.9 ± 0.8) 10^{-4}***
	1,2-dioxetanedione (peroxyoxalate)	up to 0.68 ± 0.05
Intramolecular	phenoxy-substituted 1,2-dioxetanes	0.55 ± 0.20 0.66 ± 0.33 1.0 ± 0.3
	luciferin BL	0.88

[a] Values in bold, this work. The other values are from the literature.[2]

The chemiluminescence parameters for the 1,2-dioxetanone 3, which has been first synthesized by Adam's research group[1] but for which no data on its CL properties have been reported, were studied using different activators. The initial emission intensities obtained in the presence of ACTs show a linear free-energy correlation with the oxidation potential ($E_{p/2}^{ox}$) of the activator, indicating the occurrence of an electron transfer from the activator to the peroxide in the rate-limiting step of chemiexcitation, in agreement with the CIEEL mechanism.

The catalyzed decomposition of 3 has also been studied using phenolate ions as activators. Phenolate ion 4 was generated *in situ* by TBAF mediated deprotection of the corresponding trialkylsilyl ether and the emission kinetics obtained upon addition of 3 were observed on a photon-counting apparatus (Fig. 2). The observed rate constants (k_{obs}) show linear correlation with the concentration of 4, however, the singlet quantum yields obtained in these experiments are extremely low $\Phi_S < 10^{-4}$ E/mol.

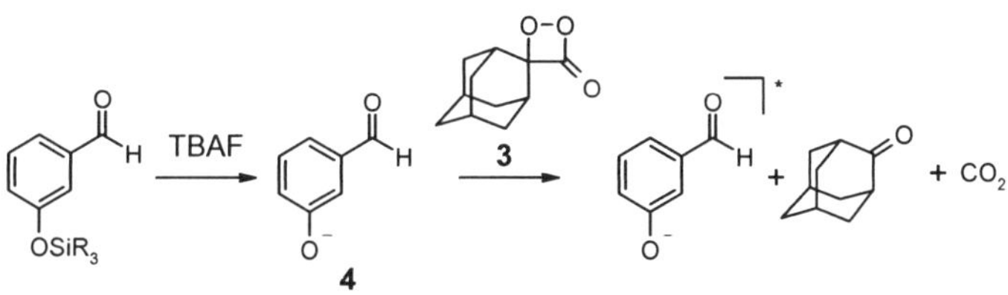

Fig. 2. Decomposition of 1,2-dioxetanone 3, catalyzed by *in situ* generated phenolate ion 4

These results, especially the low quantum yields, are of importance with respect to the mechanism of the induced 1,2-dioxetane decomposition (Fig. 1), as this system can be regarded as an intermolecular model for the latter reaction. If the induced 1,2-dioxetane decomposition would proceed by the intermolecular BET as suggested based on solvent viscosity effects,[5] the phenolate catalyzed decomposition of 3 would proceed through a very similar ion-radical pair and therefore similar quantum yields would be expected in both cases.

As the quantum yields for these transformations should be mainly determined by the efficiency of the BET (k_{BET}), the observation of very low quantum yields in the intermolecular system (3 + phenolate) clearly indicates that the induced 1,2-dioxetane decomposition should occur by an entirely intramolecular pathway (Fig. 1, lower pathway).

CONCLUSIONS

The experimental results obtained in this work indicate that the intermolecular CIEEL systems studied show much lower singlet excitation quantum yields than those obtained before for the intramolecular induced decomposition of appropriate phenoxy-substituted 1,2-dioxetanes. The CL parameters determined here for the first time in the catalyzed decomposition of spiro-adamantyl-1,2-dioxetanone indicate that also this sterically hindered and relatively stable α–peroxylactone derivative undergoes decomposition by the CIEEL mechanism, although in this case the unimolecular decomposition pathway being always the predominant one. However, the correlation of the quantum yields with the oxidation potentials of the activators indicates the involvement of an electron or charge transfer in the rate-limiting step of the chemiexcitation mechanism. Finally, we show here that phenolates can act as activators in the catalyzed decomposition of the peroxides studied, however, the singlet quantum yields obtained in this process are extremely low, indicating consequently that the induced 1,2-dioxetane decomposition should occur by an entirely intramolecular chemiexcitation pathway.

ACKNOWLEDGEMENTS

Financial support for this work has been provided by Fundação de Amparo à Pesquisa do Estado de São Paulo – FAPESP and Coordenação de Aperfeiçoamento de Pessoal de Nível Superior – CAPES.

REFERENCES

1. Schuster GB. Chemiluminescence of organic peroxides. Acc Chem Res 1979; 12: 366-73.
2. Baader WJ, Stevani CV, Bastos EL. Chemiluminescence of organic peroxides. In: The Chemistry of Peroxides; Rappoport, Z, eds; Chichester; Wiley 2006: 1211-78.
3. Catalani LH, Wilson T. Electron transfer and chemiluminescence. Two inefficient systems: 1,4-dimethoxy-9,10-diphenylanthracene peroxide and diphenoyl peroxide. J Am Chem Soc 1989; 111: 2633-9.
4. Adam W, Reinhardt D, Saha-Möller CR. From the firefly bioluminescence to the dioxetane-based (AMPPD) chemiluminescence immunoassy: a retroanalysis. Analyst 1996; 121: 1527-31.
5. Adam W, Bronstein I, Trofimov AV, Vasil'ev RF. Solvent-cage effect (viscosity dependence) as a diagnostic probe for the mechanism of the intramolecular chemically initiated electron-exchange luminescence (CIEEL) triggered from a spiroadamantyl-substituted dioxetane. J Am Chem Soc 1999; 121: 958-61.
6. Stevani CV, Silva SM, Baader WJ. Studies of the mechanism of the excitation step in peroxyoxalate chemiluminescence. Eur J Org Chem 2000: 4037-46.
7. Adam W, Blancafort L. Reaction of α-peroxy lactones with C, N, P, and S nucleophiles: Adduct formation and nucleophile oxidation by nucleophilic attack at and biphilic insertion into the peroxide bond. J Org Chem 1997; 62: 1623-9.

STUDY OF METAL BINDING TO MONOMERIC RED FLUORESCENT PROTEIN, DSRED-MONOMER

YASMEEN RAHIMI, SURESH SHRESTHA, SAPNA K. DEO
Department of Chemistry & Chemical Biology, Indiana University Purdue University Indianapolis, Indianapolis, IN 46202

INTRODUCTION

Red fluorescent protein (DsRed) is an Anthozoan fluorescent protein (28 kDa, 225 amino acid residues; ex 558 nm, em 583 nm) that was originally isolated from the *Discosoma* species from a reef coral.[1-4] The X-ray crystal structure of the protein shows that DsRed has a tetrameric configuration.[5] Each DsRed monomer has an 11-stranded β-can tertiary structure similar to that of the green fluorescent protein (GFP). The monomer units interact with each other via hydrophobic and hydrophilic interactions to form a tetrameric unit.[5] The chromophore of DsRed is formed from post-translational modification of the tripeptide sequence (Gln66-Tyr67-Gly68) similar to GFP.[4,6] The observed red emission in DsRed is believed to be due to the extended conjugation in the chromophore through dehydrogenation of the αC-N bond of Gln66.[4] DsRed tetramer has a slow folding rate, and forms intracellular aggregates, which limits its use as a fluorescent reporter.[7] To overcome these problems, several mutagenesis studies have been performed.[2,8-11] A monomeric protein (mRFP1) has been created by Tsien's group that folds better than the DsRed tetramer.[8] A monomeric DsRed is also available commercially as DsRed-Monomer Fluorescent Protein from Clontech, CA, USA. This monomeric DsRed (DsRed-Monomer) was created by substituting 45 amino acids in the native DsRed. The excitation and emission wavelength maximum of DsRed-Monomer are 557 nm and 585 nm, respectively (Fig. 1). Recently, Sumner *et al.* reported that tetrameric DsRed binds to copper ions and the binding event caused the fluorescence of DsRed to be quenched.[12] DsRed was thus employed as the biosensing element for copper detection. On the basis of this observation, we performed a study to investigate if DsRed-Monomer can bind copper ions. Since tetrameric DsRed has a tendency to aggregate and takes longer to mature, a monomeric form of the protein with efficient folding and reduced tendency to aggregation should be better suited as the biosensing element in the development of copper biosensor. Our study showed that upon binding of copper, the fluorescence of DsRed-Monomer was decreased. In addition, the protein did not respond to other divalent metal ions, indicating high selectivity for binding of copper ions. Here we describe a metal binding study performed on DsRed-Monomer.

EXPERIMENTAL

Protein Expression and Purification. *E. coli* JM107 cells were transformed with pDsRed-Monomer (Clontech, CA, USA) plasmid. The cells were grown in Luria Bertani (LB)-broth containing ampicillin to an optical density of 0.6 measured at

600 nm. Protein expression was induced using IPTG (Isopropyl ß-D-1-thiogalactopyranoside). The cells were harvested by centrifugation and then sonicated. The crude protein was obtained by centrifugation and then purified using an affinity chromatography protocol. Purity was determined by SDS-PAGE analysis and the concentration was determined using the Bradford assay.

Copper Binding Study. A fluorescence intensity of various dilutions of purified DsRed-Monomer protein was measured by exciting the sample at 558 nm and recording the resulting emission at 586 nm. Fluorescence readings were taken using the Cary Eclipse Fluorescence Microtiter Plate Reader from Varian Inc. (Walnut Creek, CA, USA). For the copper binding study, 100 μL of different concentrations of copper solution was added to a 100 μL of 10 μM protein in an individual microtiter well, then a final concentration of protein in each well was 5 μM. A fluorescence reading was taken, and a calibration curve for copper was generated.

Selectivity Study. Stock solutions of Fe^{2+}, Co^{2+}, Ni^{2+}, Zn^{2+}, Ba^{2+}, Ca^{2+} or Mg^{2+} were added to the DsRed-Monomer protein of concentration 5 μM to give a final concentration of metal ion to 25 μM. Ca^{2+} and Mg^{2+} were added to get 1 mM final concentration as these ions are found in high concentration in environment and in biological samples. The fluorescence intensity was monitored at emission wavelength of 586 nm.

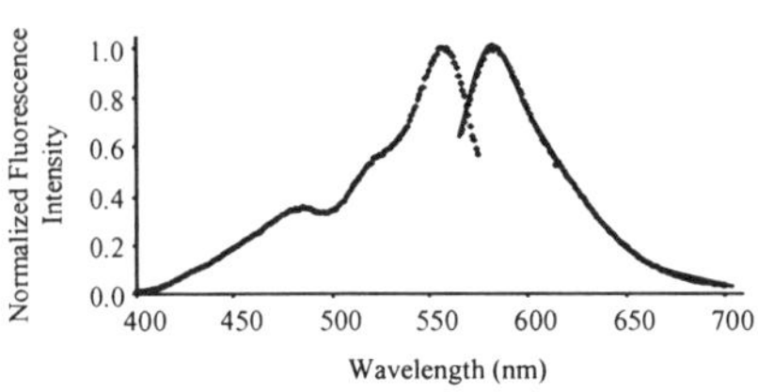

Fig. 1. Excitation and emission scans of DsRed-Monomer

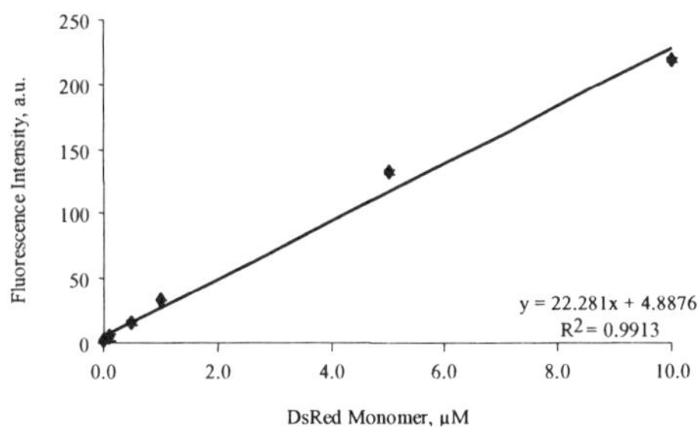

Fig. 2. Fluorescence intensity plotted against dilutions DsRed-Monomer

RESULTS AND DISCUSSION

The plasmid that encodes the gene for the DsRed-Monomer protein (obtained from Clontech) was employed to express the protein. The expressed protein was purified using the affinity chromatography protocol. Fluorescence excitation and emission spectra were obtained for the purified protein (Fig. 1). The excitation wavelength maximum of 558 nm and an emission wavelength maximum of 586 nm correspond to the literature reported values. A calibration curve was generated for DsRed-Monomer to determine the lowest protein concentration that can be used for copper binding studies (Fig. 2). A protein concentration of 5 μM was selected from this study. A copper binding study was performed by adding various concentrations of copper to a protein solution in microtiter wells so that the final protein concentration in each well was 5 μM. Addition of copper to the DsRed-Monomer resulted in the quenching of protein's fluorescence, similar to that observed in the case of DsRed tetrameric protein by Sumner, *et al.* A calibration curve was generated for copper by plotting percent fluorescence quenching against copper concentrations (Fig. 3).

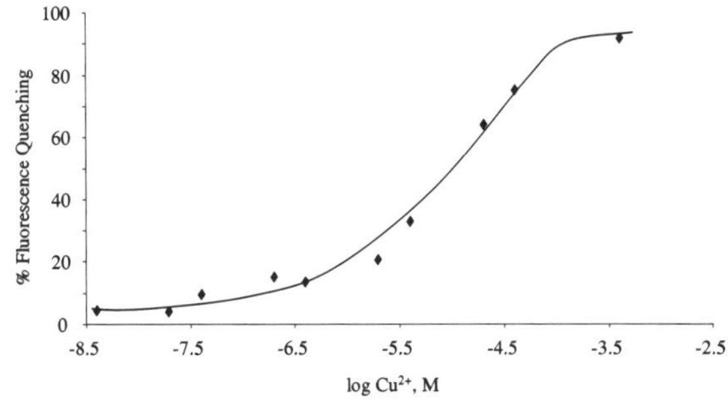

Fig. 3. Increase in fluorescence quenching of DsRed-Monomer with increase in copper concentration

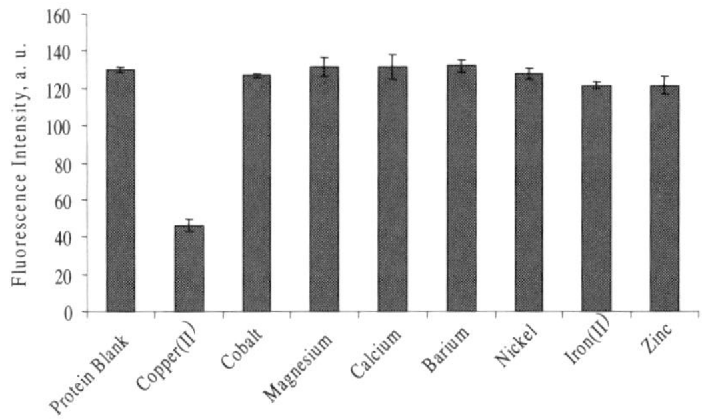

Fig. 4. Selectivity of DsRed-Monomer for copper ions

A copper concentration of 5 μM showed 33% quenching of DsRed fluorescence. However, when the copper concentration was increased to 500 μM, more than 92% of the fluorescence was quenched. The detection limit for copper (calculated using the S/N of 3) was 7.9 x 10^{-7} M. As the detection limit depends on the concentration of the protein used, an even lower detection limit can be obtained by decreasing the protein concentration. In addition, use of a microtiter plate reader limits the sensitivity of detection. Nevertheless, to mimic the application of the DsRed-Monomer as the sensing element for Cu^{2+}, all the studies were performed in the microtiter well plate. The sub-micromolar detection limit obtained using this system indicates that the DsRed-Monomer can be used in the development of Cu^{2+} sensor. Specificity of DsRed-Monomer for copper ions was also evaluated. The DsRed-Monomer fluorescence was unaffected by the addition of other metal ions, namely Ca^{2+}, Mg^{2+}, at a final concentration of 1 mM, and Fe^{2+}, Co^{2+}, Ni^{2+}, Zn^{2+}, and Ba^{2+} at a final concentration of 25 μM (Fig. 4). Addition of Cu^{2+} at this same concentration quenched more than 64% fluorescence of DsRed-Monomer. These results indicate that the quenching observed is not a collisional quenching, but rather quenching that is very specific to copper binding. Hence, the DsRed-Monomer may be the good alternative to the tetrameric DsRed in the development of Cu^{2+} sensor.

ACKNOWLEDGMENTS

We would like to thank American Cancer Society (IRG 84-002-22) for funding.

REFERENCES

1. Matz MV, Fradkov AF, Labas YA, Savitsky AP, Zaraisky AG, Markelov ML, et al. Fluorescent proteins from nonbioluminescent Anthozoa species. Nat Biotechnol 1999;17: 969-73.
2. Verkhusha VV, Lukyanov KA. The molecular properties and applications of Anthozoa fluorescent proteins and chromoproteins. Nat Biotechnol 2004;22: 289-96.
3. Baird GS, Zacharias DA, Tsien RY. Biochemistry, mutagenesis, and oligomerization of DsRed, a red fluorescent protein from coral. Proc Natl Acad Sci U S A 2000; 97: 11984-9.
4. Gross LA, Baird GS, Hoffman RC, Baldridge KK, Tsien, RY. The structure of the chromophore within DsRed, a red fluorescent protein from coral. Proc Natl Acad Sci U S A 2000; 97: 11990-5.
5. Yarbrough D, Wachter RM, Kallio K, Matz MV, Remington SJ. Refined crystal structure of DsRed, a red fluorescent protein from coral, at 2.0-A resolution. Proc Natl Acad Sci USA 2001; 98: 462-7.
6. Wiehler J, von Hummel J, Steipe B. Mutants of *Discosoma* red fluorescent protein with a GFP-like chromophore. FEBS Lett 2001; 487: 384-9.
7. Mizuno H, Sawano A, Eli P, Hama H, Miyawaki A. Red fluorescent protein from *Discosoma* as a fusion tag and a partner for fluorescence resonance energy transfer. Biochemistry 2001; 40:2502-10.
8. Campbell RE, Tour O, Palmer AE, Steinbach PA, Baird GS, Zacharias DA, et al. A monomeric red fluorescent protein. Proc Natl Acad Sci USA 2002;99: 7877-82.
9. Shaner NC, Campbell RE, Steinbach PA, Giepmans BN, Palmer AE, Tsien RY. Improved monomeric red, orange and yellow fluorescent proteins derived from *Discosoma* sp. red fluorescent protein. Nat Biotechnol 2004;22:1567-72.
10. Terskikh AV, Fradkov AF, Zaraisky AG, Kajava AV, Angres B. Analysis of DsRed mutants. Space around the fluorophore accelerates fluorescence development. J Biol Chem 2002; 277: 7633-6.
11. Bevis BJ, Glick BS. Rapidly maturing variants of the *Discosoma* red fluorescent protein (DsRed). Nat Biotechnol 2002; 20: 83-7.
12. Sumner JP, Westerberg NM, Stoddard AK, Hurst TK, Cramer M, Thompson RB, et al. DsRed as a highly sensitive, selective, and reversible fluorescence-based biosensor for both Cu(+) and Cu(2+) ions. Biosens Bioelectron 2006; 21: 1302-8.

TIME-DEPENDENT SPECTRAL CHANGE OF CHEMILUMINESCENCE FROM FIREFLY LUCIFERIN IN DEOXYGENATED DIMETHYL SULFOXIDE

RIKUO SHIBATA,[1] YASUHIKO YOSHIDA,[1] NAOHISA WADA[2]

[1] Department of Applied Chemistry, Faculty of Engineering, Toyo University, Kawagoe, Saitama 350-8585, Japan; [2] Faculty of Life Sciences, Toyo University, Itakura, Gunma 374-0193, Japan, E-mail:kuri@eng.toyo.ac.jp

INTRODUCTION

The molecular mechanism and light emitters of the chemiluminescence (CL) of firefly luciferin (Ln) have been interpreted as outlined in Fig.1.[1-3] The oxygenation of Ln in the presence of a strong base (e.g., *t*-BuOK) produces a dioxetanone, which successively decomposes to form an electronically-excited oxyluciferin (Oxyln*).[4,5] Oxyln* emits green- or red-colored light depending on its ionized state: a monoanion of Oxyln* in the keto form emits red-colored light, while its mono- and/or di-anion in the enolic form emits green-colored light at high base concentrations.[6,7]

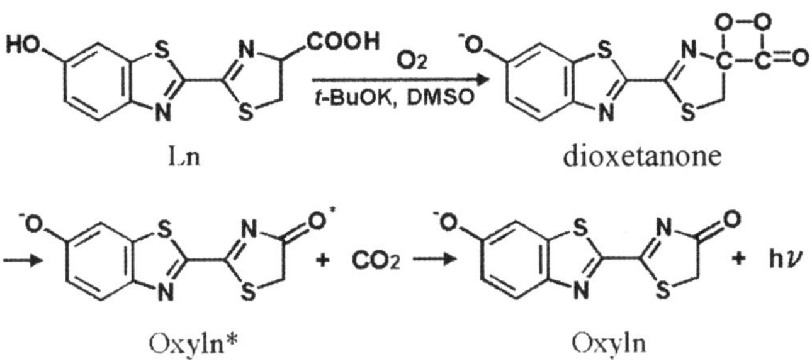

Fig. 1. The reaction scheme of Ln CL

We have investigated the process of formation of the dioxetanone and light emitters of Ln CL. Firstly, the intermediates were produced by stopping the reaction of Ln with oxygen. At high *t*-BuOK concentrations, the stable intermediate (M_{420}) of Ln (absorption maximum 420 nm) was produced[8] (M_{420} deprotonated at OH, COOH and the C_4 site of Ln).[9] Next, M_{420} CL was initiated by reacting M_{420} with oxygen. The time course of M_{420} CL could be measured with a photomultiplier R376 (Hamamatsu Photonics Co.) combined with band-pass filters, but not using multi-channel detector with a photo diode array for the very weak M_{420} CL.[9] Consequently, a photomultiplier module (PM-module) attached to 32 stripes of band-pass filters (Hamamatsu Photonics Co.) was applied to this problem. We report the first measurement of the time-resolved spectra of M_{420}.

MATERIALS AND METHODS

The M_{420} sample comprised a solution of 0.25 mM Ln in deoxygenated dimethylsulfoxide (DMSO) with 5 mM *t*-BuOK. Then, oxygen gas was dissolved in the M_{420} solution at a rate of 200 mL/min using the apparatus described previously,[9] except for the PM-module being placed on the side face of the reaction cell. The PM-module collects light emission every 0.1 sec, over the range 400 to 710 nm at an interval of 10 nm. Analytical grade Ln, *t*-BuOK, and DMSO (0.2% w/w) were purchased from Sigma Chemical Co, Merck Co. and Kanto Chemical Co., respectively. These reagents were treated in an atmosphere of argon to prevent Ln from changing to dehydroluciferin (drLn) in the presence of oxygen.[10] Absorption and fluorescence (FL) spectra were recorded with a Hitachi U-3200 and a F-4010 spectrophotometer, respectively.

RESULTS AND DISCUSSION

Figure 2 shows the time resolved spectral change of M_{420} CL for the time duration of dissolving oxygen, 30 seconds, a part of the total 120 second measurement. A green-colored emission with a strong intensity and a fast decay in the time interval $0=t_d=60$, especially $0=t_d=18$ (called $t_d^{(1)}$), and a red- colored emission in the time interval $0= t_d = 120$, especially $18=t_d=120$ (called $t_d^{(2)}$) were observed. In $0=t_d=10$, the very weak intensity of red-colored emission showed a fast rise and decay. In $10=t_d=120$, it rose slowly to the maximum at $t_d=30$, and then decayed steadily. Meanwhile, in a solution of *t*-BuOK in DMSO the superoxide anion of O_2^- was detected[11] and was observed to emit light. However, the CL intensity attributed to O_2^- was too weak to be taken into consideration.

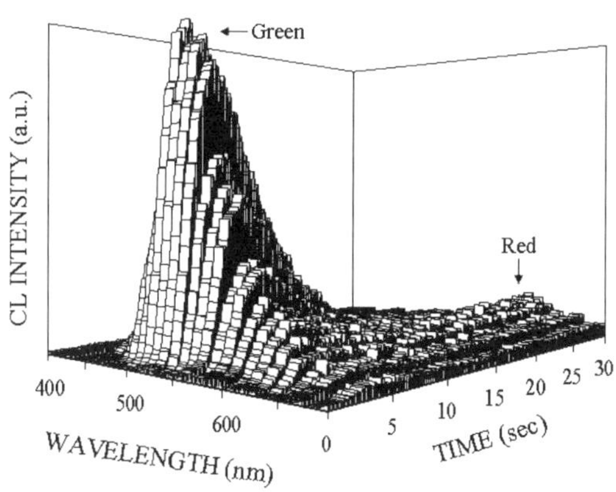

Fig. 2. The time-resolved spectral change of M_{420} chemiluminescence. Green- and weak red-colored light emissions with the wavelength maxima of 510 nm and 620 nm are observed especially in the time duration $0 \leq td \leq 18$ and $18 \leq td \leq 30$, respectively

Fig. 3 shows the spectra of the green- and red-colored light emissions integrated over the ranges of $t_d^{(1)}$ and $t_d^{(2)}$. The green-colored light emission shows a maximum at 510 nm with the broad half band width of 55 nm (*c.f.,* FL peak of the light emitter Oxyln in a DMSO solution containing added *t*-BuOK was 557 nm).[6] Here, it should be noted that the large difference, 47 nm between these wavelength maxima may be beyond experimental error.

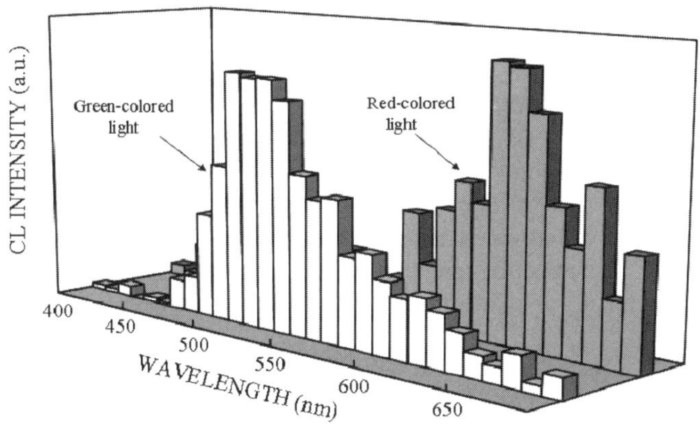

Fig. 3. The integrated spectra of the green- and red-colored light emissions. The intensity at each of the bars in the figure is obtained by summation of the intensities measured in $t_d^{(1)}$ and $t_d^{(2)}$ at 0.1 sec. intervals, respectively

On the other hand, the red-colored light emission shows a wavelength maximum at 620 nm with half band width of 40 nm. The wavelength maximum, 620 nm corresponds to that of Ln CL in the presence of *t*-BuOK, 632 nm.[2] The wavelength maxima from our preliminary measurement with a photomultiplier combined with band pass filters, 520 nm and 630 nm[8] were in agreement with the maxima obtained here. Absorption and FL spectra of the products formed after these emissions were measured for $0{=}t_d{=}120$ seconds to investigate the emitters of M_{420} CL. The product formed in $t_d^{(1)}$ (N_{420}) had an absorption and FL maxima at 420 nm and 560 nm (*c.f.*, Oxyln in a basic solution).[7] However, there was a large difference (50 nm) between the FL maximum of N_{420} and the CL maximum of the green-colored light emission. During the time $t_d^{(2)}$ when red-colored light predominates, the absorption and FL spectra showed maxima at 535 nm (product P_{535}) and 620 nm, respectively. The FL maximum of P_{535} (620nm) was coincident with the maximum of the red-colored light emission. We suggest that P_{535} might be a keto form of Oxyln monoanion and N_{420} (the green emitter) is an enolic form of Oxyln anion. Further, work to identify the products N_{420} and P_{535} is in progress.

In conclusion, a) the intermediate M_{420} is a candidate for the precursor of dioxetanone because the reaction of M_{420} with oxygen produces CL, b) it was for the first time observed that green- and red-colored light emissions were simultaneously produced under the chemical conditions adopted without the enzyme, and c) the light emitters can be assigned to anions in the enolic and keto forms of Oxyln.

ACKNOWLEDGEMENT

Supported in part by a Grant-in-Aid for Scientific Research (C) for N.W. from the Ministry of Education, Culture, Sports and Science and Technology of Japan (No.16540371).

REFERENCES

1. McCapra F, Chang YC, Francois VP. The chemiluminescence of a firefly luciferin analogue. J Chem Soc Chem Commun 1968:22-3.
2. White EH, Rapaport E, Seliger HH, Hopkins TA. The chemi- and bioluminescence of firefly luciferin. An efficient chemical production of electronically excited states. Bioorg Chem 1971;1:92-122.
3. Wada N. The molecular mechanism of firefly chemiluminescence from a physicochemical perspective. Rec Res Devel Photochem Photobiol 2001;5: 19-28.
4. White EH, McCapra F, Field GF. The structure and synthesis of firefly luciferin. J Am Chem Soc 1963; 85: 337-43.
5. Koo J-Y, Schmidt SP, Schuster GB. Bioluminescence of the firefly: Key steps in the formation of the electronically excited state for model systems. Proc Natl Acad Sci USA 1978; 75: 30-3.
6. White EH, Roswell DF. Analogs and derivatives of firefly oxyluciferin, the light emitter in firefly bioluminescence. Photochem Photobiol 1991;53:131-6.
7. Suzuki N, Sato M, Okada K, Goto T. Studies on firefly bioluminescence I synthesis and spectral properties of firefly oxyluciferin*. A possible emitting species in firefly bioluminescence. Tetrahedron 1972;28:4065-74.
8. Wada N, Shibata R. Absorption spectral changes of firefly luciferin in deoxygenated dimethyl sulfoxide. J Phys Soc Jpn 1997;66:3312-3.
9. Shibata R, Yoshida Y, Wada N. Filter-photometry of chemiluminescence from firefly luciferin intermediate M_{420} in deoxygenated demethyl sulfoxide. J Photoscience 2002;9:290-2.
10. McCapra F. Alternative mechanism for dioxetan decomposition. J Chem Soc Chem Commun 1977:946-7.
11. Wada N, Mitsuta K, Kohno M, Suzuki N. ESR studies of firefly D-(-)-luciferin chemiluminescence in dimethyl sulfoxide. J Phys Soc Jpn 1989;58:3501-4.

A QUANTUM-CHEMICAL APPROACH TO THE AMINO ANALOGS OF FIREFLY LUCIFERIN

NAOHISA WADA, HIROAKI FUJII, HIRONORI SAKAI
Department of Life Sciences, Toyo University, 1-1-1 Izumino Itakura-machi, Gunma 374-0113, Japan, E-mail: bhwada@itakura.toyo.ac.jp

INTRODUCTION

Various analogs of the substrate luciferin (Ln-OH) were used in order to study how *in vitro* firefly bioluminescence has the high quantum yield and wide spectral distribution of emission through the characteristic interaction with the wild-type luciferase (PpyLuc) from North American firefly (*Photinus pyralis*). White *et al.* have synthesized the amino analogs of Ln-OH substituting its 6'-OH group by amino derivatives, namely aminoluciferin (Ia), acetylaminoluciferin (Ib), trifluoroacetyl aminoluciferin (Ic).[1]

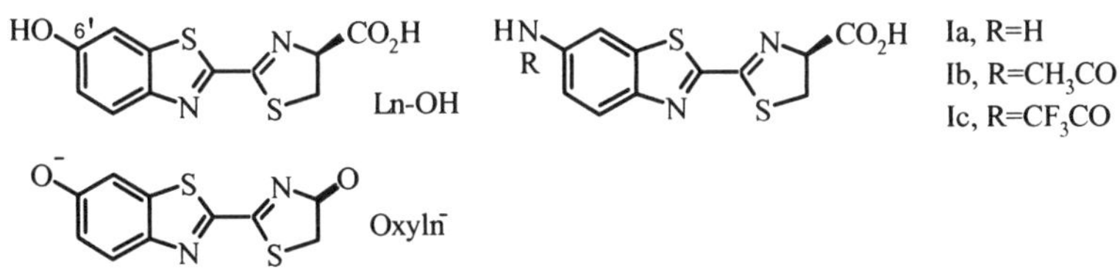

Fig. 1. The molecular structures of luciferin (Ln-OH), oxyluciferin anion (Oxyln$^-$) and amino analogs (Ia,Ib,Ic)

In this report, the optical properties of Ln-OH, the light emitter of oxyluciferin (Oxyln) and their amino analogs, and their potential energy curves (PECs) of the ground and the excited states were obtained quantum-chemically. This data was related to their spectroscopic characteristics, with the assumption that the chemical excitation process contains the transition state of dioxetanone (DOX), an intermediate generated enzymatically from the oxygenation of Ln-OH.

METHOD OF CALCULATION

The optical absorptions of Ln-OH and its amino analogs were obtained by the following procedures: After the geometries of these molecules were optimized by *ab initio* calculations within the density function theory (B3LYP) using 6-311+G (2d,p) basis set in the Gaussian 03 program,[2] the optical absorptions of the energy-minimized molecules were calculated by differential overlap for spectroscopy (INDO/S) method in MOS-F (Fujitsu Ltd., Tokyo, Japan, 2004, V6.0), where the configuration interactions (CIs) were taken into account. In *ab initio* calculations, energy minimization of molecules interacting with ethanol could be processed by the polarizable continuum model (PCM) method, and the solvent effect on the optical absorptions of the compounds could be also considered explicitly by the Onsager model in the INDO/S method. Although White *et al.* observed no absorption spectra of Oxyln and its amino analogs, the optical absorptions of these molecules could be calculated by the manner described above. In order to calculate the PECs of the ground- and the singlet excited-states (S_0, S_1, S_2) of DOXs along their decomposition from transition states to the products, the intrinsic reaction coordinate (IRC) method was applied as described above.[3]

RESULTS AND DISCUSSION

Table 1 shows the optical absorptions of Ln-OH, Oxyln and their analogs together with the absorption maxima of Ln-OH and its analogs measured by White *et al.*,[1] and the quantity ΔQ_{CT} defined by electron transfer from the benzothiazole to the thiazoline ring in the first excited state, which is related to the first absorption band of each molecule. The absorption maxima qualitatively correspond to the absorption peaks of respective molecules, and ΔQ_{CT} values of Oxyln, its analogs and Oxyln$^-$ are larger than those of Ln-OH, its analogs and Ln-O$^-$, respectively. The comparatively small ΔQ_{CT} value of Ic can be attributed to the effect of a strongly electron attracting center at the amino group.

Table 1. Experimental and calculated spectroscopic characteristics of Ln-OH, Oxyln and their analogs together with ΔQ_{CT} (a.u.) of electron transfer from the benzothiazole to the thiazoline ring in the first excited state of each molecule

	Ln-OH	Ia	Ib	Ic	LnO$^-$
	Oxyln	Oxyla	Oxylb	Oxylc	OxylnO$^-$
The position of absorption maximum (nm) [logε]	330[4.26] 269[3.85]	363[4.19] 280[3.78]	327[4.33] 225[4.44]		
The calculated absorption peak (nm) [f]	341[0.59] 247[0.36]	364[0.64] 244[0.21]	354[0.64] 256[0.34]	343[0.64] 249[0.36]	415[0.71] 250[0.26]
	393[0.70] 263[0.31]	442[0.80] 214[0.48]	410[0.74] 215[0.58]	384[0.69] 264[0.47]	518[0.99] 265[0.25]
ΔQ_{CT}	0.308	0.340	0.327	0.257	0.318
	0.466	0.464	0.469	0.399	0.359

White *et al.* observed that only Ia reacts with PpyLuc and ATPMg^{2+} among the amino analogs of Ln-OH to produce red light regardless of pH, contrary to the reaction with Ln-OH emitting in the yellow-green spectrum at neutral and basic pH and in the red spectrum at acid pH.[1] In general, the change of bioluminescence color might reflect both variation in the charged state of the emitter itself and in the specific interaction of the excited emitter with certain amino acid residues in the active center of PpyLuc. According to X-ray diffraction studies of wild-type Japanese Genji-botaru (*Luciola cruciata*) luciferase (LcrLuc) and its red mutant complexed with a high-energy intermediate analog, the color-shift of emission from yellow-green to red is caused by lowering the degree of molecular rigidity of the excited Oxyln$^-$* induced by a significant conformational change in the wild-type LcrLuc, but not in the red mutant.[4] Therefore, red-color emission, regardless of pH change on mixing the analog Ia with PpyLuc and $ATPMg^{2+}$, implies that the charged state of the emitter does not change and/or the specific interaction between Ia and the active center in PpyLuc does not induce a large conformational change of PpyLuc. On the other hand, the quantum yield of bioluminescence might be largely affected by the following factors: 1) the actual conformational adaptation of Ln-OH and its analogs to the pocket in the active center of PpyLuc in order for the forward reaction to proceed forming DOX, 2) the probability of non-adiabatic transition from the reactant in the ground state to the product in the singlet excited-state through DOX along the reaction path, 3) the difference in the fluorescence yield of light emitters among Oxyln, its analogs and Oxyln$^-$. As far as factor 1) is concerned, the bulk of the substituted groups at the C_6'-side chain of Ln-OH, its analogs and Ln-O$^-$ is depicted in Fig. 2 by the space filling representation.

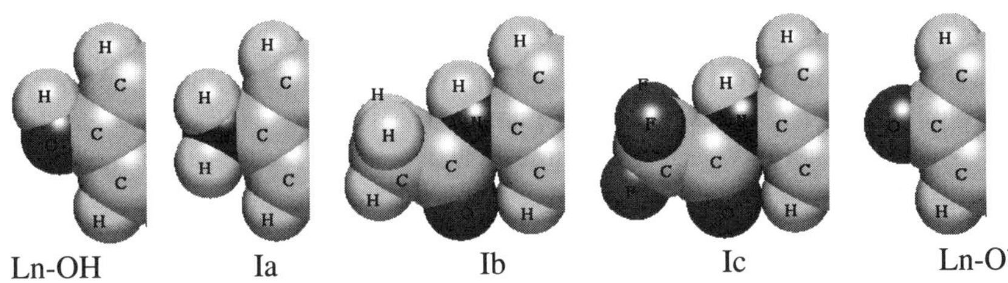

Fig. 2. The bulk of the substituted groups at C6' side chain of Ln-OH, Ia, Ib, Ic and Ln-O$^-$ measured in the same scale

By considering the spread of filling space along the long axes of Ib and Ic, the formation of DOXs from Ib and Ic might be blocked due to mismatch of location between the thiazoline moieties of these molecules toward the open side of the pocket of PpyLuc and certain amino acid residues catalyzing the oxygenation reaction. Concerning factor 2), each of the PECs along the reaction coordinate from the transition state to product is shown in Fig. 3. Except in the case of Ln-O$^-$, there is no difference in dependence of PECs on the reaction coordinate in spite of replacement at the C_6'-side chain of Ln-OH; carrying forward the reaction over the concave region of these molecules, the S_1-PEC approaches the S_2-PEC. With regard to factor 3), Oxyln$^-$ might have the large fluorescence ($S_1 \rightarrow S_0$) quantum yield judging from the calculated oscillator strengths in Table 1 if the nonradiative transition process does not compete with the radiative one.

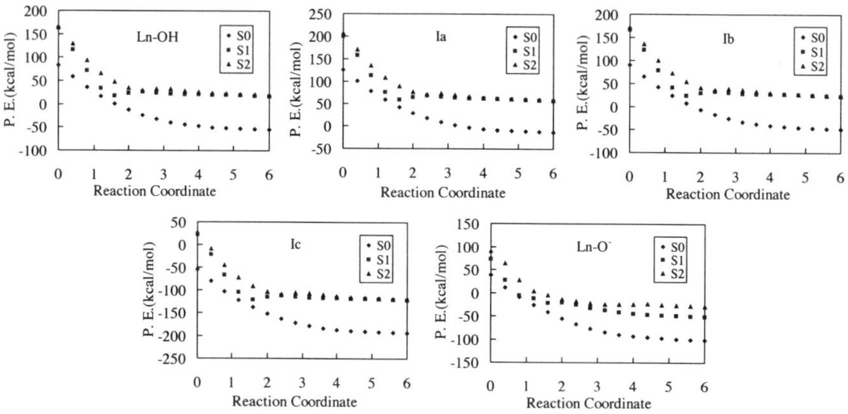

Fig. 3. The PECs of S_0, S_1, S_2 states calculated by IRC along the reaction coordinate from the transition state of DOX to product in case of Ln-OH, its analogs and Ln-O$^-$. P.E. means the potential energy of the system

It is, therefore, presumed that the quantum yield of bioluminescence and its color distribution must result from the multiplier effect of many factors on the applicable emitter. The results reported in this study must be analyzed by much more precise methods, but the quantum-chemical approach is a useful procedure to clarify the problems concerning bioluminescence phenomena that have not been investigated by experimental techniques.

ACKNOWLEDGEMENTS

This work was partly supported by grants for H.S. from Toyo University. We thank Research Center for Computational Science (RCCS) for using its super computer.

REFERENCES

1. White EH, Worther H, Seliger HH, McElroy WD. Amino analogs of firefly luciferin and biological activity thereof. J Am Chem Soc 1966; 88: 2015-9.
2. Frisch MJ, et al. Gaussian 03 Revision C.02 Pittsburgh, PA: Gaussian Inc. 2003.
3. Wada N, Sakai H. Decomposition feaction of dioxetanone in firefly. bioluminescence by computer experiment. J Biol Phys 2005; 31:403-12.
4. Nakatsu T, Ichiyama S, Hiratake J, Saldanha A, Kobayashi N, Sakata K, et al. Structural basis for the spectral difference in luciferase bioluminescence. Nature 2006; 440: 372-5.

FIREFLY-LUCIFERIN AND ITS ANALOGS: A SOURCE OF NEW LUMINESCENCE DYES AND LIGANDS

D WEISS,[1] S TRUPP,[1] LFML CISCATO,[2] R BECKERT,[1] E ANDERS[1]
[1] *Inst. Org. Chem., Friedrich-Schiller-University Jena, Humboldtstr. 10, D-07743 Jena, Germany; E-mail: c5diwe@rz.uni-jena.de;*
[2] *Instituto de Quimica, Universitade de Sao Paulo-Av. Lineo Prestes, 748 Bloco 12S CEP 05508-900, Sao Paulo, Brazil*

INTRODUCTION

The bioluminescence of beetles is one of the most impressive phenomena in nature. The discussion about the mechanism of this reaction about 30 years ago[1] and still continues.[2] Besides the general biological aspects we are interested in how excited states may be formed by chemical reactions. One of the main problems in this topic is the question how really good emitter molecules may be constructed. The overall quantum yield of the firefly bioluminescence is more than 90%. This means that the light emitter of this reaction must do its work very well and must be able to emit different colors. One of the most popular hypotheses says that in the excited state of the emitting oxyluciferin exist a quaistautomeric equilibrium between the red emitting thiazolinone 1 and the yellow/green emitting thiazole 2. Beside this there may exist a pH dependent mesmeric equilibrium in 2, as well as a number of rotamers to the central single bond.[3]

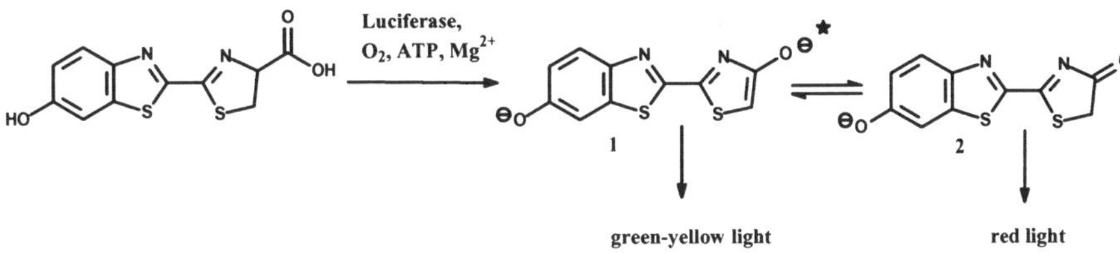

Fig. 1. Luminescence of firefly luciferin

MATERIALS AND METHODS

The thiazoles 10a-e were synthesized by a method described in the literature[5] and the ethers by a classic Williamson ether synthesis. All substances were characterized by spectroscopic methods and elemental analysis. The structures of the complexes were determined by X-ray crystallography. The fluorescence was measured with a JASCO FP 6500 in ethanol as solvent.

RESULTS AND DISCUSSION

To learn more about the photophysical properties of this very complex light emitting system we have synthesized some new model substances. Starting from the nitrile 3 it was possible to synthesize the two side protected oxyluciferin derivative 7. We were not able to observe the thiazolone 5. The endproduct of the condensation between 4 and mercaptolactic acid always was 6. If there is a tautomeric equilibrium between the keto-form 5 and the enol-form 6 this equilibrium lies clearly on the side of 6. This corresponds with the general chemical principles that the formation of aromatic ring systems is always favored thermodynamically.

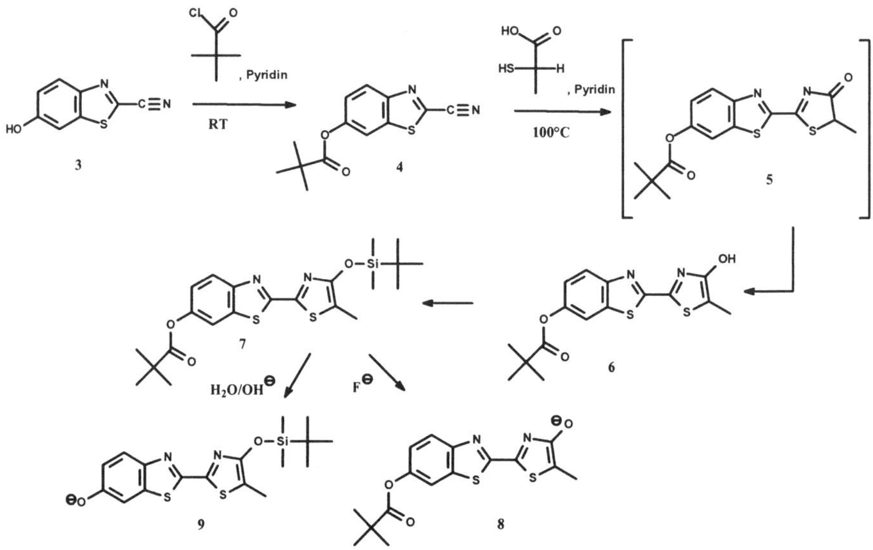

Fig. 2. The synthetic pathways to new firefly luciferin derivatives

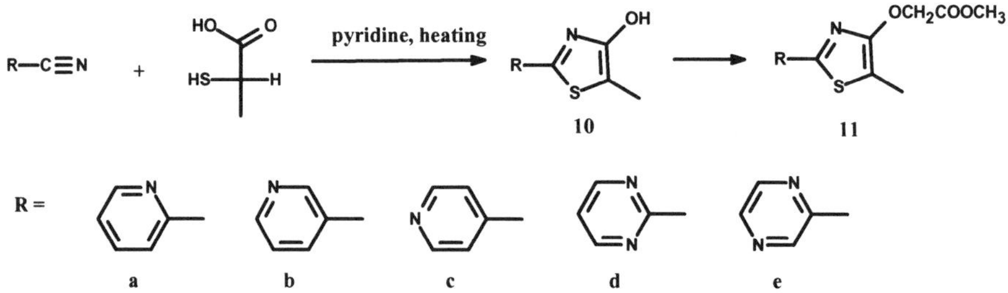

Fig. 3. The synthesis of new thiazoles

To our surprise the two side protected molecule 7 shows an intensive blue fluorescence (λ_{em} = 435 nm) with a quantum yield more than 90 % and a Stokes shift of .90 nm. This luminescence changes drastically when one of the protective groups is removed. With fluoride ions the anion 8 is formed having a red fluorescence. With hydroxide ions the anion 9 is formed showing the typical yellow/green luminescence of fireflies. In summary it was a great surprise for us that such a relative small molecule is able to emit light in such a broad spectral range.

These results led us to synthesize new thiazoles with different substituents. Starting from the corresponding nitriles we have got the hydroxythiazoles 10a-e in good yields. These substances are sparingly soluble, yellow needles without any fluorescence. This changes dramatically when the ethers 11a-e are formed. The substances become colorless and show a blue fluorescence when excitated with ultraviolet light. The quantum yields rise up to 95 %. Just like luciferin and oxiluciferin the molecules 10a, d and e have an azadien –substructure which should be able to form complexes with metal ions. X-ray structures of the crystals show that in the solid state the N-atoms are in a transoid position.[1] Nevertheless it is possible to prepare complexes with different metal precursors.

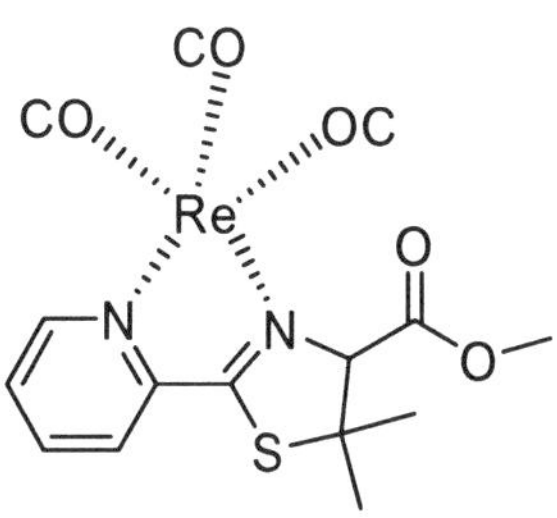

12

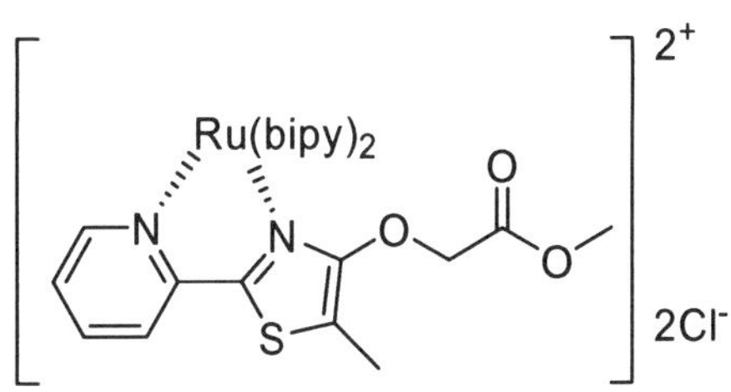

13

Fig. 4. General structures of the rhenium- and ruthenium complexes

The figures shows a rhenium (0) Complex 12 of the luciferin like molecule and the ruthenium (II) complex 13 of the oxyluciferin like molecule 11a. The X-ray structures show that in both cases the N atoms are in the cisoid position and therefore able to act as donor atoms in heterocyclic ligands.

ACKNOWLEDGEMENTS

We thank the Fonds der Chemischen Industrie, the DAAD, the FAPESP and the CAPES for financial support.

REFERENCES

1. White EH, Steinmetz MG, Miano JD, Wildes PD, Morland R. Chemi- and bioluminescence of firefly luciferin. J Am Chem Soc 1980; 102; 3199-208.
2. Barros P, Bechara JH. Bioluminescence as a possible auxiliary oxygen detoxifying mechanism in elaterid larvae. Free Rad Biol Med 1998;24;767-77.
3. Weiss D, Beckert R, Lamm K, Baader WJ, Bechara EJH, Stevani CV. Playing with luciferine – new results on the luminescence of a well known molecule. J Biolumin Chemilumin 2000;15:229.
4. Trupp S, Diploma Thesis, Jena, Germany, 2004.

CHEMILUMINESCENCE ENHANCEMENT EFFECT OF TRACE METAL ELEMENTS

YUMIKO YOSHIKI
Department of Food Science, Ishikawa Prefectural University,
1-308 Suematsu, Nonoichi-machi, Ishikawa, 921-8836, Japan

INTRODUCTION

Some trace metal elements (TME), such as Mn, Fe, Cu and Zn, are indispensable micronutrients in the preservation of the growth and health of humans, and are widely distributed in nature. They are constituents of enzymes and vitamins, and play an important role in oxygen metabolism. Either insufficient or excess intakes of TME may impair various biological reactions or cause serious disease through free radical formation. For example, in rheumatoid arthritis, there is significant oxidative stress and decrease of reactive oxygen species-related enzymes (SOD, GSH and catalase etc), showed fluctuating TME levels in serum fluctuate from normal levels. The decomposition of H_2O_2 catalyzed by a variety of TME has been widely studied.[1] During the H_20_2 decomposition reaction with TME, a weak light emission occurs and this involves formation of a peroxo-metal complex, and this undergoes auto-decomposition. The H_2O_2/gallic acid/$KHCO_3$/MeCHO system is one of the chemiluminescence (CL) systems involving H_2O_2 decomposition.[2] The CL and the H_2O_2 decomposition in the H_2O_2/gallic acid/$KHCO_3$/MeCHO system are enhanced by the addition of TME. In this study, we focused on the CL enhancement effect of TME and we report its role in the decomposition of H_2O_2 in the H_2O_2/gallic acid/$KHCO_3$/MeCHO system.

MATERIALS AND METHODS

Chemicals. Hydrogen peroxide was obtained from Santoku Co. (Tokyo, Japan). Gallic acid, $FeCl_2$, $CuCl_2$, $ZnCl_2$, $MnCl_2$ and $KHCO_3$ were purchased from Nakarai Tesque (Tokyo, Japan). Acetaldehyde was from Merck (Darmstadt, Germany). **Chemiluminescence measurements.** The chemiluminescence (CL) intensity was measured using a CL detector model CLD-110 (Tohoku Electronic Co., Ltd.) connected to a Waters model 515 pump. The wavelength was set in the range of 300-650 nm. The measurement was carried out under flow conditions using distilled water, with a flow rate of 1 mL/min. Each 10 μL of 196 mmol/L H_2O_2, 5 mmol/L gallic acid, 1 mol/L $KHCO_3$ in 356 mmol/L MeCHO was injected with 10 μL of $FeCl_2$, $CuCl_2$, $ZnCl_2$ and $MnCl_2$ (total 40 μL). **Determination of emission spectra.** The CL spectra were measured by a simultaneous multiwavelength analyzer model CLA-SP2 (Tohoku Electronic Co., Ltd.). The wavelength range of the spectroscope was 400-850 nm. In the experiment, 0.5 mL of 196 mmol/L H_2O_2, 5 mmol/L gallic acid, 1 mol/L $KHCO_3$ in 356 mmol/L MeCHO was injected into the stainless steel aperture with 0.5 mL of

$FeCl_2$, $CuCl_2$, $ZnCl_2$ and $MnCl_2$ (total 2.0 mL). The light emission was determined for 3 min. The spectral energy was calculated with 1 Einstein ($2.86\times10^4/\lambda$, kcal/mol) using the maximum on the spectrum curve.

RESULTS AND DISCUSSION

The addition of trace metal elements (TME), especially Mn^{2+}, Fe^{2+}, Cu^{2+} and Zn^{2+}, enhanced the CL from the H_2O_2/gallic acid/$KHCO_3$/MeCHO system (standard system CL in this study). TME also increased H_2O_2 decomposition 2-14 fold compared with the standard CL system. The rate of H_2O_2 decomposition increased dependent on the CL intensity ($Mn^{2+} > Fe^{2+} > Cu^{2+} > Zn^{2+}$). For the CL enhancement effect, TME had an optimum concentration (Fe^{2+}: 0.78-3.12 mmol/L, Zn^{2+}: 0.19-0.78 mmol/L, Mn^{2+}: 0.024-0.095 mmol/L, Cu^{2+}: 0.012-0.095 mmol/L) (Fig. 1).

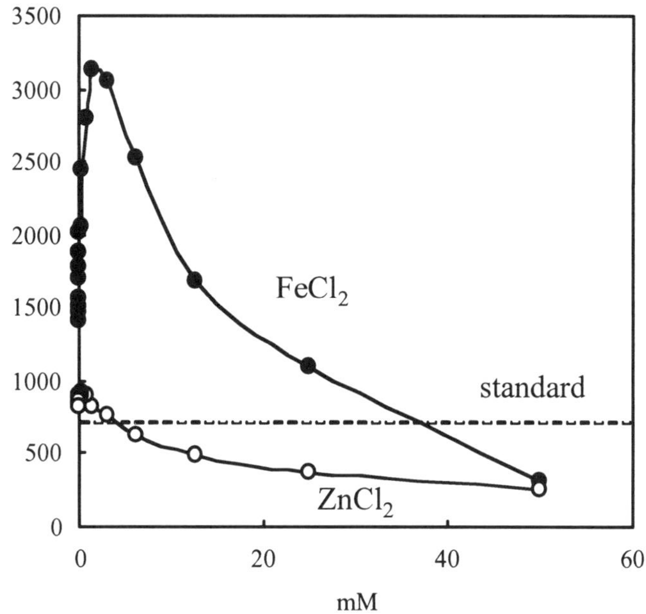

Fig. 1. Effects of $FeCl_2$ and $ZnCl_2$ concentration on photon flux

The emission spectrum of the standard CL system (Emax: 650 nm) shifted to 510 and 530 nm as shoulder peaks with a main peak around 650 nm with the addition of Fe^{2+} and Cu^{2+} respectively (Fig. 2). A definitive difference in spectrum from the standard CL system was observed in the Mn^{2+} addition system (Emax: 530 nm). Emission energy calculated from the spectrum increased from 44 kcal/mol to 54 kcal/mol in order of $Mn^{2+} > Fe^{2+} > Cu^{2+} > Zn^{2+}$. This experimental evidence suggests that electron population of 3d and 4s of TME is related to the emission spectra.

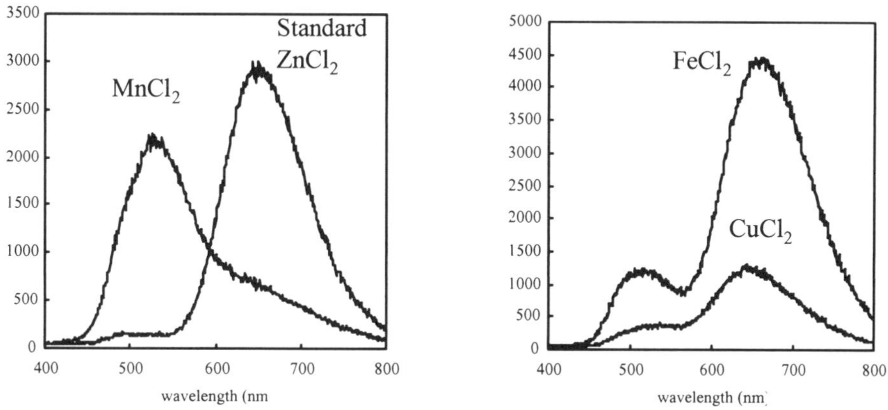

	Emax (nm)	Ee (kcal/mol)
Standard	650	44.0
$ZnCl_2$	650	44.0
$CuCl_2$	650 (530)	44.0 (54.0)
$FeCl_2$	650 (510)	44.0 (56.1)
$MnCl_2$	530	54.0

Fig. 2. Emission spectra and Ee observed from TME additive system

Free radical analysis using ESR showed the generation of gallic acid$^{\cdot}$, $CH_3^{\cdot}$, $CH_3O^{\cdot}$ from the CL standard system. These radicals increased and decomposed rapidly with the TME addition. $HO^{\cdot}$ spectra, generated from Fenton reaction ($H_2O_2 + M \rightarrow OH^- + HO^{\cdot} + M^+$), were not observed in any TME addition systems. A correlation of free radical intensity and photon intensity suggested that the CL enhancement effect of TME resulted from a catalytic effect on the CL reaction rate.

REFERENCES

1. Hanaoka S, Lin J-M, Yamada M. Chemiluminescence behavior of the decomposition of hydrogen peroxide catalyzed by copper(II)-amino acid

complexes and its application to the determination of tryptophan and phenylalanine. Anal Chim 2000; 409: 65-73.

2. Yoshiki Y, Iida T, Akiyama Y, Okubo K, Matsumoto H, Sato M. Imaging of hydroperoxide and hydrogen peroxide-scavenging substances by photon emission. Luminescence 2001; 16:1-9.

PART 6

BL/CL OCEANS, LAND & SPACE

BIOLUMINESCENCE IN THE SPACEFLIGHT AND LIFE SCIENCE TRAINING PROGRAM AT THE KENNEDY SPACE CENTER

VA KRATASYUK,[1] SM GUSEV,[1] NN REMMEL,[1] OA OSIPENKO,[1]
EN ESIMBEKOVA,[2] DM SHOEMAN,[3] TW DRESCHEL,[4] PV CHETIRKIN,[4]
MJ CORRELL,[5] RA BUCKLIN,[5] VE RYGALOV[6]

[1] Krasnoyarsk State University, 660041 Krasnoyarsk, Russia; [2] Institute of Biophysics, SB RAS, 660036 Krasnoyarsk, Russia; [3] Merritt Island High School, Science, FL, USA; [4] Kennedy Space Center, FL, 32899 USA; [5] University of Florida, Fl, USA; [6] University of North Dakota, ND, USA

INTRODUCTION

Bioluminescence studies have been incorporated into the six-week Kennedy Space Center Spaceflight and Life Science Training Program (STSTP). The SLSTP program is designed to attract the "best and brightest" US and Canadian students to careers in space science and engineering. This six-week program was sponsored by NASA and implemented at the Kennedy Space Center (KSC). Dynamac and Bionetics Corporations provided the research projects that were integrated in a program designed and assessed by a consortium of post-secondary institutions.

The SLSTP mission was: “to provide a summer pre-professional training program to introduce undergraduates to spaceflight and life sciences associated with NASA’s space exploration mission. The students worked with a project counselor, summer program director, and Principal Investigator. The Principal Investigator provided research projects and experience while mentoring the trainees in a field or laboratory work environment, evaluated the performance of the trainee each week using attendance, performance levels in laboratory projects, and oral/written presentations of laboratory findings. Trainees were expected to be cooperative, productive, and contribute to lab discussions. The Space Life Science Laboratory (SLSL) at KSC supported the life science experiments that were launched into low-Earth orbits from the Kennedy Space Center. Resident scientists designed and managed biological experiments for the Space Shuttle and International Space Station. Resident and guest scientists were investigating the feasibility of growing plants on Mars and developing numerous life support systems necessary for “long-duration space travel” including sensors, low-energy lighting systems, waste treatment systems, and advanced hydroponics systems.

The selected research project at the SLSL designed by Krasnoyarsk State University PIs was “Bioluminescent Biosensors for Space Biotechnology”.[1-3] The overall goal of our experiments is the design of the bioluminescent biosensors for a simulated flight experiment in environmental chambers and Mars green house at the Kennedy Space Center.

The research project had two components:

Design and test bioluminescent sensors for *in-situ* monitoring the stress responses in plants growing in Mars greenhouse and environmental chambers.

Enzymes of luminous bacteria – luciferase and NADH:FMN-oxidoreductase - were used to develop bioluminescent assay methods (NADH, FMN, aldehyde or other metabolites) in leaf, stem and root tissue extracts and to evaluate effects of controlled stress on the plants' metabolism. Radish, onion, lettuce, and parsley leaves, growing under stress conditions were tested. The results will be used for developing immobilized chemicals (enzymes and substrates) for bioluminescent biosensors.[4]

Design and test bioluminescent sensors to monitor and control air, water and soil contaminations in closed ecological systems associated with extreme environments, including space.

The luminous bacteria and their enzymes were used in this study. This research should clarify whether bioluminescent bioassays can be used as a quick response alarm test in controlling acute toxicity from various sources such as air, water or soil samples in Closed Ecological Systems (Environmental Chamber, Mars Greenhouse, Shuttle etc.) at Kennedy Space Center.

This project was performed by a team of biologists, physicists and engineers. Members of the Glowing Experiments Team conducted ground-based, preflight experiments and characterizations of various flight experiments in Environmental chambers and Mars green houses using luminous bacteria and their enzymes – luciferases. They characterized the adaptation of biological systems (plants in greenhouse) to flight stress conditions and control of the environments in chambers with bioluminescent integral toxicity bioassays and other devices. During this period of experiment development, the team members worked in collaboration with other SLSTP members and as a group to collectively monitor the progress of each task. The development tasks culminated in the engineering design of bioluminescent biosensors for Space applications. The data formed the basis of the final reports and SLSTP presentations. The team included two students who have been conducting independent pre-flight experiments with bioluminescent biosensors application under the direction of PI and Co-PI. Both students have done experiments together but from two different sides: control of toxicity in environments and control of stress metabolites inside the plants. Students prepared posters, presentations and abstracts for a student conference. Briefly, their results are:

Monitoring plant stress with bioluminescent biosensors.[5]

Closed ecological systems are required for long duration space missions. Pertinent to such systems is a sensor capable of monitoring plant health accurately and cost effectively. A method for *in-situ* monitoring of plant stress has been developed. This method utilizes bacterial bioluminescent reactions to convert plants' metabolite stress signals into quantifiable light emissions. The concentrations of coupled enzymatic system substrates in plants (NADH, FMN and other stress metabolites) may infer the stress levels of those plants. Because they are reagents in bacterial bioluminescent reactions, these plant substrates can influence light emissions in a predictable manner. *In-situ* monitoring requires a biologically friendly water-based metabolite-extraction method. Such an extraction method has been developed. Of the metabolites tested—NADH, flavin mononucleotide (FMN), aldehydes—the bioluminescent assay had the greatest sensitivity for FMN (concentrations at the ten nanomole level were detectible). Onion and radish grown at CO_2 levels ranging from 350-4000 ppm in the Kennedy Space Center SLS Lab's experimental growth chambers were obtained for sampling. Results show that plants' FMN levels vary depending on plant type, environmental stress, and plant cell type. Radish extracts indicated FMN levels four times higher than those of onion extracts. The concentration of FMN in radishes is not uniformly distributed throughout the plant: tubers appear to contain five times more FMN than leaves; stems contain three times more. An increase in environmental CO_2 levels by a factor of five increases FMN production by 20%. This correlation may be attributed to higher photosynthetic rates in high CO_2 environments. The finding that plant FMN concentration varies with stress, in conjunction with our ability to convert that dependence into a quantifiable luminescent signature, is a considerable step towards developing biosensors for use in the closed ecological systems of tomorrow's space endeavors.

Development of bioluminescent sensors to control contaminants in the air, water, and soil of closed ecosystem.[6]

The environmental control of contaminants in closed ecological life support systems is crucial for the long term success of human space exploration. It will be important to develop biological sensors to monitor environmental conditions inside closed ecosystems. The purpose of this research is to develop highly sensitive and low cost bioluminescent sensors that are that capable of monitoring multiple aspects of the environment in space life support systems. Bioluminescent assays were used to detect contaminants in the air, water, and soil taken from environmental growth chambers at the SLSL at KSC. Two methods were developed to monitor the levels of toxicity in closed ecological systems: (1) *in vivo* assay using luminous recombinant *E. coli* bacteria with LUX-gene from *Photobacterium phosphoreum*, (2) *in vitro* assay using the coupled enzyme system: NADH:FMN–oxidoreductase–luciferase. The biological components of the assay were injected into a microplate. Then test samples were injected into the biological solution. A PerkinElmer Victor 2 bioluminometer measured the emission of light. The conditions to conduct environmental toxicity assays were determined, including the volume of bacteria required and the standard reaction mixture for the coupled enzyme system. The sensitivity of bioluminescent systems to model pollutants: ethanol and butanol was determined. The biological systems were found to be highly sensitive to liquid ethanol 1 μL and buthanol 2 μL. The bioluminescent assays developed for testing liquids were adapted to test for toxicity in the air. A new bioluminescent method for control of the process of purification of bacteria from water was developed with NanoCeram filters. These preliminary investigations of bioluminescent systems show the advantages of this approach for applications in ecological space life support monitoring.

The following educational objectives were met using bioluminescence in SLSTP:

- To give students an opportunity to learn scientific laboratory techniques as well as lab safety practices and government regulations.
- To develop and test methods of using bioluminescence as a bioassay with known results while evaluating the procedures developed for student use.
- To allow students to design and set up experiments with the intent to model pure science, while developing and refining methods for testing bioluminescence as a bioassay within environmental growth chambers.
- To give students an opportunity to work cooperatively and communicate effectively in a scientific research environment.
- To train students to collect and analyze data in preparation of a formal presentation and abstracts.

ACKNOWLEDGEMENTS
This research were funded by the National Aeronautics and Space Administration, Fulbright Program, Russian Humanitarian Science Foundation (grant N06-06-00229a), Award No. RUX0-002-KR-06 and Y1-B-02-19 of the U.S. CRDF and RF Ministry of Education", BRHE Programs; "Molecular and Cellular Biology" program of the Russian Academy of Science. The thoughtful contributions and assistance of SLSTP Trainees Frank H. Mycroft and Li Yang; SLS researchers Michelle Birmele, Michael Roberts have benefited from this research considerably.

REFERENCES
1. Kratasyuk VA, Kuznetsov AM, Gitelson JI Bacterial bioluminescence in ecological education. In: Hastings JW, Kricka LJ, Stanley PE, eds. Bioluminescence and Chemiluminescence: Molecular Reporting with Photons, Chichester: Wiley 1997: 177-80.
2. Kratasyuk VA, Kudinova IJ Practical enzymology course based on bioluminescence. Luminescence 1999; 14: 189-92.
3. Gitelson JI, Kratasyuk VA. Bioluminescence as an educational tool. In: Stanley PE, Kricka LJ. eds. Bioluminescence and Chemiluminescence: Progress and Current Applications. Singapore: World Scientific, 2002: 175-82.
4. Kratasyuk VA, Esimbekova EN. Polymer immobilized bioluminescent systems for biosensors and bioinvestigations. In: Arshady R. ed. Polymeric Biomaterials, The PBM Series, London: Citus Books, 2003; 1: 301-43.
5. Mycroft FH, Kratasyuk VA Monitoring plant stress with bioluminescent biosensors. In: Gravitational and Space Biology (ASGSB 2005 Twenty-first Annual Meeting Abstracts of American Society for Gravitational and Space Biology, November 1-4, 2005), 2005; 19:6.
6. Yang L, Kratasyuk VA. Development of bioluminescent sensors to control contaminants in the air, water, and soil of closed ecosystems. In: Gravitational and Space Biology (ASGSB 2005 Twenty-first Annual Meeting Abstracts of American Society for Gravitational and Space Biology, November 1-4, 2005), 2005; 19:15.

A COMPARISON OF BIOLUMINESCENCE AND CHLOROPHYLL FIELDS OF THE WORLD OCEAN

SA PIONTKOVSKI,[1,2] YU N TOKAREV,[2] LA LEVIN[3]
[1] *MSRC, Stony Brook University, Stony Brook 11794-5000,*
Email: spiontkovski@notes.cc.sunysb.edu;
[2] *Institute of Biology of the Southern Seas, Sevastopol 99011, Ukraine;*
[3] *Institute of Biophysics, Akademgorodok, Krasnoyarsk, Russia*

INTRODUCTION

Bioluminescence is recognized to be widespread in the marine environment.[1] Primary sources of spontaneous or stimulable bioluminescence in all oceans are phytoplankton dinoflagellates, zooplankton copepods, and bacteria[2] which are capable of producing effects visible even from satellites.[3] On the global scale, the bioluminescence intensity acts as a sensitive indicator of plankton biomass,[4] toxicity in the marine environment, and the functional state of planktonic ecosystems. It is important in the global mapping of the oceanic biota,[5,6] upper layer biology based on remote sensing[7] and fishery.[8] The plankton bioluminescence however has never been analyzed in a form of global distribution patterns. The objective of our study was to create an annual bioluminescent map of the world ocean, applicable to the upper 100 m layer based on direct field measurements.

METHODS AND DATA

The database from the vertical profiles of bioluminescence obtained by the Former Soviet Union (FSU) expeditions in the Atlantic, Pacific, Indian Oceans and the Mediterranean Sea basin was assembled, to produce the world ocean map. Data included about 24,000 oceanographic stations with bioluminescence profiles obtained in the upper 100 m layer:

Basin	# Data sets (cruises)	# Stations with bioluminescent profiles
Arctic Ocean	4	1034
Atlantic Ocean	22	7098
Indian Ocean	4	5801
Pacific Ocean	10	6129
Mediterranean and Black Sea	30	4684

A breakdown of oceanographic stations with bioluminescence profiling, CTD casts and plankton sampling over regions is exemplified by Fig. 1.

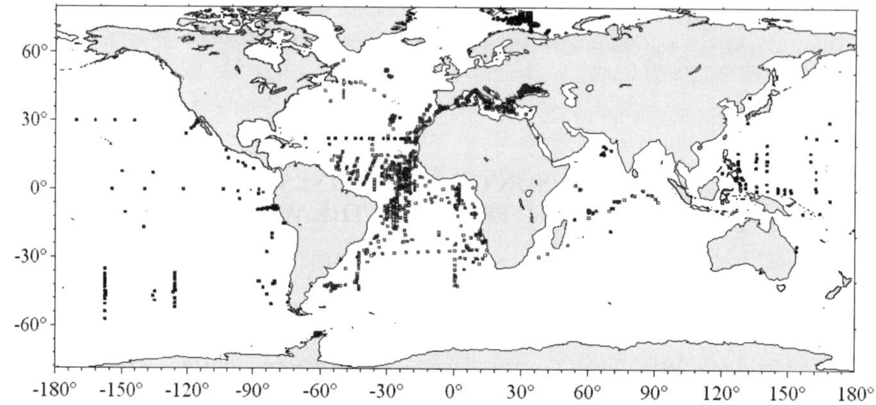

Fig. 1. Sampling sites in the World Ocean (1966-2002)

The technical schemes of the two major FSU bathyphotometers (SALPA and ROMASHKA) were quite similar. The first one was designed and used predominantly by the Institute of Biology of the Southern Seas (Sevastopol, Ukraine) while the ROMASHKA was designed by the Institute of Biophysics (Krasnoyarsk, Russia). Both Institutes have been working on the basis of joint programs and information exchange. At the same time ROMASHKA has a slightly different range for measuring the intensity of bioluminescence: from 10^{-12} to 10^{-4} W cm^{-2}, within 5 subranges. Bathyphotometer SALPA has a dynamic band of 57 db and the sensitivity of the bioluminescence sensor from 10^{-13} to 10^{-8} W cm^{-2} with a band of spectral sensitivity from 160 to 600 nm.

To map the spatial structure of the bioluminescent field, the "Surfer" Golden Software package was applied. The Kriging gridding method and the smoothing of the matrix of the grid, with the weight of matrix center and a set of the distance weighting power was used to create a two-dimensional field. The Coastal Zone Color Scanner (CZCS) chlorophyll *a* data (seasonally averaged from January 1978 through March 1986) were used, to compare global patterns of bioluminescence and phytoplankton (represented by the chlorophyll field). The CZCS data set is a collection of monthly composites of ocean chlorophyll concentration derived from the CZCS flown aboard the Nimbus-7 satellite from 1978 through 1986. Although the CZCS data are fairly old compared to the latest (SEWIFS, MODIS, etc) remotely sensed global chlorophyll images, the CZCS were selected for comparison, because they matched the years of the FSU expeditions (mostly from 1975 to 1990).

RESULTS AND DISCUSSION

Our attempt to reveal global scale trends of bioluminescent field and to compare bioluminescent maps with the maps of the remote sensing of ocean color is exemplified in Fig. 2. The bioluminescence data in the upper 50 m layer were averaged from November through April.

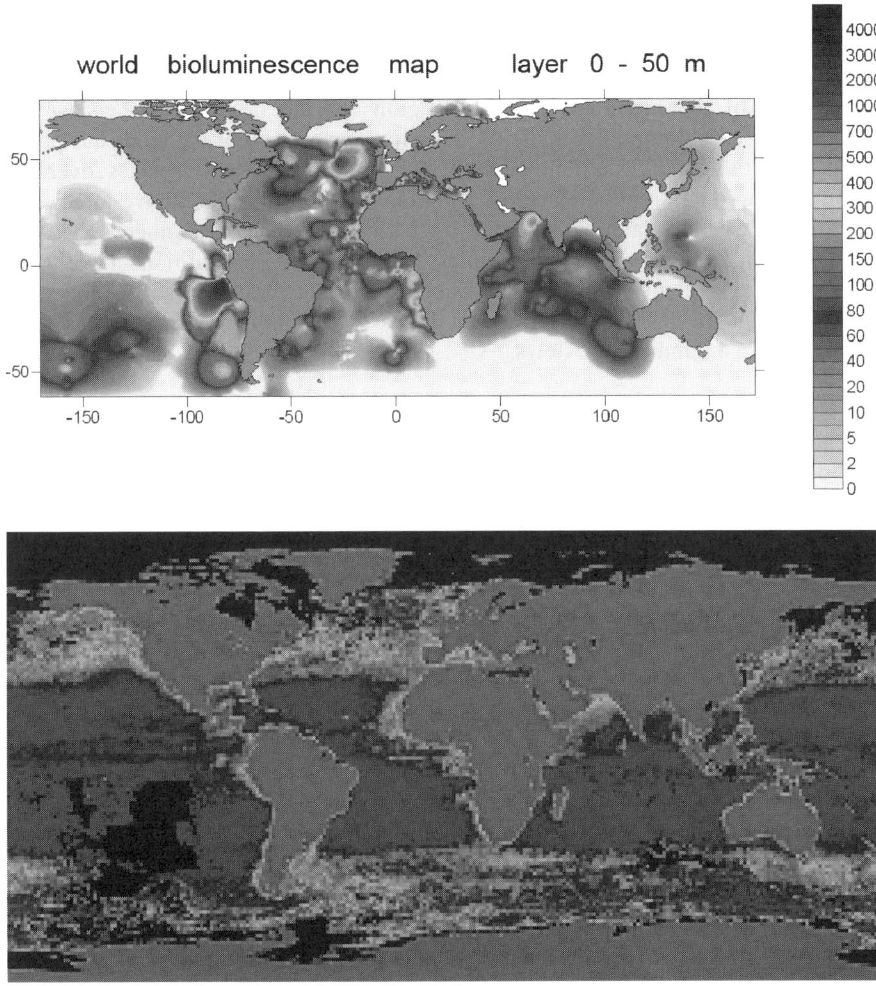

Fig. 2. Bioluminescent and chlorophyll fields of the global ocean. Upper panel: Bioluminescence intensity (0-50m): 10^{-12} W cm^{-2} l^{-1}. Lower panel: The CZCS chlorophyll a data. http://disc.gsfc.nasa.gov/interdisc/readmes/czcs.shtml#ref3

Comparing maps of bioluminescent and chlorophyll fields, one might note elements of similarity, as well as discrepancy. For instance, tongues of enhanced bioluminescence and chlorophyll concentration directed from African coastal upwellings towards the open ocean waters are well pronounced in both maps. In some cases (like Peruvian upwelling in the Pacific Ocean or the north-eastern Atlantic), zones of enhanced bioluminescence occupy huge areas –not supported that much by the chlorophyll *a* concentration. This means that in some areas, local maxima of bioluminescence might be formed by the bioluminescent species comprising relatively small portion of the total phytoplankton. On the other hand, we understand a certain distortion of the global bioluminescence pattern, caused by the absence of measurements in many regions of the ocean (not the case for remotely sensed chlorophyll a pattern). The first comparison of the maps has posed an issue of the ratio of bioluminescent to non-bioluminescent species over the regions. Using data on phytoplankton sampling carried out in the above mentioned expeditions, we are planning to come up with appropriate assessments, because phytoplankton, and zooplankton samples from the FSU expeditions have been treated to the level of species. The abundance of the bioluminescent, and non-bioluminescent species will be defined on the basis of a bioluminescence species list summarized in a number of reviews.[4,9,10,11]

REFERENCES

1. Rohr J, Schoonmaker J, Latz M, Hyman M. 2000. Flow visualization in the ocean: Implications of laboratory bioluminescence experiments. Future Directions in Bioluminescence Research. ONR Workshop. San Diego.
2. Herring PJ. Sources of bioluminescence in the ocean. In: Future Directions in Bioluminescence Research. 2000. An ONR Sponsored Workshop. San Diego, CA, USA.
3. Miller SD, Haddock SHD, Elvidge CD, Lee TF. Detection of a bioluminescent milky sea from space. PNAS 2005;102:14181-4.
4. Gitelzon II, Levin LA, Utyshev RN, Cherepanov OA, Chugunov YV. 1992. Ocean bioluminescence. Gidrometeoizdat, St. Petersburg, 284pp (in Russian).
5. Neilson DJ, Latz MI, Case JF. Temporal variability in the vertical structure of bioluminescence in the North Atlantic Ocean. J Geophys Res 1995: 100: 6591-603.
6. Piontkovski SA, Tokarev YN, Bitukov EP, Williams R, Kiefer D. The bioluminescent field of the Atlantic Ocean. Mar Ecol Progr Ser 1997: 156: 33-41.
7. Ondercin DG, Atkinson CA, Kiefer DA.. The distribution of bioluminescence and chlorophyll during the late summer in the North Atlantic: Maps and a predictive model . J Geophys Res 1995:100:6575-90.
8. Butler MJA, Mouchot MC, Barale V, LeBlanc C. The application of remote sensing technology to marine fisheries: an introductory manual. FAO Fish Tech Pap 1988:295:165.
9. Clarke GL, Conover RJ, David CN, Nicol JA. Comparative studies of luminescence in copepods and other marine animals. Limnol Oceanogr 1962:4:163-80.
10. Evstigneev PV, Bitukov EP. 1990. Bioluminescence of marine copepods. Naukova Dumka, Kiev, 144 pp (In Russian).
11. Lapota D, and Losee JR. Observations of bioluminescence in marine plankton from the sea of Cortez. J Exp Mar Biol Ecol 1984:77:209-240.

BRAZILIAN SPECIES OF BIOLUMINESCENCE FUNGI

CASSIUS V. STEVANI, ANDERSON G. DE OLIVEIRA,
LUIZ F. MENDES, OLIVIA DOMINGUES
Laboratório de Bioluminescência de Fungos Instituto de Química - Universidade de São Paulo,C. P. 26077/05599-970, São Paulo, S.P., Brazil

INTRODUCTION

The fungi bioluminescence (BL) history begins with Aristotle (384-322 BC), who was the first to observe light emission from decomposing wood and to differentiate this phenomenon from fire.[1] Afterwards, bioluminescent mushrooms have been cited sporadically by several authors. One of the most interesting descriptions was made by Georg Everard Rumph (1637-1706), a Dutch physician, merchant and Indonesian consul in his *Herbarium Amboiense*. Rumph reported that natives used to wander at night carrying bioluminescent mushroom in their hands in order to illuminate their way in the forest. Two centuries later, Micronesian natives still use bioluminescent mushrooms on their heads as an ornament or spread them on their faces to scare people. Very often, however, shining mushrooms were destroyed by natives as they were considered a sign of bad luck.[1] Although many botanists have described bioluminescent fungi worldwide, much more attention was paid to the glowing rotten wood than to the organisms responsible for the light emission. A connection between decomposing wood and mycelium BL was evidenced only in the beginning of the 19th century when scientists were able to correlate BL with oxygen.

Until recently, 42 species in 9 genera, of bioluminescent fungi were known throughout the planet.[2] More than a half occur in Australia and Asia and belong to the genus *Mycena*. Certainly, however, a much larger number of bioluminescent fungi in tropical forests still awaits description. Despite the fact that Brazilian forests have the largest plant and animal biodiversities in the world, only two bioluminescent fungi were known till 2002. *Mycena rorida* was first described in 1948, but its bioluminescence "had not been strictly observed, but 'surmised".[2] In 1988, a yellow mushroom identified as *Pleurotus olearius* was found in the State of Goiás (Center-West of Brazil).[3] Unfortunately, neither identification is reliable, and no material was preserved in a herbarium for further studies. It is noteworthy to mention that some authors report a blue, sometimes white glow from fruiting bodies[2], but the attribution of different colors can be explained by the low BL intensity and the inability of human eyes to distinguish colors under dim light. In fact, the spectral distribution of BL displayed by fruiting bodies and mycelium peaks in the green region, at about 520 nm.[3-5]

During 2002-2006, our research group was able to collect 11 new luminous mushroom species, 8 of them collected in the Ribeira River State Park (PETAR) near the municipality of Iporanga, State of São Paulo, Brazil. Addition of these species to the literature descriptions of bioluminescent fungi increased the reported number of taxonomically valid luminous fungus species to an astounding 54 species, representing 22% of all species described in the world so far. The present work is the first report on new luminous mushrooms since Wassink's publication in 1978.[2] Elucidation of the chemistry of firefly bioluminescence lead to the development of several analytical tools such as an ATP-based bioassay for food microbial contaminants and the use of the *luc* reporter gene in molecular biology. Likewise, unraveling the bioluminescence mechanism in fungi may also lead to novel analytical tools. Our group is currently investigating the mechanism of bioluminescence in fungi as well as the development of a bioluminescent ecotoxicological bioassay based on the acute toxicity response against toxicants applied on the surface of *Gerronema viridilucens* mycelium cultivated in agar.

MATERIALS AND METHODS

All fruiting bodies were collected at night in near total darkness, and were detected in the field as sources of green luminescence. Specimens were placed in plastic collecting boxes in the field for transportation to the field laboratory. All fresh specimens were then photographed with a Speedlight flash gun and subsequently in dark conditions using 20–50 minute exposures. Photographs were obtained with a Nikon N80 camera equipped with a Nikkor AF Zoom 28-105 mm lens, and carrying Fuji Superia ASA 400 or ASA 800 color print films, at f5.5 and f11. Fresh specimens were maintained under cool conditions until the following morning when macromorphological data were registered under daylight conditions.

From 2002 until 2006 11 different species of bioluminescent fungi were collected. Cultures of *G. viridilucens*[3] and *Mycena lucentipes*[5] were obtained initially by inoculation of fruiting bodies collected in December 2002 at PETAR in a potato dextrose agar (PDA) containing medium with 500 mg/L of benomyl (Sigma-Aldrich) and chloramphenicol (Sigma-Aldrich). Small fragments of fruiting bodies were first washed with previously sterilized deionized water, immersed in a 5.0% (v:v) solution of common household bleach for 1 min, washed again in the water and deposited on the PDA medium, near a flame. After one month, separation of mycelia in PDA media (with no benomyl and chloramphenicol) allow us to observe light emission from the cultures with dark-adapted eyes. Cultures were then transferred to a enriched medium [2.0% agar (Oxoid), 1.0% sugar cane molasses (food grade), and 0.10% yeast extract (Oxoid)] and cultivated ever since for short-term usage at 20-25°C and pH 6.0 ± 0.5.

RESULTS AND DISCUSSION

Collecting specimens at night is the only way to document the enormous diversity of bioluminescent fungi in the woods. Our work in PETAR disclosed seven new species of *Mycena* spp and *Gerronema viridilucens* from remnant Atlantic Forest grown over a limey soil (Fig. 1).

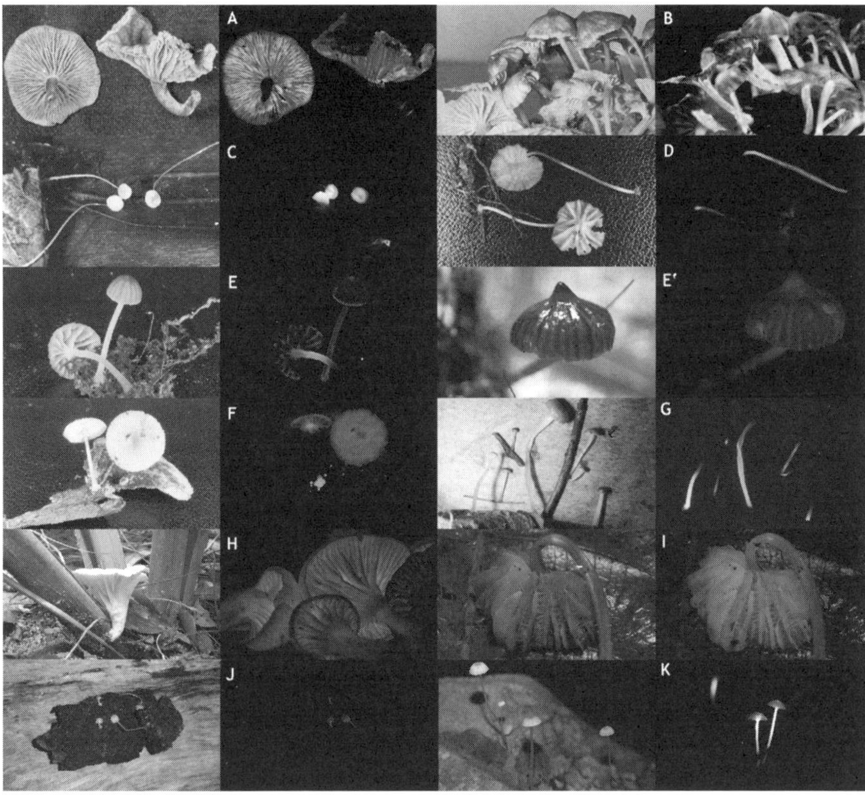

Fig. 1. Eleven species of bioluminescent fungi found in Brazil (A-K) during 2002-2006. A: *Gerronema viridilucens*, B: *Mycena lucentipes*, C: *M. asterina*, D: *M. singeri*, E: *M. fera* (white variation), E': *M. fera* (black variation), F: *M. discobasis*, G: Mycena sp.1, H: Unknown species from Piauí State, I: Unknown species from PETAR, J: Unknown species from Paraná State and K: Unknown species from Amazonas State.

We believe that there are many more species of mycenoid fungi with bioluminescent properties to be described. When taxonomists report new species of euagarics they rarely record macromorphological data at night, so their luminescent properties keep undetected. We found so far three species of *Mycena* that heretofore were unknown as being luminescent organisms, *M. discobasis*, *M. fera* and *M. singeri*.[5]

The biological roles of fungi bioluminescence have been recently reviewed.[6,7] Sivinski hypothesized that luminescence may attract predators of fungivores, or may serve as a warning signal to repel nocturnal fungivores. Sivinski's hypotheses are difficult to be proved as fungivores could be attracted by fruiting bodies odor instead by their bioluminescence.[7] In order to prove that fungivores really see the light emitted by the fungus, electroretinographic data should be obtained.[8]

The greater number of *Mycena* (33 species) spread worldwide raises the question whether bioluminescence evolved in the mycenoid fungi independently numerous times, or if it is a primal feature subsequently lost numerous times. We suspect that the metabolic pathway leading to bioluminescence in fungi may have evolved to counteract the oxygen toxicity and that it appeared once early in the evolutionary history of the lineage, and that the final light-emitting steps were lost a number of times. Bioluminescence has been suggested as an advantageous biochemical process for several organisms as an auxiliary antioxidant protection against deleterious effects of reactive oxygen species produced mainly by mitochondria during respiration.[5] Ongoing molecular phylogenetic and biochemical studies in our laboratory are in progress to evaluate this hypothesis.

ACKNOWLEDGEMENTS
We thank the FAPESP for financial support.

REFERENCES

1. Harvey EN. Bioluminescence. New York: Academic Press, 1952.
2. Wassink EC. Luminescence in fungi. In: Herring PJ. ed. Bioluminescence in Action. New York: Academic Press, 1978: 171-97.
3. Desjardin DE, Capelari M, Stevani CV. A new bioluminescent agaric from São Paulo, Brazil. Fungal Divers 2005; 18: 9-14.
4. O'Kane DJ, Lingle WL, Porter D, Wampler JE. Spectral analysis of bioluminescence of *Panellus stypticus*. Mycologia 1990; 82: 607-16.
5. Desjardin DE, Capelari M, Stevani CV. Bioluminescent *Mycena* species from São Paulo, Brazil. Mycologia 2006; accepted.
6. Bermudes D, Petersen RH, Nealson KH. Low-level bioluminescence detected in *Mycena haematopus* basidiocarps. Mycologia 1992; 84: 799–802.
7. Sivinski JM. Arthropods attracted to luminous fungi. Psyche 1981; 88:383–90.
8. Lall AB, Järvilehto MVA, Cronin TW, Carvalho AA, de Souza JM, Hill A, et al. Vision in click-beetles (Coleoptera: Elateridae): structure, visual and screening pigments, and spectral correspondence between species visual sensitivity and bioluminescence emission. J Comp Physiol 2006; in press.

MOLECULAR PHYLOGENY OF CYPRIDINID OSTRACODES AND THE EVOLUTION OF BIOLUMINESCENCE

ELIZABETH TORRES AND VANESSA L. GONZALEZ
Department of Biological Sciences, California State University, Los Angeles, 5151 State University Drive, Los Angeles, CA 90032, USA

INTRODUCTION

The ostracode family Cypridinidae contains approximately 200 species, of which about half are bioluminescent. Both cypridinid sexes are luminescent and all luminescent species can presumably use the light they produce to deter predation. Males of more than 60 cypridinid species from the Caribbean use their light for courtship.[1] The courtship displays occur nightly on coral reefs throughout the Caribbean and are species specific. Cohen and Morin[1] reconstructed the phylogenetic relationships within the Cypridinidae based on 58 morphological characters from 46 species and all known genera within the Cypridinidae. They proposed that bioluminescence evolved just once in the Cypridinidae. Bioluminescent signaling for courtship appears to have evolved subsequently, and just once, in the Caribbean cypridinids. Here, we constructed a molecular phylogeny to independently test the hypotheses of Cohen and Morin.[1] Analyses of sequences from the 16SrRNA and 12SrRNA mitochondrial genes from 36 species, representing 19 genera, were performed to reconstruct relationships within the Cypridinidae. The patterns of evolution of bioluminescence and signaling were examined by mapping behavioral characters onto the molecular phylogeny. The specific objectives of this study were to (1) assess the monophyly of bioluminescent taxa and the homology of cypridinid bioluminescence, (2) assess the monophyly of Caribbean taxa and the homology of luminescent signaling in the Caribbean, and (3) assess the monophyly of newly named genera within the Cypridinidae.

METHODS

Total genomic DNA was extracted from individual ethanol preserved ostracodes with the QIAamp® DNA mini kit (QIAGEN, Valencia, CA, USA) following the tissue protocol. Primers 16Sbr[2] and 16SOA1 were used to amplify a fragment of the mitochondrial 16S rRNA gene, and primers 12sai and 12sbi[2] and forward primer 12sd to amplify a fragment of the mitochondrial 12S rRNA gene. PCR products were sequenced with the Big Dye terminator v3.1 Cycle Sequencing Kit (Applied Biosystems, Foster City, CA) and analyzed on an ABI 3100-Avant Genetic Analyzer. Sequences were aligned in Clustal X.[3] A 340 bp fragment of the 16SrRNA gene was sequenced for 36 species representing 19 genera within the Cypridinidae and combined with an incomplete data set of 399 bp of 12S rRNA for 24 species from 12 genera. Simultaneous analyses were performed with maximum parsimony (MP) in PAUP*[4] (random addition of taxa, 100 replicates, gaps treated as fifth character state) and Bayesian analysis in Mr. Bayes 3.0[5] (ngen=50000) under the best-fit model (GTR+I+G) selected by AIC in MrModeltest 2.2.[6]

RESULTS AND DISCUSSION

The mitochondrial phylogenies support the hypothesis that bioluminescence evolved once within the Cypridinidae, which suggests that bioluminescent systems are homologous within the Cypridinidae. In both the MP (Fig. 1) and Bayesian consensus trees, all luminescent taxa form a monophyletic clade, but with low bootstrap and Bayesian consensus values. *Vargula tubulata,* a species that has not yet been directly tested for luminescence, falls within the luminescent clade. It has a pair of large tusks with nozzles on the upper lip, a feature that is only found in all other bioluminescent cypridinids.[1] Its luminescence capability remains ambiguous, and its position in the morphological phylogeny is uncertain because it shares other features with non-luminescent cypridinids.[1]

There is weak support for the hypothesis that Caribbean signaling cypridinids are a monophyletic group, suggesting that luminescent signaling may be homologous and is not convergent. In the MP tree (Fig. 1), the Caribbean signaling species form a monophyletic group, but with low bootstrap support. In the Bayesian tree (not shown), the Caribbean signaling taxa form two well-supported clades, but their relationship within the luminescent clade is unresolved. A single non-signaling but luminescent species, "Cypridina (Vargula)" *tsujii*, from eastern Pacific temperate waters off of California and Mexico, clusters within the Caribbean signaling clade, a result consistent with the morphological cladograms of Cohen and Morin.[1] This species has presumably been separated from its Caribbean relatives at least since the emergence and final closure of the Isthmus of Panama approximately 3.1-2.8 million years ago.[7] It may have lost the signaling condition, or its signals have not yet been detected in the field. As in the morphological analysis of Cohen and Morin,[1] the genus *Vargula* is polyphyletic. However, the relationship of *Vargula norvegica* (the type species of the genus *Vargula*) to the other luminescent cypridinids is unresolved in the molecular analyses presented here. The monophyly of the newly named signaling genera, *Photeros* (=Phot Group)[1] and *Kornickeria*,[8] is supported in the molecular trees. Many luminescent signaling species remain incorrectly assigned to the genus *Vargula*. Cohen and Morin[1] have proposed at least 5 additional genera of signaling species, designated as Groups H, R, C, Z and U. The H and C groups are supported by the molecular results in this study, but the monophyly of the R, Z, and U groups are uncertain. More signaling species need to be included in the molecular data set to assess the monophyly of newly proposed luminescent signaling genera, and examine the patterns of signaling within genera.

The molecular phylogeny presented here is largely congruent with the morphological cladograms of Cohen and Morin[1]. Other authors have included molecular data from cypridinids to examine the evolution of bioluminescence in myodocopid ostracodes.[9,10]

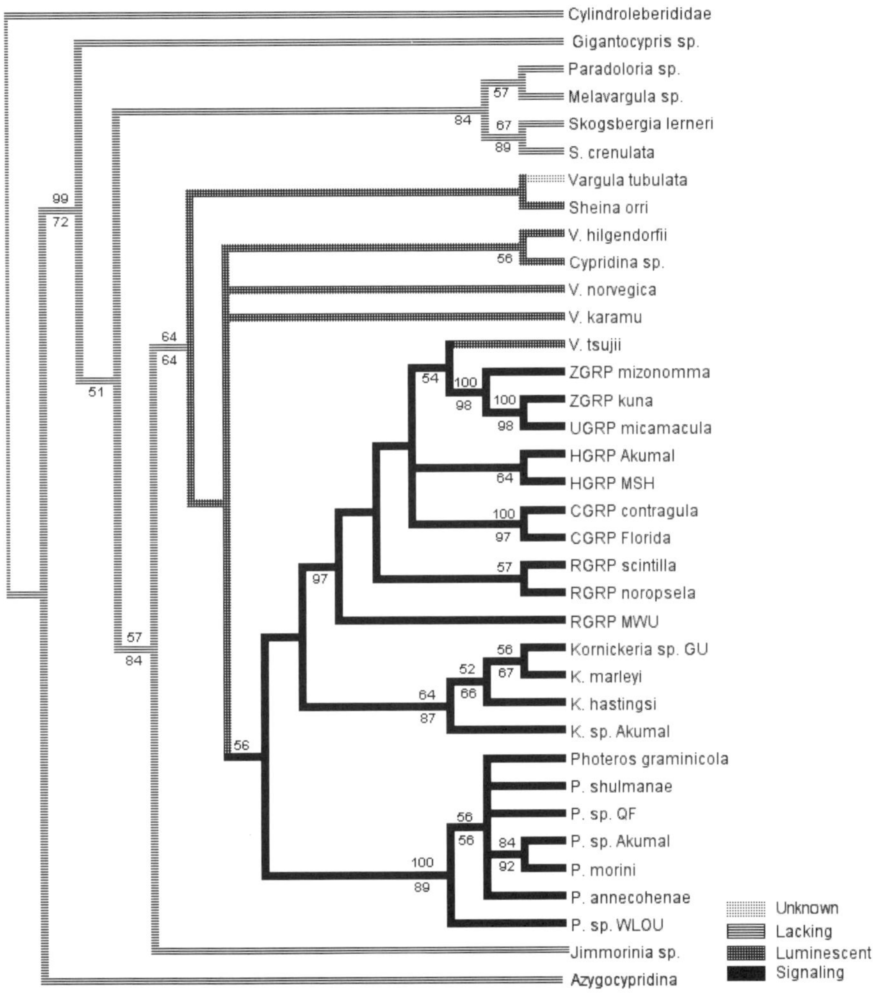

Fig. 1. MP strict consensus of 54 equally parsimonious trees (L=1451, CI=0.4313, RI=0.5170, 276 parsimony informative characters); bootstrap values from MP analysis (100 replications) indicated above branches. Fifty percent majority consensus values from the Bayesian analysis indicated below branches.

Ours is the first to include several species and genera within the Cypridinidae, allowing us to test specific hypotheses about the evolution of bioluminescence and signaling behavior within the Cypridinidae.

ACKNOWLEDGMENTS

We are grateful to S. Haddock, K. Abe, M. Grygier, A. Parker, A. Heger, and M. Bennett for providing samples; many specimens were identified and provided by A.C. Cohen and J.G. Morin. This work was supported by the NSF (DEB-9412813, DEB-0002692 and CEA-CREST). V.L.G. was supported through an NSF LSAMP Bridges to the Doctorate fellowship (NSF HRD-0331537).

REFERENCES

1. Cohen AC, Morin JG. Sexual morphology, reproduction and the evolution of bioluminescence in Ostracoda. Bridging the gap: Trends in ostracode biological and geological Sciences. Paleontol Soc Papers 2003: 37–53.
2. Palumbi SR. Nucleic acids II: the polymerase chain reaction. In: Hillis DM, Moritz C, Mable BK. eds. Molecular Systematics, Second edition. Sunderland, MA: Sinauer Associates, 1996: 205-47.
3. Thompson JD, Higgins DG, Gibson TJ. CLUSTAL W: improving the sensitivity of progressive multiple sequence alignment through sequence weighting, position specific gap penalties and weight matrix choice. Nucleic Acids Res 1994; 22:4673-80.
4. Swofford DL. PAUP*. Phylogenetic Analysis Using Parsimony (*and Other Methods). Version 4. Sunderland, MA: Sinauer Associates, 2000.
5. Huelsenbeck JP, Ronquist F. MRBAYES: Bayesian inference of phylogeny. Bioinformatics 2001; 17:754-5.
6. Nylander JAA. MrModeltest 2.2. Program distributed by the author. Uppsala University: Evolutionary Biology Centre, 2004.
7. Coates AG, Obando JA. The geologic evolution of the Central American Isthmus. In: Jackson J, Budd A, Coates AG, eds. Evolution and Environment in Tropical America. Chicago: Univ Chicago Press, 1996: 31-56.
8. Cohen AC, Morin JG. The cypridinid copulatory limb and a new genus Kornickeria (Ostracoda: Myodocopida) with four new species of bio-luminescent ostracods from the Caribbean. Zool J Linn Soc 1993, 108:23-84.
9. Oakley TH. Myodocopa (Crustacea: Ostracoda) as models for evolutionary studies of light production and vision: multiple origins of bioluminescence and extreme sexual dimorphism. Hydrobiologia 2005:179-82.
10. Wakayuma N, Abe K. The evolutionary pathway of light emission in Myodocopid Ostracoda. Biol J Linn Soc 2006;87: 449-55.

INDEX

1,2-dioxetanes 223, 231
activation parameters 225
17-β Estradiol 126
2-hydroperoxycoelenterazine 7
4-coumarate:CoA ligases 21
6'-amino-*D*-luciferin 107
9,10-diphenylanthracene 216
9-[[(3-Carboxypropyl)[(4-methyl-phenyl)sulfonyl]amino]-carbonyl]-10-(3-sulfopropyl)acridinium inner salt 59, 63
ab initio calculations 243
ABLAMSKI D 119
Acetylaminoluciferin 243
Acetylcholine 191
Acridan esters 171
Acridinium
ester 171
labels 139
ACS Centaur 60
Activators 215, 221
Acyl-CoA synthetase 19
ADAMCZYK M 59, 63
Adenylate kinase 79, 87
Adenylate-forming enzymes 21
Aequorin 103, 159, 189
Air contaminants 259
AK rapid® technology 87
ALLAIN B 75
Alpha-phenyl-N-t-butylnitrone 216
AMBRETTI S 201
Amino acid sequence 27
Aminoluciferin 243
Analogues
amino 243
Cypridina luciferin 168
ANDERS E 247
Antioxidants 168
capacity 83
Antiradical capacity 83, 85
Apoaequorin 159
Aq 103
Arabidopsis thaliana 20
ARAKAWA H 67, 103
Aristotle 265
ARTS R 171
ASAI T 177
Assay 57–174
ATP 41, 151
cell-based 119
dual luciferase 99
enumeration 87
glucose 230
H_2O_2 229
immunoassay 103, 147
multicolor 177, 175–212
multiplexed 119
proteolytic 107
pyrophosphate 67
RNA 143
ATP 79, 87
ATPase 151
ATP-metry 16
ATP-releasing reagent 77
ATP-sulphurylase 95
bioluminescence 75
food pathogens 79
microbial 41
non-microbial 41
Autotaxin 63
BAADER WJ .83, 215, 219, 223, 231
Back electron transfer 215, 232
BacLite® RapiTECT GN™ 87
Bacterial
cell count 91
meningitis 185
quorum sensing 111
Bacteriophage 75
BP1 75
BP6 75
BARALDINI M 123
BART 95, 96

BARTOLONI FH 215, 219
Barton-Kellog olefination reaction 224
BASTOS EL 83, 215
Bathyphotometers
ROMASHKA...........................262
SALPA...................................262
BECKERT R...............................247
Beetle 23
luciferase............................20, 197
BERTHOLD A 177
BERTHOLD F...............................41
BET...215
Bile acids.....................................120
Biofilm..111
Bioluminescence
activation.................................165
bacteria....................................261
beetles 247
biosensors................................258
copepods 261
dinoflagellates..........................261
emission spectra.......................120
evolution 269
fungi...................................52, 265
imaging 191
immunoassay....................103, 104
inhibition.................................165
land....................................255–72
oceans................................255–72
phytoplankton 261
space..................................255–72
species.....................................264
zooplankton.............................261
Bioluminescent fields..................263
Bioluminescent map
Arctic Ocean 261
Atlantic Ocean 261
Black Sea 261
Indian Ocean............................261
Mediterranean Sea 261
Pacific Ocean...........................261
world ocean.............................261
Biosensor............................ 111, 123
bis(2,4,6-trichlorophenyl)oxalate 215
Blotting....................................... 173
b-Luc 103
Bootstrap values 21
BORISOVA VV 3
BRANCHINI B 119
BRASHEAR RJ 59, 63
BRET (Bioluminescence Resonance Energy Transfer)............. 123, 167
BROVKO LY......................... 71, 75
BUCKLIN RA............................ 257
C57BL/6-*Min (Min*/+) mice 209
Ca^{2+}... 115
Caged ATP 151
CAMPBELL AK.......................... 35
Campyloxenini 25
Carbachol 191
CARDWELL T 135
Caribbean cypridinids................. 269
CARNE C..................................... 95
CASANOVA S 201
Caspase-3 71
CBG99.. 181
CCD (Charge coupled device)....... 48
camera 46, 53, 172, 201
imaging...................................... 51
cDNA 143
Ceftazidime 89
CellTiter-Glo™ Reagent............... 102
Cerebral cortex 189
c-fos.. 55
promoter 177
Chagas' disease 139
CHANG CD 139
Charge transfer–induced luminescence 14
Chemically initiated electron exchange luminescence 215
Chemiexcitation 12, 215, 219, 231
Chemiluminescence
enhancement........................... 251
flow injection.................. 135, 227

land....................................255–72
oceans..................................255–72
probes.......................................155
reaction mechanism 13
space...................................255–72
CHEN N..............................181, 197
CHENG K...................................139
CHETIRKIN PV..........................257
Chlamydia trachomatis 95
Chlorophyll.................................261
CHO...116
Cholesterol.................................119
7 alpha-hydroxylase.................119
Choline...................................59, 63
oxidase.......................................64
CHURCH VL 95
CIEEL...................................215, 231
mechanism...............................219
Circadian rhythmicity 47
CISCATO LFML. 215, 219, 223, 247
Clock gene 45
Coelenterazine . 3, 7, 115, 123, 131, 159, 189, 197
hydroperoxycoelenterazine..........7
Colistin..128
Contaminants
soil...259
water.......................................259
Copepod...3
Copper..236
CORAZZA S 115
CORRELL MJ 257
Cortical network..........................189
Corynebacterium diphtheriae.......173
CRE-induction pathway.................99
Crohn's disease............................111
Cryogenic CCD camera 48
CUTTER KL...............................185
CV-1 197
CyBi®Lumax...............................118
Cylindroiulus distinctus 16
CYP27A1...................................119
CYP7A1.....................................119
Cypridina (Vargula) 11, 167
luciferase 167
luciferin 11, 167
ostracodes............................... 269
Cypridinidae 269
Cysteine.. 15
DAUNERT S...................... 111, 131
DE OLIVEIRA AG.................... 265
Deep sea fish 38
DEO SK 111, 131, 235
Dimethyl sulfoxide..................... 239
Dioxetanone 239
Diphtheria.................................. 173
toxin....................................... 174
Direct somatic cell count.............. 91
Dixon plots 28, 29
DOMINGUES O 265
DOSAKA S 53
DRESCHEL TW 257
DRISCOLL M........................ 71, 79
DROBAC E................................ 189
Drosophila melanogaster......... 19, 20
DsRed..........................205, 206, 235
Monomer Fluorescent Protein . 235
Dual plasmid strategy................. 100
E. coli 36, 43, 80, 81, 88, 89, 119, 185, 194, 205, 209
JM107.................................... 235
ECKERT CR 83
ECL 202
ECV-304.................................... 185
Egfp 193
Elateridae...................................... 23
phylogeny 25
Electron transfer 231
Electroporation........................... 194
Emission spectra..............49, 50, 132
Enhanced yellow fluorescent protein 123
ENSOR M 131
Enumeration assay........................ 87
Enzyme embedded organic monolith 227

Epithelial cells 127
EPR spectra 217
EREMEEVA EV 7
ESIMBEKOVA EN 257
ESTEVES CH 83
Estrogen 123
receptor 123
EYFP 123
Fab fragment 103
FAN F 99
Farnesoid X receptor 119
Fatty acyl-CoA synthetase 19
Fenchone 223
Filtravettes™ 91
Firefly 151, 243
Firefly luciferase 15, 19, 71, 103
biotinylated 103
evolution 21
Luciola cruciata 67
thermostable mutants 95
Firefly luciferin 239, 243, 247
synthetic pathways 248
Fish 203, 261
deep sea 38
symbiont 35
FITC labelled casein 74
FLAK T 115
FLAMMERSHEIM HJ 223
Flavin monooxygenase 27
Flavonoids 84
FLIPR^TETRA 118
FLOMENHOFT D 111
Flow cytometry 197
Flow injection analysis 135
Fluorescence resonance energy transfer 123
Fluorescent reporters 193
Fluoro-Opto-Electronic Count 91
Food industry 41
Foodborne infections 75
Foodborne pathogens 71, 79
FOOTE N 87
Four-color labeling 197
FRANK LA 3, 7
FRET 123
FRUNKZHYAN VG 91
FUJII H 243
Fungi 265
Fusion genes 181
Fusion protein 189
GABA 189
GADDIPATI SR 193
Gallic acid 251
GANDELMAN OA 95, 107
GARVIN D 99
Gastrointestinal disorders 111
Gaussia 123
GAVRILOV GB 91
G-coupled protein receptors 99
GEISSINGER U 197
Gene
constructs 100
duplication 19, 22
expression 45, 47, 53, 193
recruitment 19, 22
GeneMorph II Random Mutagenesis kit 115
Gentamicin 89
Gerronema viridilucens 266, 267
GFP (Green fluorescent protein) 45, 193, 197, 235
aequorin 189
GFP-Ruc 197
GI-101A 181
Global mapping 261
Glowells ™ 49, 50, 52
Glucose oxidase 227
GLV-1h84 181
GLV-1h86 181
GOLOVLEV V 143
GOLZ S 3
GOMI K 103
GONZALEZ VL 269
G-protein coupled receptors 115
Gram-negative bacteria 87, 111
GREEN M 87

GREENMAN J ... 185
GREGORY S ... 87
GRIFFITHS MW ... 71, 75
GRUMMT F ... 197
GUARDIGLI M ... 201
GUSEV SM ... 257
GZ-11 ... 171
Haemolysin ... 193
HAFNER K ... 41
HAJDOK C ... 71
HALLER A ... 139
Hammett plot ... 12
HAMZINK M ... 171
HATTA-OHASHI Y ... 45
HAWKINS EM ... 99
HAYASAKA N ... 45
hCL1/hPEST ... 101
HEK 293 ... 117
HEK 293 cells ... 101
HeLa cells ... 45, 46, 54
Hemin ... 83
HepG2 ... 119
Herbarium Amboiense ... 265
HERRING PJ ... 35
HICKEY PC ... 49
High throughput instruments ... 42
High throughput screening ... 63, 115, 119, 167
HILL PJ ... 35, 193, 205
HIRANO T ... 11
HIV
  recombinant proteins ... 148
hly ... 193
HMB-45 ... 204
Horbaschek C ... 197
Horner-Wadsworth-Emmons ... 223
Horseradish peroxidase ... 171, 227
Hotaria papariensis ... 16
Hotaria parvula ... 16
Hotaria tsushimana ... 16
Hotaria unmunsana ... 16
hPEST ... 101
HPV ... 203
HSP70B promoter ... 46
Human Immunodeficiency Virus 147
Hydrogen peroxide ... 63
Hydrogen sulphide ... 135
Hydroperoxycoelenterazine ... 7
Hygiene assays ... 80
Hygiene standards ... 75
IGARASHI R ... 151
IKEDA H ... 11
Imaging ... 45, 49
  assay ... 229
  optical ... 175–212
  system ... 45
Imidazole ... 228
Imidazopyrazinone ... 12
Immunoassay ... 103, 147
  bioluminescent ... 103, 104
Immunocapture ... 90
Immunohistochemistry ... 201
In situ hybridization ... 201
In vivo imaging ... 49, 52
INDO/S ... 243
Inflammatory bowel disease ... 111
Injectors ... 42
INOUYE S ... 19, 103
Instrumentation ... 39–56
ITO K ... 103
JACKSON SK ... 127
JACOBSON KB ... 143
JALAL H ... 95
Jasmonic acid ... 20
JATHOUL AP ... 107
JIANG L ... 139
KAMAHORI M ... 67
KAMBARA H ... 67
KAWAMOTO H ... 227
KAWASAKI K ... 167
KAYKOVA EV ... 27, 31
Kennedy Space Center ... 257
Klebsiella pneumoniae ... 88
Klenow DNA polymerase ... 67
KONDO H ... 11
KOOIJMAN M ... 171

Kornickeria ... 270
KRATASYUK VA ... 257
KUMARI A ... 111
LAMBOLEZ B ... 159, 189
Lampyridae ... 23
Laser
  Nd:YAG ... 151
LAW GHE ... 107
LC-MS/MS ... 66
LE TH ... 197
LEGOTSKY SA ... 15
LEVIN LA ... 261
LIAWRUANGRATH S ... 135
Life Science Training Program ... 257
Light organs ... 35, 38
Listeria monocytogenes ... 193, 194
LLIS™ ... 51
LOMAKINA GY ... 15
Loop-Mediated Amplification (LAMP) ... 95
Low light imaging ... 198
Low Light Imaging Standard™ ... 49
Luc ... 107, 119
luc2 ... 100
luc2P ... 100
Luciferase ... 1–38, 45, 259
  bacterial ... 27, 31
  dual-luciferase assays ... 99
  enhanced for thermostability ... 107
  firefly ... 15, 19, 103
  FMN-oxidoreductase ... 163
  Gaussia ... 123
  green-emitting ... 178
  inhibition ... 28
  Metridia longa ... 3
  orange-emitting ... 178
  Photobacterium leiognathi ... 27
  red-emitting ... 178
  reporter ... 53
  thermal stability ... 16, 31
  thermoinactivation ... 17
  water-organic solvents ... 31
Luciferin ... 23, 45, 243
  Cypridina ... 167
  firefly ... 239, 243, 247
Luciola cruciata ... 16, 19, 20
Luciola italica ... 119
Luciola lateralis ... 16
Luciola mingrelica ... 15, 16
Luciola noctiluca ... 16
Luminescence
  biochemistry ... 213–54
Luminescence imaging ... 201, 210
  calibration ... 49
  single cells ... 53
Luminescence microscope ... 45, 54
Luminol ... 83
Luminometer ... 41, 45, 177
  BG-1 tube ... 72
  bioluminometer ... 28
  BLM 8801 ... 4, 163
  chemiluminometer, NightOWL NC100 ... 228
  custom flow-through ... 136
  Hamamatsu R5929 ... 11
  LUM-1 ... 91
  Luminoscan Ascent ... 96
  Luminoskan ... 120
  Microplate luminometer XS4 ... 42
  MPC-1 ... 147
  PhL microplate ... 159
  photon-counting ... 53
  PMT ... 91
  Sirius ... 145
  Tecan GENios Pro ... 128
  Turner ... 35
  Uni-Lite® luminometer ... 80
Luminoview ... 46
Luminoview LV-100 ... 53
Lumiphos Plus ... 202
Lumorase™ ... 107
Lumtek test system ... 91
Lux ... 127
luxABCDE cassette ... 207
luxABCDE operon ... 207
luxA-E ... 206

luxCDABE 185
Lysophospholipase D 63
Lysostaphin 90
M13 145
M3R 99, 102
MAEDA M 67, 103
MAFFT program 19, 25
MAGLIULO M 123
MAKARUSHIN AA 91
MAKI S 11
MANDEVILLE R 75
MARKOVA SV 3, 7
Marlene DeLuca Prize iii
MASTROIANNI N 115
MATTINGLY PG 59, 63
MCLA 155
McMurry 223
MENDES LF 265
Meningococcal disease 172
Metal binding 235
MIC - D 228
Mice 197, 209
Michaelis constant 29
MICHELINI E 119, 123
Microarray 145
Microbial ATP 41
Microbial contamination 41
Microparticles 139
Microplate luminometer
XS4 .... 42, 43
Microplates 41, 42, 64
Microscope, luminescence 45, 53
Milk 71, 91
assay 93
Min mice 210
MIRASOLI M 201
Mithras LB940 .... 59, 63
Monophyletic clade 270
MOORE CA 95
MOS-F 243
mPer2::luc knock-in mice 45
MRP2 201
MRSA screening 90
Multicolor assays 175–212
MUNAKATA E 67
MUNESUE M 227
MURRAY JAH 95, 107
Muscarinic chloride 102
MUSIANI M 201
Mutagenesis 159
Mycena 265
Mycena asterina 267
Mycena discobasis 267, 268
Mycena fera 267, 268
Mycena lucentipes 266, 267
Mycena rorida 265
Mycena singeri 267, 268
Myocardial ischemia 59
N-acyl homoserine lactones 111
NADH:FMN-oxidoreductase 163, 259
NASA 257
N-butyryl-DL-homoserine lactone 112
N-dodecanoyl-DL-homoserine
lactone 112
Neisseria meningitidis 185
NELSON SM 127, 185
Neutral density filter 49
N-hexanoyl-DL-homoserine lactone 112
Ni-chelate chromatography 19
NightOWL NC100 228
NISHIMURA W 103
Nitrocellulose 172
NIWA H 11
N-octanoyl-DL-homoserine lactone 112
NOZAKI O 227
N-tetradecanoyl-DL-homoserine
lactone 112
Nucleic acids 95
Nude mice 184
OBA Y 19, 23
Obelia longissima 7, 131
Obelin 8

binding sites 7
cysteine residues 131
mutant 8, 131
OBINATA R 67
Ocean 261
OHASHI-HATTA Y 53
OHGIYA S 167
ÔHIRA H 23
OHMIYA Y 167, 177
OLIVEIRA MA 231
Omalisidae 23
On-board reagent addition 42
On-board vortexer 43
Optical imaging 175–212
Orthopoxvirus 181
OSIPENKO OA 257
Oxalyl chloride 220
Oxyluciferin 239, 247
$p16^{INK4a}$ 202
p53 gene 69
PAGUIO A 99
PAP 104, 105
Paralogous genes 20
Parsimonious trees 271
PASINI P 111
Patch-clamp 190
PAUP program 19, 25
PCR 19, 23, 95, 194
real time quantitative 143
p-cyanophenol 85
PEREHINEC TM 193
Peroxyl radicals 83
Peroxyoxalate 215, 219
reaction 216
Peruvian upwelling 263
pGEX-6P-2 119
Phengodidae 23
Phenol 85
Phloroglucinol 85
Phosphatidyl phospholipase C 193
Phospholipase A2 63
Photeros 270
Photinus pennsylvanica 16
Photinus pyralis .. 15, 16, 20, 109, 243
Photobacterium 36
Photobacterium kishitanii 35, 36
Photobacterium leiognathi 27, 31
Photobacterium phosphoreum. 35, 36, 259
Photolysis 151
Photomultiplier tube 136
Photon counters 41
Photon-counting luminometer 53
Photorhabdus luminescens .. 130, 185
Photuris pennsylvanica 108
Phrixothrix hirtus 178
Phylogenetic analysis 20, 23
Phylogenetic tree 21
elateridae 24
Physiculus 35, 36
Phytoplankton 264
Pilocarpine 102
PIONTKOVSKI SA 261
Plant stress 258
Plaque assay 76
plcA 193
Pleurotus olearius 265
p-nitrophenol 85
Point-of-care 97
PorA 173
Potassium permanganate 136
Poxviridae 181
PPi 144
PrfA 193
Primary biliary cirrhosis 201
PRISM® 139
PROKHOROVA EV 163
Pronase E 73
Prostatic acid phosphatase 105
Proteinase 71
PSA (Prostate-specific antigen) .. 104, 105
Pseudomonas aeruginosa 89, 112, 127
Pseudomonas sp. 20
Pseudomonas spp. 71
pSLR-cfos 177

PVDF 172
Pyrocoelia rufa 16
Pyrophorini 25
Pyrophosphate 67
Pyruvate phosphate dikinase 67
QAZI SNA 193
qPCR 98
Quantum yield 11, 232
Quercetin 84
Quorum sensing 111
RAAB V 181, 197
RACE 19
RAHIMI Y 235
RAMSAY C 71, 79
Random mutagenesis 159
Rapid hygeine monitoring 79
Rapid microbiology 41
RapiTECT GN™ 87
Real time quantitative PCR 143
Real-time monitoring 181
Real-time reporter 185
REES CED 35, 193
Refolding 4
REMMEL NN 257
Remote sensing 261
Renilla luciferase ... 99, 101, 123, 197
Reporter assay 53, 177
Resorcinol 85
Reverse-transcription quantitative bioluminescence assay 143
RFP (red fluorescent protein) 197, 235
RFP-CBG99 197
Rhagophthalmidae 23
Rhagophthalmus ohbai 178
Rhenium complexes 249
Rheumatoid arthritis 251
rhlR 112
Ribeira River State Park 265
Ribosomal DNA 23
Rluc 123
ROBINSON GM 127, 185
RODA A 119, 123, 201
RODA E 201
ROMANOVA NA 75
ROMOFF P 83
ROTHERT A 131
ROWE L 131
Ruc-GFP 181
Rumph, Georg Everard 265
Ruthenium complexes 249
Rutin 84
RYGALOV VE 257
SAGEGAMI-OBA R 23
SAKAI H 243
SALBILLA V 139
SALISBURY VC 127, 185
Saliva 113
Salmonella typhimurium ..75, 77, 209
Sanitizers 75
Saponin 128
SATIENPERAKUL S 135
SCHOCHETMAN G 139
SCHREIBERS M 41
SET-oxygenation process 12
SEUBERT C 197
SHAH S 139
SHASHIDHAR H 111
SHIBATA R 239
SHIGEYOSHI Y 45
Shining mushrooms 265
SHOEMAN DM 257
SHRESTHA S 235
Single Nucleotide Polymorphism.. 67
Sirius C 177
Site-directed mutagenesis 119
Site-specific mutagenesis 16
SloDK 159
SMITSMAN C 171
SNPs analysis 67
Somatic cell count 91
SOUTHWORTH TL 119
Spaceflight 257
Spectroscopic characteristics 244
Standard Plate Count 91
Steady-state 27

STECHA P.....99
Sterol 27-hydroxylase.....119
STEVANI CV.....265
Stool.....113
STRITZKER J.....197, 205
SUGIYAMA T.....53
SUKOVATAYA IE.....27, 31
Sulfide.....135
Sulfopropyl acridinium.....141
Sulfur.....163
SUN X.....147
SUN YE.....143
Superoxide.....158
Superoxide anions.....155
Suprachiasmatic nucleus slices.....45
SUZUKI H.....45, 53
SUZUKI S.....67
SWANSON B.....99
SZALAY AA.....181, 197, 205, 209
TAKAHASHI T.....45
TAKAHASHI Y.....11
TANAKA M.....151
TAPP C.....71
TARKKANEN V.....41
Tauroursodeoxycholic.....121
TCPO.....215
Tecan GENios Pro.....128
Tenebrio molitor.....19, 20
TERANISHI K.....155
Tet-luciferase.....56
Thiazoles.....248
THP-1.....185
Throscidae.....23
TIETZE CJ.....197
Time-dependent spectra.....239
Time-lapse bioluminescence imaging.....47
TISHKOV VI.....15
TISI LC.....95, 107
TK promoter.....101
TOKAREV YN.....261
TORRES E.....269
Trace metal.....251
  elements.....251
TRICOIRE L.....159, 189
Trifluoroacetyl aminoluciferin.....243
Tritium light sources.....49
Troponin I.....59
  cardiac-specific.....59
TRUPP S.....247
Trypanosomas cruzi.....139
TSATSOS PH.....63
TSUZUKI K.....159
Tumors
  benign.....210
  metastases.....209
  solid tumors.....205, 209
Tumor-specific targeting.....205
TYULKOVA NA.....27, 31
UGAROVA NN.....15, 91
Ultra-Glo™.....107
Ultra-High Temperature.....71
Uni-Lite® luminometer.....80
Ursodeoxycholate.....121
Vaccine.....172
Vaccinia virus.....181, 197
Vargula.....270
Varian Eclipse.....83
VENTUROLI S.....201
VETROVA EV.....163
Vibrio cholerae.....209
Victor® Multilabel Plate Reader.....72
Virulence.....127
Viscosimetric Assay.....91
VV- CBG99-RFP.....197
VV Ruc-GFP.....197
VYSOTSKI ES.....3, 7
WADA N.....151, 239, 243
WATKIN J.....79
WEIBEL S.....205
WEISS D.....223, 247
Western blot.....147, 171
WHITAKER S.....35
WHO (World Health Organization).....147, 181
Williamson ether synthesis.....247

Wittig 223
WOOD KV 99
World ocean 262
WORSCHECH A 197
WU C 167
X-ray diffraction 151
YANG X 147
YOSHIDA Y 239
YOSHIKI Y 251
YU YA 181, 197, 209
ZELLNER E 197
ZHANG Q 181, 197
ZOMER G 171